Community-based Water Law and Water Resource Management Reform in Developing Countries

Community-based Water Law and Water Resource Management Reform in Developing Countries

Edited by

Barbara van Koppen

Mark Giordano

and

John Butterworth

www.cabi.org

CABI is a trading name of CAB International

CABI Head Office
Nosworthy Way
Wallingford
Oxfordshire OX10 8DE
UK

Tel: +44 (0)1491 832111
Fax: +44 (0)1491 833508
E-mail: cabi@cabi.org
Website: www.cabi.org

CABI North American Office
875 Massachusetts Avenue
7th Floor
Cambridge, MA 02139
USA

Tel: +1 617 395 4056
Fax: +1 617 354 6875
E-mail: cabi-nao@cabi.org

A catalogue record for this book is available from the British Library,
London, UK

Library of Congress Cataloging-in-Publication Data

Community-based water law and water resource management reform in
developing countries / edited by Barbara van Koppen, Mark Giordano and
John Butterworth.
     p. cm. -- (Comprehensive assessment of water management in
agriculture ; 5)
   Includes bibliographical references and index.
   ISBN 978-1-84593-326-5 (alk. paper)
  1. Water--Law and legislation--Developing countries. 2. Water resources
development--Developing countries. I. Koppen, B. C. P. van (Barbara C. P.)
II. Giordano, Mark. III. Butterworth, John. IV. Series: Comprehensive
assessment of water management in agriculture series ; 5.

  K3496.C66 2008

  346.04'691--dc22   2007021936

ISBN-13: 978 1 84593 326 5

Produced and typeset by Columns Design Ltd, Reading, UK
Printed and bound in the UK by Biddles Ltd, King's Lynn

Contents

Contributors

Seleshi Bekele Awulachew, *International Water Management Institute (IWMI), ILRI-Ethiopia campus, PO Box 5689, Addis Ababa, Ethiopia; e-mail: s.bekele@cgiar.org*

Mukand Singh Babel, *School of Civil Engineering, Asian Institute of Technology, PO Box 4, Khlong Luang, Pathumthani 12120, Thailand; e-mail: msbabel@ait.ac.th*

Rutgerd Boelens, *General Coordinator of the WALIR Program and Researcher with Wageningen University and Research Centre, Wageningen, Netherlands; e-mail: rutgerd.boelens @wur.nl*

Bryan Bruns, *Consulting Sociologist, Santa Rosa Beach, Florida, USA; e-mail: bryanbruns@ bryanbruns.com*

Rocio Bustamante, *Coordinator for WALIR in Bolivia and Researcher with Centro AGUA, San Simon University, Cochabamba, Bolivia; e-mail: vhrocio@entelnet.bo*

John Butterworth, *IRC International Water and Sanitation Center, Delft, Netherlands; e-mail: butterworth@irc.nl*

Bill Derman, *Professor of Anthropology, Michigan State University, East Lansing, Michigan, USA and Fulbright Visiting Professor, Department of International Development and Development Studies (NORAGRIC), Norwegian University of Life Sciences, Norway; e-mail: derman@msu.edu or bill.derman@umb.no*

Alan B. Dixon, *Department of Geography, University of Otago, PO Box 56, Dunedin, New Zealand; e-mail: alan.dixon@geography.otago.ac.nz*

Hugo de Vos, *Freelance Researcher on Institutional Aspects of Natural Resource Management in Latin America; e-mail: voswiz@versatel.nl*

Desalegn Chemeda Edossa, *PO Box 19, Haramaya University, Ethiopia; e-mail: dchemeda @yahoo.com*

Anne Ferguson, *Department of Anthropology, Michigan State University, East Lansing, Michigan, USA; e-mail: fergus12@msu.edu*

Mark Giordano, *Head: Institutions and Policies Research Group, International Water Management Institute (IWMI), Colombo, Sri Lanka; e-mail: mark.giordano@cgiar.org*

Ashim Das Gupta, *School of Civil Engineering, Asian Institute of Technology, PO Box 4, Khlong Luang, Pathumthani 12120, Thailand; e-mail: adg@ait.ac.th*

Anne Hellum, *Faculty of Law, University of Oslo, 0130, Oslo, Norway; e-mail: anne.hellum@ jus.uio.no*

Bruce Lankford, *School of Development Studies, University of East Anglia, Norwich, UK; e-mail: b.lankford@uea.ac.uk*

Rose Machiridza, Department of Soil Science and Agricultural Engineering, University of Zimbabwe, Box MP 167, Mount Pleasant, Harare, Zimbabwe; e-mail: roma877@yahoo.co.uk

Emmanuel Manzungu, Department of Soil Science and Agricultural Engineering, University of Zimbabwe, Box MP 167, Mount Pleasant, Harare, Zimbabwe; e-mail: manzungu@mweb.co.zw

Everisto Mapedza, Researcher in Policies and Institutions, International Water Management Institute (IWMI), Southern Africa Regional Programme, PBag X813, Silverton 0127, South Africa; e-mail: e.mapedza@cgiar.org

Abraham Mehari, PhD Research Fellow in Land and Water Development, UNESCO-IHE, Netherlands; e-mail: abrahamhaile2@yahoo.com or a.meharihaile@unesco-ihe.org

Ruth Meinzen-Dick, International Food Policy Research Institute (IFPRI), 2033 K Street NW, Washington, DC 20006, USA; e-mail: r.meinzen-dick@cgiar.org

Wapulumuka Mulwafu, History Department, Chancellor College, University of Malawi, PO Box 280, Zomba, Malawi; e-mail: wmulwafu@chanco.unima.mw

Albert Mumma, Faculty of Law, University of Nairobi, Parklands Campus, PO Box 30197, Nairobi, Kenya; e-mail: cepla@nbnet.co.ke

Willie Mwaruvanda, Rufiji Basin Water Office, Ministry of Water and Livestock Development, Iringa, Tanzania; e-mail: rufijibasin@yahoo.co.uk

Regassa Ensermu Namara, Economist, International Water Management Institute (IWMI), PMB, CT 112, Cantonments Accra, Ghana; e-mail: r.namara@cgiar.co.org

Leticia Nkonya, Department of Sociology, Anthropology and Social Work, Kansas State University, 204 Waters Hall, Manhattan, Kansas 666502-4003, USA; e-mail: letinkonya@yahoo.com

Leah Onyango, Lecturer in Urban and Regional Planning, Maseno University, Private Mail Bag, Maseno, Kenya and Graduate Attachment, World Agroforestry Centre; e-mail: leahonyango@yahoo.com

Jessica L. Roy, former PhD Student at the University of California–Santa Cruz, USA and Graduate Attachment, World Agroforestry Centre (ICRAF); she died in August 2004 while conducting field research reported in this book.

Bart Schultz, Professor of Land and Water Development, UNESCO-IHE; Top Advisor, Rijkswaterstaat, Civil Engineering Division, Utrecht, Netherlands; and President Honorary of the International Commission on Irrigation and Drainage (ICID); e-mail: b.schultz@unesco-ihe.org

Tushaar Shah, Principal Scientist, International Water Management Institute (IWMI), South Asia Program, Anand Office, Anand, Gujarat, India 388001; e-mail: t.shah@cgiar.org

Pinimidzai Sithole, Centre for Applied Social Sciences, University of Zimbabwe, Harare, Zimbabwe; e-mail: spinimidzai@yahoo.com

Brent Swallow, Theme Leader for Environmental Services, World Agroforestry Centre (ICRAF), PO Box 30677, Nairobi, Kenya; e-mail: b.swallow@cgiar.org

Barbara van Koppen, Principal Scientist, International Water Management Institute (IWMI), Southern Africa Regional Programme, PBag X813, Silverton 0127, South Africa; e-mail: b.vankoppen@cgiar.org

Frank van Steenbergen, MetaMeta Research, Paarskerkhofweg, 5223 AJ's-Hertogenbosch, Netherlands; e-mail: fvansteenbergen@metameta.nl

Adrian P. Wood, Centre for Wetlands, Environment and Livelihoods, University of Huddersfield, Queensgate, Huddersfield, HD1 3DH, UK; e-mail: a.p.wood@hud.ac.uk

Preface

Water resource management reform today emphasizes user participation. However, in developing country contexts the water laws and institutions which have followed from this reform have consistently ignored how people actually manage their water. Informal rural and peri-urban water users have managed their water resources for centuries and continue to respond to new opportunities and threats, often entirely outside the ambit of formal government regulation or investment. The community-based water laws which guide this informal management in fact govern water development and management by significant numbers of water users, if not the majority of citizens and the bulk of the poor, who depend on water for multiple uses for fragile agrarian livelihoods. These community-based arrangements tend to have many of the people-based, pro-poor attributes desired in principle, if not always found in practice in current water management reform agendas – they are typically robust, dynamic and livelihood-oriented, and often encompass purposeful rule-setting and enforcement and provide incentives for collective action. At the same time, they can also be hierarchical and serve to entrench power and gender disparities.

Ignoring community-based water laws and failing to build on their strengths, while overcoming their weaknesses, greatly reduce the chance of new water management regimes to meet their intended goals. In contrast, when the strengths of community-based water laws are combined with the strengths of public sector contributions to water development and management, the new regimes can more effectively lead to sustainable poverty alleviation, gender equity and overall economic growth. Indeed, the challenge for policy makers is to develop a new vision in which the indispensable role of the public sector takes existing community-based water laws into full account.

This book contributes to this new vision. Leading authors analyse living community-based water laws in Africa, Latin America and Asia and critically examine the interface between community-based water laws, formal water laws and a variety of other key institutional ingredients of ongoing water resource management reform.

Most chapters in the book were selected from papers presented at the international workshop 'African Water Laws: Plural Legislative Frameworks for Water Management in Rural Africa', held in Johannesburg, South Africa, 26–28 January 2005, co-organized by the International Water Management Institute (IWMI), the Department of Water Affairs and Forestry (DWAF) South Africa, the National Resources Institute UK (NRI), and the Faculty of Law, University of Dar es Salaam, Tanzania (www.nri.org/waterlaw/workshop). The support given to this workshop by the Comprehensive Assessment on Water Management in Agriculture, the Water Research Commission, South Africa, EU, DFID and CTA is gratefully acknowledged.

The completion of this volume has been made possible, first of all, by the willing and punctual contributions of the authors of the fifteen chapters. Kingsley Kurukulasuriya carefully and promptly edited all chapters. The maps were designed by Simon White. Mala Ranawake, Pavithra Amunugama, Nimal Attanayake and Sumith Fernando provided further indispensable editorial support. The editors are grateful for these contributions.

The Editors

Series Foreword: Comprehensive Assessment of Water Management in Agriculture

There is broad consensus on the need to improve water management and to invest in water for food to make substantial progress on the Millennium Development Goals (MDGs). The role of water in food and livelihood security is a major issue of concern in the context of persistent poverty and continued environmental degradation. Although there is considerable knowledge on the issue of water management, an overarching picture on the water–food–livelihoods–environment nexus is required to reduce uncertainties about management and investment decisions that will meet both food and environmental security objectives.

The Comprehensive Assessment of Water Management in Agriculture (CA) is an innovative multi-institute process aimed at identifying existing knowledge and stimulating thought on ways to manage water resources to continue meeting the needs of both humans and ecosystems. The CA critically evaluates the benefits, costs and impacts of the past 50 years of water development and challenges to water management currently facing communities. It assesses innovative solutions and explores consequences of potential investment and management decisions. The CA is designed as a learning process, engaging networks of stakeholders to produce knowledge synthesis and methodologies. The main output of the CA is an assessment report that aims to guide investment and management decisions in the near future, considering their impact over the next 50 years in order to enhance food and environmental security to support the achievement of the MDGs. This assessment report is backed by CA research and knowledge-sharing activities.

The primary assessment research findings are presented in a series of books that form the scientific basis for the Comprehensive Assessment of Water Management in Agriculture. The books cover a range of vital topics in the areas of water, agriculture, food security and ecosystems – the entire spectrum of developing and managing water in agriculture, from fully irrigated to fully rainfed lands. They are about people and society, why they decide to adopt certain practices and not others and, in particular, how water management can help poor people. They are about ecosystems – how agriculture affects ecosystems, the goods and services ecosystems provide for food security and how water can be managed to meet both food and environmental security objectives. This is the fourth book in the series.

The books and reports from the assessment process provide an invaluable resource for managers, researchers and field implementers. These books will provide source material from which policy statements, practical manuals and educational and training material can be prepared.

The Comprehensive Assessment of Water Management in Agriculture calls for Institutional Reform to address issues of equity, sustainability and efficiency in water resource use for agriculture. The assessment recognizes that effective reform has been elusive, and that reform is needed in the reform process itself. This book focuses on the critical issue of institutional and legal water arrangements that can strengthen poor rural women's and men's access to water and, thus, contribute to poverty reduction and gender equity. The book envisions a new role for the state in informal rural economies in developing countries in which community-based water laws also play their full roles. The book assesses legal and institutional challenges based on in-depth empirical analyses of community-based water laws in Africa, Latin America and Asia.

The CA is carried out by a coalition of partners that includes 11 Future Harvest agricultural research centres supported by the Consultative Group on International Agricultural Research (CGIAR), the Food and Agriculture Organization of the United Nations (FAO) and partners from over 200 research and development institutes globally. Co-sponsors of the assessment, institutes that are interested in the results and help frame the assessment, are the Ramsar Convention, the Convention on Biological Diversity, FAO and the CGIAR.

Financial support from the governments of The Netherlands and Switzerland, EU, FAO and the OPEC foundation for the Comprehensive Assessment for the preparation of this book is appreciated.

David Molden
Series Editor
International Water Management Institute
Sri Lanka

Foreword

Barbara Schreiner

Deputy Director General, Department of Water Affairs and Forestry, South Africa

From space, our world is a blue planet, bathed in vast blue oceans, wrapped in water. It is, to all intents and purposes, a very wet planet. Like all things, however, the devil is in the detail. As you move closer to the blue planet, the picture changes. You see that most of the vast rolling waters are salty, unfit for human use. On the land, rivers, aquifers, lakes, dams and wetlands and in the atmosphere, clouds contain the tiny proportion of water on which humans, many animals and plants survive. As humans, we are dependent on this fresh water for our survival. If you move even closer, however, you will see how unevenly distributed the water resources are. There are areas of land abundantly endowed with water. There are vast areas of land surface where there is little, or no, water. As you move even closer, you might see how much water some sections of the population have and how little others have. You might see areas of green-watered yards and swimming pools; you might see jumbles of shacks tightly packed in dusty, dry and barren areas. If you have the right kind of telescope you might notice that it is the poor, in rural and urban areas, who truly experience water scarcity. It is the poor who have little, or no, access to water and the poor who suffer the worst impacts of water pollution, droughts and floods.

It is at this point that you might realize that the management of our precious water resources is deeply political, deeply influenced by issues of access to power. In the context of many developing countries, it is also influenced by the juxtaposition of different water management paradigms. In many developing countries, the 'official' management systems for water stem from colonial and post-colonial formal systems. At the local level, however, customary practices are still in place, managing water according to systems and practices that may be many, many decades old.

Understanding customary water management practices is important because they often define the de facto institutional environment of the rural poor far more than the formal institutional arrangements determined by legislation and government administration. Water is key to agriculture-based livelihoods. Over time, rural smallholders have devised many ways to develop and manage local water resources, through wells, tanks, water harvesting, river diversions and small dams. The social capital manifest in customary arrangements embodies creativity, resilience, local appropriateness, broad compliance and ownership: in sum, the experience of centuries.

It would, however, be inappropriate to assume that local and customary systems are without problems. Local communities have their own power dynamics, and are often divided by clan allegiances, by gender and by levels of wealth. Such divisions may filter through into access to water and may perpetuate inequities at the local level.

There are thus often two parallel water-management paradigms – customary practices and formalized legal approaches. Both may have their strengths and their weaknesses. The two may, in some cases, be directly contradictory. For example, customary arrangements may entrench inequities in access to water, such as gender inequities which relegate women to a secondary legal status, while the formal paradigm may require gender equity. Ethnic divides, often exploited for colonial divide and rule, and clan-based access to power are other retrogressive elements of customary systems that need to be changed. On the other hand, the formal legal arrangements may not support the needs of localized people on the ground. As some authors in this book have suggested, formalized legal approaches may even, unintentionally, disadvantage the poor who do not have the necessary access to formal structures to make them work to their advantage.

Africa, in particular, shares a common history of externally imposed water legislation that have systematically marginalized existing customary arrangements, including access to water and ownership of land. This history urgently needs to be written from an African perspective, challenging the ever-enduring dominance of the European perspective in the history of African water laws. The famous Kenyan author, Ngugi wa Thiongo, challenged Africans to decolonize their minds – a challenge that pertains equally in the arena of water legislation and management paradigms.

As Africans and as citizens of developing countries, the challenge lies with us to take the best of customary practices, the best of formal systems and to create, in the interests of the poorest of our citizens, a water-management paradigm that is located in the needs and culture of our own societies. This book brings together a wide range of experience to examine the benefits and challenges of customary practices and their relationship to formal systems. It articulates, in particular, an African perspective on managing water in the interests of the poor, the rural and the marginalized people of Africa and other developing countries. It is an important contribution to the discourse on integrated water resources management, bringing the perspective of developing countries clearly to the fore.

Abbreviations and Acronyms

$ – US dollar/s
ACTS – African Centre for Technology Studies
BATNA – Best alternative to a negotiated agreement
BFA – Beneficiary farmer association (India)
BIS – Bureau of Indian Standards
BPL – Below poverty line
BSAC – British South Africa Company
CADA – Command Area Development Agency (India)
CAPRI – Collective action and property rights
CASS – Center for Applied Social Studies (Zimbabwe)
CBNRM – Community-based natural resources management
CBO – Community-based organization
CBS – Central Bureau of Statistics (Kenya)
CC – Catchment Council (Zimbabwe)
CDE – The Centre for Development and Environment (Switzerland)
CEDAW – Convention on the Elimination of All Forms of Discrimination Against Women
CESCR – Convention on Economic, Social and Cultural Rights
CMA – Catchment Management Authority (Malawi)
CNA – Comisión Nacional del Agua (Mexico)
CoE – Council of Elders (India)
CONAIE – The Confederation of Indigenous Nationalities of Ecuador
CONIAG – Consejo Interinstitucional del Agua (Bolivia)
COTAS – Comités Técnicos de Aguas Subterráneas (Mexico)
CRC – Convention on the Rights of the Child
CWUA – Catchment Water User Association (Tanzania)
DANIDA – Danish International Development Agency
DEAP – District Environment Action Plan (Kenya)
DFID – Department for International Development (UK)
DSDO – District Social Development Officer (Kenya)
DSE – German Foundation for International Development
DWD – Department of Water Development (Kenya)
ECLAC – United Nations Economic Commission for Latin America and the Caribbean
ELWDP – Eastern Lowland Wadi Development Project (Eritrea)

EMCA – Environmental Management and Coordination Act (Kenya)
ESAP – Economic Structural Adjustment Programme
ESCOM – Electricity Supply Commission of Malawi
EWRP – Ethiopian Wetlands Research Programme
FAO – Food and Agriculture Organization of the United Nations
FTLP – Fast-track Land Reform Program (Zimbabwe)
GLC – Governments' Land Commission (Pakistan)
GOI – Government of India
GOM – Government of Malawi
GTZ – German Technical Cooperation
HIV/AIDS – Human immunodeficiency virus/a(cquired) i(mmune) d(eficiency) s(yndrome)
IAR – Irrigation Allocation Ratio
IAs – Institutional arrangements
ICESCR – International Covenant on Economic, Social and Cultural Rights
ICRAF – World Agroforestry Centre (Kenya)
IDI – Infrastructure Development Institute (Japan)
IE – Institutional environment
IFAD – International Fund for Agricultural Development
IFPRI – International Food Policy Research Institute
IIED – International Institute for Environment and Development
IIP – Irrigation Improvement Project (Yemen)
ILO – International Labour Organization
IMF – International Monetary Fund
IMT – Irrigation management transfer
ISI – Indian Standards Institution
IWMI – International Water Management Institute
IWRM – Integrated Water Resources Management
IWUA – Irrigation Water User Association (Tanzania)
KA – Kebele administration (Ethiopia)
LAB – Land Administration Body (Pakistan)
LIFCA – Legal infrastructure framework for catchment apportionment
LVEMP – Lake Victoria Environment Management Programme (Kenya)
MAS – Movimiento al Socialismo (Andes)
MCP – Malawi Congress Party (Malawi)
MOW – Ministry of Water (Tanzania)
MOWLD – Ministry of Water and Livestock Development (Tanzania)
MYP – Malawi Young Pioneers (Malawi)
NEAP – National Environmental Action Plan (Kenya)
NEMA – National Environmental Management Authority (Kenya)
NEPAD – New Partnership for Africa's Development
NGO – Non-governmental organization
NIB – National Irrigation Board (Kenya)
NIE – New Institutional Economics
NNMLS – Northern New Mexico Legal Services
NRM – Natural Resources Management
NSSO – National Sample Survey Organization (India)
NUFFIC – Netherlands Organization for International Cooperation in Higher Education
NWCPC – National Water Conservation and Pipeline Corporation (Kenya)
ODG – Overseas Development Group (UK)
PA – Peasant association or kebele (Ethiopia)
PEAP – Provincial Environment Action Plan (Kenya)
PIM – Participatory irrigation management

PIU – Provincial Irrigation Unit (Kenya)
RBMSIIP – River Basin Management and Smallholder Irrigation Improvement Project (Tanzania)
RBO – River Basin Office (Tanzania)
RBWO – Rufiji Basin Water Office (Tanzania)
RDC – Rural District Council (Zimbabwe)
RDP – Rural Development Project (Malawi)
RIM – Registry Index Map (Kenya)
RLA – Registration of Land Act (Kenya)
RO – Reverse osmosis (India)
RSA – Republic of South Africa
SADC – Southern Africa Development Community
SCC – Sub-catchment Council (Zimbabwe)
SEB – State Electricity Board (India)
SFT – Settlement Fund Trustee (Kenya)
SIDA – Swedish International Development Cooperation Agency
SLSA – Sustainable Livelihoods in Southern Africa
SMC – Scheme Management Committee (Malawi)
SMUWC – Sustainable Management of the Usangu Wetland and its Catchment (Tanzania)
SSP – Sardar Sarovar Project (India)
SWMRG – Soil Water Management Research Group (Tanzania)
TA – Traditional authority (Malawi)
TCE – Transaction cost economics
TOEB – Tropical Ecology Support Programme (Germany)
TTL – Tribal Trust Lands (Zimbabwe)
UNDP – United Nations Development Programme
URT – United Republic of Tanzania
USAID – United States Agency for International Development
VSA – Village Service Area (India)
WALIR Program – Water Law and Indigenous Rights Program
WFP – World Food Programme
WHO – World Health Organization
WMI – Wetland Management Institution (Ethiopia)
WRMA – Water Resources Management Authority (Kenya)
WRUA – Water Resources User Association (Kenya)
WSB – Water Service Board (Kenya)
WSP – Water Service Provider (Kenya)
WSRB – Water Service Regulatory Board (Kenya)
WTF – Water Trust Fund (Kenya)
WUA – Water User Association
ZCTU – Zimbabwe Congress of Trade Unions
ZINWA – Zimbabwe National Water Authority

1 Community-based Water Law and Water Resource Management Reform in Developing Countries: Rationale, Contents and Key Messages

Barbara van Koppen,[1]* Mark Giordano,[2]** John Butterworth[3]*** and Everisto Mapedza[1]****

[1] *International Water Management Institute, Southern Africa Regional Programme, Silverton, South Africa;* [2] *Head: Institutions and Policies Research Group, International Water Management Institute, Colombo, Sri Lanka;* [3] *IRC International Water and Sanitation Center, Delft, Netherlands; e-mails: *b.vankoppen@cgiar.org; **mark.giordano@cgiar.org; ***butterworth@irc.nl; ****e.mapedza@cgiar.org*

Abstract

Water resources management reform in developing countries has tended to overlook community-based water laws, which govern self-help water development and management by large proportions, if not the majority, of citizens: rural, small-scale water users, including poor women and men. In an attempt to fill this gap, global experts on community-based water law and its interface with public sector intervention present a varied collection of empirical research findings in this volume. The present chapter introduces the rationale for the volume and its contents. It further identifies key messages emerging from the chapters on, first, the strengths and weaknesses of community-based water law and, second, the impact of water resources management reform on informal water users' access to water and its beneficial uses.

Impacts vary from outright weakening of community-based arrangements and poverty aggravation or missing significant opportunities to better water resource management and improved well-being, also among poor women and men. The latter interventions combine the strengths of community-based water law with the strengths of the public sector. Together, these messages contribute to a new vision on the role of the state in water resources management that better matches the needs and potentials of water users in the informal water economies in developing countries.

Keywords: community-based water law, water reform, developing countries, IWRM, public sector.

Rationale for This Volume

Since the late 1980s, an unprecedented reform of water resource management has taken place across the globe, as heralded by events such as the declaration of the Dublin Principles in 1992. Worldwide, this reform has radically redefined the role of the public sector, with the state's conventional primary role as investor in water infrastructure being questioned. Partly as a result of these policy changes, public investments in water have declined in the expectation that the private sector would step in to fill the gap. Existing irrigation schemes have been transferred from government control to users, while privatization of the domestic water sector has been encouraged. Thus, the role of the state has shifted more towards that of regulator, promoting decentralization and users' participation.

In order to fulfil their regulatory roles, states have promoted measures such as the strengthening of formal administrative water rights systems, cost recovery and water pricing (the 'user pays' principle), the creation of new basin institutions and better consideration of the environment (the 'polluter pays' principle). Together, this set of regulatory measures is usually referred to as 'Integrated Water Resources Management' (IWRM).

Although the emphasis on users' participation suggests otherwise, water resources management reform has paid little attention to community-based water laws in rural areas within developing countries. Community-based water law is defined as the set of mostly informal institutional, socio-economic and cultural arrangements that shape communities' development, use, management, allocation, quality control and productivity of water resources. These arrangements, anchored in the wisdom of time, are embedded in local governance structures and normative frameworks of kinship groups, smaller hamlets, communities and larger clans and groupings with common ancestry. In developing countries, they often exist only in oral form.

Reformers have tended to ignore, frown upon or even erode community-based water law as they have pushed forward the IWRM principles. This is startling, because these arrangements govern the use of water by large proportions, if not the majority, of the world's citizens: the rural women and men, often poor, who, in self-help mode, use small amounts of water as vital inputs to their multifaceted, agriculture-based livelihoods. Moreover, reforms in developing countries have often been financed by bilateral and international donors and financiers whose main aim is the use of water for improving the well-being of precisely these informal users.

Recently, the confidence with which IWRM and its redefined role of the state was promoted has started dwindling. In sub-Saharan Africa, major players like the World Bank, African Development Bank and the New Partnership for Africa's Development (NEPAD) recognize again that 're-engagement' in investments in agricultural water management, besides domestic supplies, is warranted. The private sector has not taken up this conventional public sector role. Farmers' protests against the new laws in Latin America (see Boelens *et al.*, Chapter 6, this volume) are echoed by African water lawyers concerned about the dispossession of customary water rights holders under the introduction of permit systems (Sarpong, undated).

Academic critiques are also emerging and argue that, while the typical ingredients of IWRM may work in the formalized water economies of industrialized countries, they are inappropriate in the informal water economies of the developing world (Shah and Van Koppen, 2006). As a result, there is a renewed and growing call for a new vision on a more refined role of the state and other public and civil sector entities in water resources management in the informal sectors in developing countries.

This volume seeks to contribute to developing such a vision on the role of the state in which, for the first time, community-based water arrangements play their full roles. Clearly, both the public sector and community-based water arrangements have their strengths and weaknesses, and the key question is not which one is best, but rather which combination is most appropriate to address needs in specific areas and in particular for those most at risk: rural poor women and men.

Finding an appropriate mix requires, first of all, a better understanding of community-based water law itself. Academic understanding of community-based water law has grown significantly during the past decades (cf. Von Benda-Beckmann, 1991; Shah, 1993; Ostrom, 1994; Yoder, 1994; Ramazzotti, 1996; Boelens and Dávila, 1998; Bruns and Meinzen-Dick, 2000). While the focus in this research was previously on irrigation of field crops, the scope has increasingly widened to include homestead gardening, domestic uses, livestock watering, silviculture, fisheries and even the integrated use of multiple sources for multiple purposes (Bakker *et al.*, 1999; Moriarty *et al.*, 2004; Van Koppen *et al.*, 2006).

Many questions concerning the strengths and weaknesses of community-based water arrangements are still open. To name a few: (i) how do communities induce collective action in water resources development and management and how can their systems work at scales beyond the community? (ii) How and under what conditions does spontaneous innovation, an important strength of community-based water arrangements, spread? And (iii) what are critical weaknesses of community-based arrangements where the public sector has a legitimate role in acting to enhance the human well-being through better access to water and its beneficial uses? Answers to these and other questions will be indispensable in identifying the practical implications of communities' strengths and weaknesses for the design of public policies and programmes.

The second requirement in finding a more appropriate mix of community-based water law and public sector intervention is a better understanding of the interface between these two legal systems and of the strengths and weaknesses of the public sector in meeting communities' genuine needs for improved welfare and productivity. As community-based water law has gener-

ally been ignored up till now, positive, mixed or even negative impacts of the imposition of state regulations have mostly gone unnoticed as well. A better understanding of the current interface would allow for the design of more appropriate and effective forms of public support that build upon communities' strengths, while overcoming their weaknesses.

In this volume, global experts bring rich empirical evidence together on these two core issues: community-based water law and its interface with the state and other external agencies. As shown in Fig. 1.1, the locations from which the studies draw and according to which they are organized are diverse. The first set of chapters take a broad approach, looking across low- and middle-income countries in sub-Saharan Africa, Asia and Latin America, including some comparison with high-income countries. The second set covers areas outside Africa including Latin America, India, Mexico and China, as well as the particular case of arid zones and spate-irrigation. The remaining studies, organized in alphabetical order by country,

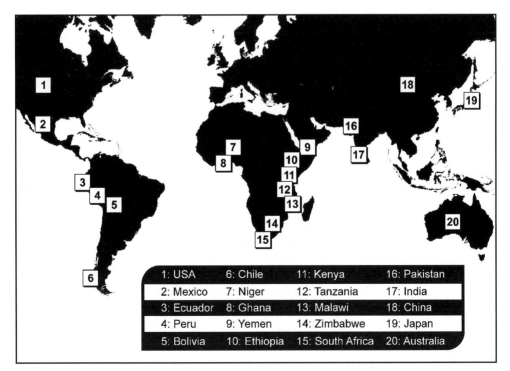

1: USA	6: Chile	11: Kenya	16: Pakistan
2: Mexico	7: Niger	12: Tanzania	17: India
3: Ecuador	8: Ghana	13: Malawi	18: China
4: Peru	9: Yemen	14: Zimbabwe	19: Japan
5: Bolivia	10: Ethiopia	15: South Africa	20: Australia

Fig. 1.1. Countries featured in this volume.

focus on Africa, the continent with the largest proportion of informal, rural, small-scale water users.

The following section provides a brief overview of the contents of all chapters. Taking the findings from the large diversity of sites with their varying foci on aspects of community-based water law and public water development and regulation together, some key messages can be derived, as presented in the third section. These messages highlight the strengths and weaknesses to be found in community-based water law and contribute to an emerging vision of the role of the state in water resources management in the informal water economies in developing countries.

Contents of the Chapters

Chapter 2, by Ruth Meinzen-Dick and Leticia Nkonya, sets the scene of pluralistic legal frameworks for water management, conflicts and water law reform, and explores the links between land and water rights. It uses examples from Africa and Asia.

In Chapter 3, Bryan Bruns focuses on the negotiation of water rights at the basin scale. He compares communities' perspectives and priorities with the assumptions that underpin current formal measures for basin-scale water allocation. The chapter identifies a set of measures for community involvement in basin management that would fit communities' own priorities and strategies significantly better.

Chapter 4, by Barbara van Koppen, discusses the entitlement dimensions of permit systems. Tracing the roots in Roman water law and the historical development of permit systems in high-income countries, she highlights differences in Europe's colonies in Latin America and sub-Saharan Africa. In the latter, permit systems were primarily introduced to serve the goal of divesting indigenous users of their prior claims. Current water law revisions promoted as IWRM in these two southern continents risk reviving dispossession of informal rural water users.

In Chapter 5, Tushaar Shah makes an encompassing analysis from the perspective of new institutional economics of the institutional environment of formal water reform and the widely prevailing institutional arrangements in informal water economies in India, also taking examples from Mexico, China and Africa. From the analysis of a range of water institutions in India it appears that the transaction costs are low and the pay-offs high in the case of six largely ignored major self-help initiatives and one potential indirect public measure. In contrast, irrigation management transfer, water policy formulation, water regulation through permits and seven other formal regulatory measures entail either excessive transaction costs or lack pay-off or both. The conclusion is that, in informal water economies, the state should: (i) support high-performing infrastructural development in a welfare mode; (ii) promote institutional innovations that reduce transaction costs and restructure incentive structures; (iii) better exploit indirect measures; and (iv) improve performance in the formalizing sectors.

In the next chapter, Rutgerd Boelens, Rocio Bustamante and Hugo de Vos discuss the interface between indigenous and formal water rights (permit systems) in Andean societies in Latin America. Evidence from a number of cases highlights the problematic 'politics of recognition' and the need for critical analysis of the power relations underpinning both legal systems.

Chapter 7, authored by Abraham Mehari, Frank van Steenbergen and Bart Schultz, compares indigenous spate irrigation arrangements in Eritrea, Yemen and Pakistan. The authors document the fair and well-enforced rules through which farmer groups have made optimal use of highly variable floods for centuries. The need for the public sector to build upon these strengths of community-based water laws is illustrated.

The first of the chapters that focus on African countries, Chapter 8, analyses the intricate collective arrangements for wise wetland use in west Ethiopia and the historically evolving interface with external landlords and government agencies. In this chapter, Alan Dixon and Adrian Wood identify effective fallback authority for rule enforcement as the greatest strength brought about by past external rulers and government, but this role is declining nowadays.

In Chapter 9, Desalegn Chemeda Edossa, Seleshi Bekele Awulachew, Regassa Ensermu

Namara, Mukand Singh Babel and Ashim Das Gupta provide a detailed analysis of the the *gadaa* system. This traditional age- and gender-based socio-political system of the Boran in South Ethiopia also influences community-based water laws, in particular conflict resolution. The authors recommend government to build upon, instead of weakening, the gadaa system.

In Chapter 10, Albert Mumma analyses the implications of Kenya's new Water Act of 2002 for the rural poor. This centralized law fails to recognize pluralistic legal frameworks. Examples include the requirements for permits for water use, which are open only for those with formal land title, so excluding the majority living under customary land tenure. Water service providers, including informal self-help groups, are required to formalize as businesses. Hence, the author expects limited effectiveness of the Act in meeting the needs of the poor.

Chapter 11, by Leah Onyango, Brent Swallow, Jessica L. Roy and Ruth Meinzen-Dick, discusses the variation in Kenya's water and land rights regimes, including women's rights, as a result of pre-colonial, colonial, and post-colonial land and water policies. Focusing on the Nyando basin, seven different land tenure systems are distinguished and documented, each with specific water rights and with varying influences of customary arrangements.

In Chapter 12, Brent Swallow, Leah Onyango and Ruth Meinzen-Dick focus on poverty trends and three pathways of irrigation and related water resources management arrangements in the same Nyando basin in Kenya. They analyse how recent state withdrawal in the top-down planning scheme led to scheme collapse and poverty aggravation. Schemes served by the centralized agency partially continued and poverty remained relatively stable, while poverty increased slowly in areas with unregulated irrigation in mixed farming.

Chapter 13, by Anne Ferguson and Wapulumuka Mulwafu, analyses the history of irrigation development and also the ongoing irrigation management transfer in two schemes in Malawi, which contributed significantly to livelihoods. Lack of clarity on new responsibilities and lack of training open the door for customary arrangements to resurface and for local elites to capture land and water resources.

Chapter 14 turns to Tanzania. Bruce Lankford and Willie Mwaruvanda elaborate a legal infrastructural framework for catchment apportionment for upstream–downstream water sharing that combines Tanzania's formal water rights system with local informal rights in the Upper Great Ruaha catchment. Various technical designs of intake structures are discussed to identify the design for proportional sharing that best fits the hydrology, users' local, transparent and fair water sharing, and also the implementation of formal water rights.

In the last chapter, Zimbabwe's customary legal systems for basic domestic and productive water uses are compared with the history of formal water law by Bill Derman, Anne Hellum, Emmanuel Manzungu, Pinimidzai Sithole and Rose Machiridza. As argued, the livelihood orientation of customary arrangements aligns well with the priority right for 'primary water uses' in national law and also with the expanding definitions of the human right to water at global levels.

Key Messages

With such a large number of chapters covering so many aspects of water resources management and so large a geographic area, exhaustive systematic comparison on the issues involved is impossible. Nonetheless, there is a remarkable consistency in the findings in a number of key areas. Here, we highlight key messages emerging from an analysis of the evidence across the chapters.

Community-based laws are both robust and dynamic

Community-based water law has shown a surprising ability to both endure and adapt. These are both key attributes of any successful institution and should be considered as the basis for, rather than impediments to, additional change and improvement. Centuries-old knowledge and institutions that are adapted to place-specific ecological characteristics of water and other natural resources and the time-tested sustainable uses of these resources have allowed communities to survive from agriculture, often in harsh ecological environments.

Community-based laws have adapted to, and driven, changing water environments. Much innovation in water development and management has occurred entirely outside the ambit of the state and state regulation, for example as a result of growing population densities, new pumping technologies and water markets, remittances from off-farm employ-ment or new output markets. Innovation occurred not only to expand water supply but also to regulate increasing conflicts over water sharing.

Community-based laws have also adapted to the influence of the state. The penetration of the state to the local level, in particular in rural areas, is generally weak but this varies around the world. In places like China, there is substantially more connection between local and national political bodies than elsewhere. For example, in sub-Saharan Africa the 'traditional' tribal author-ities that command land, water and other natural resources often exist side by side with the decen-tralized 'modern' state represented by the upcoming elected local government (Mamdani, 1996). State influence is especially strong in settlement irrigation schemes; nevertheless, customary elements continue to some extent.

Despite their robustness and dynamism, informal arrangements in rural economies on their own may be insufficient to achieve higher standards of welfare or to cope with major adverse trends, e.g. growing population density, urbanization and out-migration, adverse markets, pandemics, animal disease, civil strife or droughts and floods.

Community-based laws have application outside the community, but with limits

Unlike the widespread assumption that commu-nity-based water law is necessarily confined to restricted territories, community-based water law also operates at larger scales. The pastoralists in sub-Saharan Africa – a familiar example – whose water use agreements with each other and settled farmers cover large areas, are also cross-ing international boundaries. Community rules can also respond to today's growing water scarcity at scales beyond the community. The spontaneous groundwater recharge movement in India, for example, is massive. Similarly, in the

face of increasing abstractions from shared streams, communities in Tanzania initiated upstream–downstream rotations based on customary intra-scheme practices. Communities' methods for negotiating with distant, powerful large-scale users and for protecting their existing and new water uses, strategically soliciting state support, are also illustrated in Bolivia.

A more general pattern of how communities may deal with larger-scale water issues has been developed by Bruns (Chapter 3). He expects communities to: (i) focus on concrete problems or 'problemsheds', especially during crises; (ii) to build strategic coalitions at wider scales based on local water allocation practices and dispute resolution processes; (iii) to seek representation, and not participation by all, in multiple forums that cover the larger scales; (iv) to welcome scientific expertise that demystifies and synthe-sizes information; and (v) to seek legal support to translate their concrete demands into terms of formal law that defends their demands with state authority.

Understanding and building on these spon-taneous problem-solving alliances at large scales are indispensable, although often not sufficient, for equitable and pro-poor public intervention in water sharing across scales.

Community-based laws have livelihoods orientations, but entrench hierarchies

Community-based water law is centred on people's immediate stake in water use. It seeks to enhance members' livelihoods in a generally fair and equitable way. The absolute priority right to water of humans and animals to quench their thirst is universal. In various places, communities also prioritize water and land uses for other domestic uses and small-scale production, even if that means providing right of way over own land or handing over land to the community for water resources development. Such norms at the most local levels on how water should be used to meet basic human needs provide holistic and humane guidance for the current efforts to better define the human right to water at the highest level: the United Nations.

Notions of fairness and equity are also mani-fest in the widespread norm that those construct-ing and installing infrastructure and contributing to its maintenance in cash and kind have the strongest, although not always exclusive, rights to the water conveyed. This principle for establishing 'hydraulic property' ensures security for the fruits of investments. Also, sharing of water and its benefits under growing competition is often proportional. As water becomes scarce, each user takes a smaller share rather than some maintaining their shares while others get nothing. Norms and practices to prevent pollution also exist in community-based laws.

'Localized principles used to manage water and mitigate conflict could also provide valuable lessons for those dealing with water at the international level' (Wolf, 2000, cited in Chapter 2) – or, we would argue, at any level. This is not to say that community-based systems on their own are the best solution to the problems of water governance at all scales. Nor is it to say that some principles of different communities will not clash as the scale of the problem expands. However, also in such conditions, informal laws will be a sound basis from which to search for new possible solutions.

One significant drawback of community-based water law is that *every* community is both heterogeneous and hierarchical. Customary practices entrench gender, age, ethnicity and class differences. This is in sharp contrast to the goals, if not always the practice, of most modern states. Gender inequities are particularly pronounced. In many traditions, water governance for productive uses is strictly a male domain, excluding women from access to technologies and construction of water supplies. This handicaps women not only in using water for own productive uses, but also in meeting the disproportionate burdens of fetching water for daily domestic use that society relegates to women. The public sector has a critical role to play in removing such inequities by targeting policies and other checks and balances.

Community-based laws both confound and assist enforcement and incentives

Rule setting and enforcement are the Achilles heel of any (water) law, and community-based water laws have both strengths and weaknesses in this regard. One weakness of communities' livelihood orientation is that this also makes it morally more difficult to hold other water users, relatives and neighbours, accountable to restricting water use for livelihoods or to use the sanction of cutting water delivery to enforce agreed obligations, such as tariff payment or maintenance contributions.

However, the problem of hierarchy in community-based water law can become an advantage for law enforcement. In many instances, authoritative bodies dominated by older men and nested in multi-scale authority structures are feared but accepted because of their power to enforce behaviour in the common interest with limited transaction costs. In other cases, government can provide a useful additional influence. It will often be the case that the mere presence of such authority will be sufficient to ensure compliance, allowing the principle of subsidiarity – the devolution of decision making to the level closest to the resource – to work most of the time. Obviously, in order to meet equity goals and reduce transaction costs simultaneously, the challenge is to develop institutional devices to that end that are *not* based on gender, age or ethnic discrimination.

A clear strength of many cases of community-based water law is the crafting of the right incentive structures for those who deploy most of the effort in the common interest. For tasks like ditch watching, policing, operating infrastructure, maintenance or revenue collection, rewards are provided, even though they often remain modest. These rewards are made dependent upon the performance of the tasks.

Another advantage of community-based law is that rules are defined in terms that match the physical characteristics of water resources. Local norms related to water tend to be principles rather than rules, subject to recurring negotiation according to the ever-changing local conditions of this fugitive and variable resource. Even for spate irrigation, which captures highly unpredictable and variable floods coming from the mountain slopes, communities across countries have developed robust rules accommodating this variability.

Water permit systems and other regulations have eroded the advantages of community-based laws

One IWRM measure that risks eroding the strengths of community-based water law most directly is the promotion of permit systems. Strengthening permit systems as the single formal entitlement to water, and obliging rights holders under other water rights regimes to convert to permit systems, risks serving the same purpose for which this legal device was introduced by the colonial powers, at least in Latin America and sub-Saharan Africa: dispossession of existing prior claims to water by informal users. Community-based water law intrinsically differs from permit systems. For example, in community-based water law, water is seen as a common property resource that is to be shared, while permits stipulate individual volume-based use rights to state-owned water. It is naive to suppose that one legal system can simply be replaced by another. Moreover, vesting formal rights on the mere basis of an administrative act implicitly favours those proficient in and connected to administration.

Conditions attached to permits, e.g. formal land title or expensive registration requirements, may discriminate explicitly. Forcing permits on rural communities destroys precious social capital, creates the tragedy of the commons and favours the administration-proficient at the expense of all others, most of all poor women. A solution that is sometimes proposed is to allocate permits to collectivities, but this faces problems of defining the 'collectivity', ensuring genuine representation without elite capture and avoiding the 'freezing' of the dynamism of local arrangements. The challenge is to recognize the coexistence of plural legal entitlement systems without burden of proof.

Permits are often also expected to serve as vehicles to impose obligations on water users, for example for taxation or for imposing caps on resource use. This may work if well-resourced water departments target a limited number of formal large-scale users, but enforcing conditions on multitudes of informal water users appears unrealistic. Fiscal and other state measures or indirect measures are often more appropriate.

Regulation under the banner of IWRM can also harm informal rural communities or local entrepreneurs otherwise. Requiring sophisticated business plans for rural communities' self-help water supply risks further undermining well-functioning informal arrangements and depriving the poorest communities of indispensable financial and technical support. Water quality standards may have similar drawbacks. In these ways, regulatory IWRM measures seriously risk aggravating poverty and polarizing gender inequities.

Opportunities for taking the best from community-based law have been missed

Hard-wiring alien water-sharing rules

The key message emerging from another set of public sector interventions is that they meet communities' needs, but only partially, because critical components – either combinations of technologies and institutions or institutions on their own – fail to match communities' arrangements. Examples include state-supported improvements of intake structures for river abstraction or head works for spate irrigation. They have often succeeded in alleviating the labour required for the repeated rebuilding of traditional structures that typically wash away with strong flooding. However, these technical designs tended to hard-wire sharing rules that deviated from locally prevailing norms. This introduced new inequities. By building upon community-based and community-endorsed rules for sharing, benefits can be considerably enhanced.

Participatory irrigation management

A major missed opportunity in the past decade, leading to scheme deterioration and poverty aggravation, concerned participatory irrigation management. The institutional 'design' underpinning this move towards greater users' participation often boiled down to the assumption that it is enough to bring water users together in associations, often on paper only, irrespective of profoundly opposite interests. These newly created user associations were supposed to swiftly take over former state functions and tasks, including rule setting for water allocation, authority and enforcement, conflict resolution – e.g. between head- and tail-enders – and the creation of incentives for those who were supposed to take up, preferably on a voluntary basis, the hard work of operation, maintenance, cost recovery or conflict resolution. Especially in settlement schemes where state influence had been strongest and the numbers of small farmers largest, schemes have entirely collapsed and poverty been aggravated when government withdrew.

In other cases, the users' spontaneous participation was discouraged. State support can match local initiative well when states provide for bulk water supplies through main pipes or canals, while local users take responsibility for the connections to houses or fields. Yet, even when the latter occurred on the users' own initiative, the state can discourage this and impose newly built public distribution networks up to the field level instead.

In sum, there is a dire need for technical and institutional designs that match both farmers' initiatives and public sector abilities for construction, rehabilitation and co-management of smallholder irrigation schemes. Providing some form of fall-back authority may be the main role for the state that farmers ask for.

Basin institutions

The establishment of basin institutions is another ingredient of IWRM that risks missing important opportunities by discarding community-based water law. These costly new institutions are supposed to allow for integrated planning and implementation but they take up functions that local government, other spheres of government or communities themselves, also at larger scales, can also do and, often, more effectively.

Basin institutions entrench the bureaucratic distinction between water for productive uses – to be managed as core IWRM by basin institutions – and water for domestic uses to be left to local government or the private sector. This distinction complicates service delivery that takes people's multiple water needs from multiple sources as a starting point.

Last but not least, basin institutions are following hydrological boundaries instead of administrative boundaries, because the sharing of limited water resources in one particular basin is assumed to be the key task. Yet, water resources are often abundant but underdeveloped, in particular in sub-Saharan Africa, where less than 4% of water resources have

been developed (African Development Bank Group, 2007). For enhancing year-round storage and conveyance structures, there has rarely been any need for new, fully fledged basin institutions.

Gender

Opportunities have also been missed with regard to redressing customary inequities. Instead of reducing hierarchies intrinsic to community-based water law, public sector intervention has often reproduced or even polarized hierarchies. One example is the weakening of women's land rights in matrilineal societies during the allocation of irrigation plots. Generally, in land and water titling, women's secondary rights in the bundle of customary rights are ignored by concentrating all rights of the bundle of resource rights in men. Customary rights of way to streams, springs and other water points may also be weakened in this way. Effective targeting approaches and public sector checks and balances are still to be implemented consistently to meet the constitutional requirement of ending gender-, age- and ethnicity-based discrimination.

We can get the technologies and institutions right

This volume also documents fruitful and replicable public action in which the strengths of both community-based water law and public sector intervention are combined and lead to improved welfare and productivity, also among poor women and men. Government and other external agencies can play a direct and indirect role in enhancing access to technologies for multiple purposes year-round by providing technical support and smart subsidies or loans and by improving technical knowledge. One way to do so is by appointing engineers for advising water users and local government. This volume also entails various examples of appropriate institutional designs for rule setting and enforcement at low transaction costs and for incentive structures that ensure performance-related reward for those carrying out the legwork of collective action. Furthermore, beneficial use of water is fostered by simultaneously addressing other factors that are important for realizing the benefits of water use, including training, inputs, markets and health education.

Conclusion

Community-based water law in Latin America, sub-Saharan Africa and Asia is a precious social capital with many strengths: robust resource use is adapted to the locality; rules are dynamic and responsive to new opportunities but communities also consciously and proactively address upcoming problems at both local and larger scales; it is livelihood oriented, although hierarchical (and the latter may partially serve the goal of rule enforcement); it has nested structures for conflict resolution through rules that match notions of fairness and the physical characteristics of water resources. These characteristics are well in line with public-sector goals of enhancing well-being and productivity in rural areas, in particular among the poor. However, community-based water law is largely ignored by officialdom and professionals.

By empirically analyzing the interface between community-based water law and current IWRM measures, this volume also identifies fields in which the public sector can play an important complementary role or should, in any case, avoid eroding the strengths of community-based water law or missing opportunities to build upon those. The public sector is critical in removing gender, age and ethnic biases and in legally and factually protecting communities' small-scale water uses. Where water is still underdeveloped, the most effective way the state can develop it is by reverting to its conventional role as investor in infrastructure, but now in a genuinely participatory, inclusive and gender-equitable mode of co-development and co-management that, yet, reduces transaction costs and provides incentives. Once water resources are fully developed, equitable water allocation needs to be negotiated at larger scales.

If this volume succeeds in conveying the need for such a new vision on the role of the state in water resource management for poverty alleviation and agricultural and economic growth in developing countries and in provoking thought on how endeavours to reform the water resource management reform can be successful, it will have served its purpose.

References

African Development Bank Group (2007) *World Water Day 2007: Coping with Water Scarcity.* Press Release. African Development Bank Group, Tunis.

Bakker, M., Barker, R., Meinzen-Dick, R. and Konradsen, F. (1999) *Multiple Uses of Water in Irrigated Areas: a Case Study from Sri Lanka.* SWIM Paper 8, International Water Management Institute, Colombo, Sri Lanka.

Boelens, R. and Dávila, G. (1998) *Searching for Equity. Conceptions of Justice and Equity in Peasant Irrigation.* Van Gorcum and Co., Assen, Netherlands.

Bruns, B. and Meinzen-Dick, R. (2000) *Negotiating Water Rights.* SAGE, New Delhi, India.

Mamdani, M. (1996) *Citizen and Subject. Contemporary Africa and the Legacy of Late Colonialism.* Princeton Studies in Culture/Power/History, University Press, Princeton, New Jersey.

Moriarty, P., Butterworth, J. and van Koppen, B. (2004) *Beyond Domestic. Case Studies on Poverty and Productive Uses of Water at the Household Level.* IRC Technical Papers Series 41, IRC, NRI and IWMI, Delft, Netherlands.

Ostrom, E. (1994) *Neither Market nor State: Governance of Common-pool Resources in the Twenty-First Century.* IFPRI Lecture Series, International Food Policy Research Institute, Washington, DC.

Ramazzotti, M. (1996) *Readings in African Customary Water Law.* FAO Legislative Study 58, Food and Agriculture Organization of the United Nations, Rome.

Sarpong, G.A. (undated) *Customary Water Law and Practices: Ghana.* www.iucn.org/themes/law/pdfdocuments/LN190805_Ghana.pdf

Shah, T. (1993) *Ground Water Markets and Irrigation Development. Political Economy and Practical Policy.* Oxford University Press, Mumbai, India.

Shah, T. and van Koppen, B. (2006) Is India ripe for integrated water resources management (IWRM)? Fitting water policy to national development context. *Economic and Political Weekly* XLI (31), 3413–3421, India, 5–11 August 2006.

Van Koppen, B., Moriarty, P. and Boelee, E. (2006) *Multiple-use Water Services to Advance the Millennium Development Goals.* IWMI Research Report 98, International Water Management Institute, Challenge Program on Water and Food, and International Water and Sanitation Center (IRC), Colombo, Sri Lanka.

Von Benda-Beckmann, K. (1991) Development, law and gender skewing: an examination of the impact of development on the socio-legal position of women in Indonesia, with special reference to the Minangkabau. In: LaPrairie, C. and Els Baerends, E. (eds). The socio-legal position of women in changing society. *Journal of Legal Pluralism and Unofficial Law* 30/31, 1990–1991. Foundation for the Journal of Legal Pluralism, Groningen, Netherlands.

Wolf, A.T. (2000) Indigenous approaches to water conflict negotiations and implications for international waters. *International Negotiation* 5 (2), 357–373.

Yoder, R. (1994) *Locally Managed Irrigation Systems: Essential Tasks and Implications for Assistance, Management Transfer and Turnover Programs.* Monograph No. 3, International Irrigation Management Institute, Colombo, Sri Lanka.

2 Understanding Legal Pluralism in Water and Land Rights: Lessons from Africa and Asia

Ruth Meinzen-Dick[1] and Leticia Nkonya[2]

[1]*International Food Policy Research Institute (IFPRI), Washington, DC, USA; e-mail: r.meinzen-dick@cgiar.org;* [2]*Department of Sociology, Anthropology and Social Work, Kansas State University, Manhattan, Kansas, USA; e-mail: lenkonya@ksu.edu*

Abstract

Water rights, like the underlying resource itself, are fluid and changing; they necessarily connect people and they can derive from many sources. Much of the property rights literature has focused on rights to land but, as water rights are now receiving increasing attention from scholars and policy makers in developing countries, it is useful to examine the differences and similarities between land and water rights – as well as the linkages between the two. Without an understanding of the range and complexity of existing institutions that shape water use, efforts to improve water allocations may be ineffective or even have the opposite effects from those intended in terms of efficiency, environment, equity, empowerment and conflict reduction. Reforms need to carefully consider the range of options available. This chapter reviews the multiple sources and types of water rights and the links between land and water rights, using examples from Africa and Asia. It then examines the implications for conflict and for water rights reform processes.

Keywords: water rights, land tenure, legal pluralism, customary law, conflict management, Africa, Asia.

Introduction

Two images are often associated with the term 'property rights': (i) fixed stone walls – immobile, permanent and restricting access to the resource; or (ii) a title deed – a piece of paper with a big seal affixed in a government office. Neither of these images, deriving from the European tradition on land, is very helpful in understanding water rights, particularly in Africa and Asia. Water rights, like the underlying resource itself, are fluid and changing; they necessarily connect people and they can derive from many sources besides the government. As water rights are now receiving increasing atten-tion from scholars and policy makers in developing countries, it is useful to examine the differences and similarities between land and water rights – as well as the linkages between the two.

A starting point for this analysis is to consider why property rights matter, and why attention to water rights has lagged behind attention to land rights. Reasons given for attention to property rights are often addressed under four 'E's and a 'C': efficiency, environment, equity, empowerment and conflict reduction.

- In terms of efficiency, the arguments are often made that secure property rights are needed to provide incentives to invest in a resource. For water, this often means developing and maintaining the infrastructure, such as a well or an irrigation canal.

• Environmental arguments are closely related: property rights provide an incentive to protect the resource and, without property rights that are enforced, resources often become degraded.

• Equity relates to the distribution of the resource, and can be defined in terms of equality of access, particularly for meeting basic needs, or in terms of distribution of rights in proportion to the investments that people make, or some combination thereof. The way rights are defined determines whether people are included or excluded in the control of a vital resource for their lives. Holding property rights is thus empowering to individuals or groups, particularly control rights that recognize authority over how the resource is managed.

• Clearly defined rights are also held to reduce conflicts over resources during scarcity, which is a matter of growing concern with discussions of 'water wars'.[1]

Given this importance of property rights and of water, why has there not been more attention given to rights over water? The induced innovation hypotheses argue that establishing effective property rights is costly so, as long as a resource is abundant, there is little incentive or need to define rights over it but, with increasing demands and scarcity, there is pressure to define rights (Alchian and Demsetz, 1973). This is seen in African history, where 'frontier' areas with low population densities have generally had more loosely defined land rights than areas of high population densities and, as populations increase, land rights become more specific (Besley, 1995; Otsuka and Place, 2001).

However, while changes in land tenure institutions are more familiar, studied and debated, changes in water tenure have received less attention. Nevertheless, we also see that where water is plentiful, people often do not even know or care who else may be sharing the same river, lake or aquifer. As populations grow, demands on water rise, for household use, agriculture and industry. Those who use water are increasingly affected by the actions of other people. Coordination becomes more complex

and more crucial. In one way or another, water rights institutions, and expectations about what claims to water are socially accepted as legitimate, are constituted by such competition, influencing people's ability to obtain water.

However, water has several properties, meaning that water rights cannot be determined in exactly the same way as rights to land and other resources. Water is mobile, and most water use depends on flows. After water is diverted, some evaporates or is transpired by plants, but much water also runs back through surface channels and aquifers to be reused further downstream. Cultivation of crops, planting or cutting of trees, and other changes in land use transform the quantity and timing of water flows into and out of aquifers and rivers. While much land is dedicated to a single use, almost all water has multiple overlapping uses and users. All uses not only withdraw some water, but also add something to the water that affects the quality for users downstream, and changes in water flows affect not only human uses but also animals and the broader environment. Rights to water and the consequent patterns of use concern not only how much water is withdrawn but also water quality and the environment.

The slippery nature of water itself makes it more difficult to define water rights because of the need for so much specificity: *who* can use *how much* water from *what source*, *when* and for *what purpose*? This specificity, in turn, combined with the fugitive nature of the resource itself, increases the costs of monitoring and enforcing water laws. Instead of establishing rights once and for all, effective water rights require active management of the resource and attention to many different aspects of its use, including quality and quantity, in different places and times.

Improvements in water rights institutions can help reduce poverty, improve economic productivity and protect nature, but these lofty goals are often not achieved. Efforts to improve water allocations may be ineffective or even have the opposite effects from those intended. In this chapter, we argue that to be effective, reforms need to be grounded in a good understanding of social institutions that shape rights to water; additionally, a careful assessment of the options available for improving water

management should be made, and a willingness shown by those involved to experiment,
adapt and learn from experience. The diversity
of culture, environment, economic activities
and other conditions means there is no one best
way to improve water rights and water allocation institutions. The best route to better water
management depends on where you are starting from, with many pathways available (Bruns
and Meinzen-Dick, 2005).

From this standpoint, the increasing attention to water rights in Asia and Africa is very
encouraging, particularly those efforts that seek
to address the intricacy of rights over this
complex resource. The remainder of this chapter examines some of these complexities, and
lessons that can be drawn, not only for water
governance in those regions but for other
regions and other resources as well. We first
review the multiple sources and types of water
rights and the links between land and water
rights, before examining the implications for
conflict and water rights reform processes. Most
of the emphasis in the chapter is on how water
rights affect people, and hence we focus on the
local level, but the concluding section on reform
processes also addresses water rights at larger
levels.

Legal Pluralism in Water Rights

Property rights can be defined as: 'the claims,
entitlements and related obligations among
people regarding the use and disposition of a
scarce resource' (Furubotn and Pejovich,
1972). Bromley (1992, p. 4) points out that:
'Rights have no meaning without correlated
duties … on aspiring users to refrain from use.'
This means that property rights are not a
relationship between a person and a thing
but are social relationships between people
with relation to some object (the property).
Particularly in the case of water, rights also have
corresponding duties that apply to the holder of
those rights – usually to use the water and
dispose of wastes in a certain manner, and
often to provide money, labour or other
resources in maintaining the water supply.

The crucial point is that property rights are
effective (legitimized) only if there is some kind
of institution to back them up. In many cases,

the state is a primary institution that backs up
property rights, but this is not necessarily the
case. Irrigation or other water development
projects generate their own rules and regulations, which constitute yet another type of
'water law'. Most religions also have precepts
relating to water that can provide the basis for
entitlements or obligations regarding water.
Particularly in the case of water rights, we find
many examples of customary law (which
changes over time) that is backed by local
authority and social norms. User groups may
define their own rules for a water point.

At the other end of the scale, international
treaties such as the Ramsar convention generate yet another type of law that can provide a
basis for placing claims on water resources, e.g.
to prevent wetlands from being developed.
Particularly in Africa, where so many countries
share in international river basins, treaties and
other international laws are relevant to the allocation of these shared waters.

The pluralism of water law is further
increased because each of these types of law –
especially state, customary and religious – may
itself be plural. Government land laws often
contradict water acts. Many communities have
different ethnic groups living side by side and
using the same water, but having different traditions regarding its use. In particular, many sites
have farmers and pastoral groups, with different ways of life and ideas on water. The mix of
religions adds to this plurality. All of these types
of law will be interpreted differently in different
places, generating a plethora of local laws.

These different types of water laws are not
neatly separated; rather, they overlap and influence one another. Nor are all equally powerful
– their influence will vary. Figure 2.1 illustrates
these overlapping types of law, which can be
thought of as force fields, with variable
strengths (Meinzen-Dick and Pradhan, 2002).

For example, customary law may be very
strong and state law virtually unknown or irrelevant in a remote community with low migration
and low penetration of state agencies but, in a
heterogeneous community with high migration
rates in the capital city, customary law may be
much weaker than state law (as illustrated in the
Nyando basin, Kenya, by Onyango et al.,
Chapter 11, this volume). In the case of rural
land rights in Africa, Bruce and Migot-Adholla

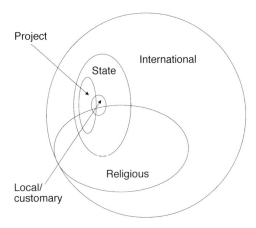

Project

State

International

Religious

Local/
customary

Fig. 2.1. Overlapping legal orders relating to water
(from Meinzen-Dick and Pradhan, 2002).

(1994) found that customary land tenure
arrangements provided just as much tenure
security as government-issued title to the
resource. Given the even higher costs of enforcing water rights (compared with land rights) and
the limitations of government agency capacity,
especially in most rural areas, customary law,
backed by local norms and community sanctions, may also be as effective as state law as a
basis for claiming water rights in many parts of
Asia and Africa.

Bundles of Rights

As with rights over land or trees, water rights are
not usually homogeneous 'ownership' rights
that permit one to do anything with the
resource, but they may rather be considered as
bundles of rights that may be held by different
parties. Indeed, because of the complex inter-
relations between these individual rights and
rights-holders, they could even be considered
as a 'web of interests' (Arnold, 2002, cited in
Hodgson, 2004). The exact definition of these
bundles varies, but they are often grouped into
two broad categories: (i) *use* rights of *access*
and *withdrawal*; and (ii) *decision-making* rights
to regulate and control water uses and users,
including the rights to *exclude* others, *manage*
the resource or *alienate* it by transferring it to
others (Schlager and Ostrom, 1992). To these
may be added the rights to *earn income* from a

resource, which Roman legal traditions have
referred to as *usufruct* rights (see also Alchian
and Demsetz, 1973). Rights to earn income
from a resource (even without using it directly)
can be separate from the use and management
of the resource, as when government departments collect revenue from water users or when
individuals or communities collect a charge
from others who use water – a factor that is
increasingly important in the context of water
transfers.

An example from Kiptegan, a spring protection site in the Nyando basin of Kenya, illustrates this:

● Because of strong local norms that no one
should be denied basic water needs, anyone
has the right to withdraw water from the
pipe below the spring for drinking.
● People may also use water for their cattle,
but only from the cattle trough, and they are
expected to help keep the trough clean.
● Those community members who have paid
some of the cost of developing the spring
protection are entitled to a higher level of
service, including, if hydrologically feasible
and they have paid for it, a piped water
supply to meet domestic needs and some
small garden uses at their homestead, and to
have a say in selecting committee members.
● The members of the committee, who provide additional time and labour, also have
decision-making or control rights, including
decisions on who can join or who is excluded
from the user group, and how the spring and
its infrastructure will be managed. They also
collect fees from the group members, but do
not earn income from this themselves.

These represent a blend of customary law,
'project law' (in the form of rules developed
with external assistance when the spring was
protected) and rules developed and modified
by the user group.

While the exact definition of these bundles of
rights varies from place to place, we find several
common elements in many water laws in Africa:

● The state generally claims some kind of ultimate 'ownership' rights over water, which
may not be felt at all at the local level, or it
may require that individuals or groups who
want to use or develop a water source need

to get some kind of permission from the state.

- There are widespread notions that anyone is entitled to water for 'primary uses', which are usually interpreted as basic domestic needs, as well as needs for household gardens, but may include other productive livelihood needs. Islamic law has formalized this as a 'right to thirst' for people and animals. Indeed, many African societies recognize water needs of animals as well as people. As one Kalengin proverb in Kenya says: 'Even the hyena is entitled to water', with the implication that no one can be denied water (Onyango et al., Chapter 11, this volume).
- While basic use rights are strong, they are also usually quite flexible. Rather than being clearly defined in terms of who can draw how much water, access rights are socially negotiated, either individually or by groups, depending on changing local circumstances (Witsenburg and Adano, 2003). In range-lands, Ngaido (1999) discusses the impor-tance of *access options* for people to use another individual's or group's land and water resources under conditions like drought, which provide a measure of resilience against ecological stress. Cleaver (1998, p. 351) reports a similar pattern for domestic water in Zimbabwe: 'As a precaution against drought, women rarely rely on one source of water but maintain access to a number of different supplies, often through reciprocal social networks. Incentives to cooperate may therefore be indirect and relate to the need to maintain good relations with neigh-bours and kin in a more general sense.'
- Control rights of management and exclusion are often held by the local chiefs, groups or individuals who developed the source. The effectiveness of these management authori-ties in setting and enforcing the rules and in maintaining the source varies greatly, as does the extent to which they are participatory or autocratic. Indeed, effectiveness and deci-sion-making practices are related. In Burkina Faso, McCarthy et al. (2004) found that where the chiefs made decisions in collab-oration with community members, rather than by themselves, there was a significantly higher cooperative capacity, which led to

better resource outcomes. Similarly, in Zimbabwe, Cleaver (1998, p. 355) reported: 'Critical decisions about the rationing of water from particular sources are only successfully enforced in those communities where the decision has been taken at a meet-ing of the whole community rather than a committee alone. Consensus may enhance collective management since it reduces the need for compulsion, monitoring, and sanction.'

- Most state, customary and religious laws do not grant alienation rights (to sell, give away or otherwise transfer one's rights to some-one else).[2] More people can be allowed in, but there is no profit to individuals in giving up their rights to water.

Broad patterns of water rights in Asia show a number of similarities, with the state claiming ownership of water. Customary and religious laws also emphasize that all people should be given water for basic domestic needs, although water use for even basic garden irrigation is often more restricted. The state or local farmer groups exercise control rights over how the irri-gation systems and their water are used. However, even within systems that have highly formalized rules, access to water is socially nego-tiated, either among communities or between communities and government agency staff (for examples, see Pradhan et al., 1997; Bruns and Meinzen-Dick, 2000). There are some informal water markets, especially for groundwater, by which those with wells can sell water to other farmers or to industries, but these are generally 'spot markets', not long-term transfer or alien-ation of the underlying water rights (Easter et al., 1998).

Types of Water Rights

As with other types of property rights, water rights can be broadly classified as public, common or private property, according to who holds the rights and, particularly, the decision-making rights of allocation, which lie at the heart of water rights (Bruns and Meinzen-Dick, 2000, 2005; Paul, 2003).

Public water rights are rights held by the state where the government allocates rights to users. Many countries adhere to some form of

Public Trust Doctrine, a principle dating back at least to Roman law, which maintains that the state holds navigable waters and certain other water resources as common heritage for the benefit of the people. Under this doctrine, control over water is an aspect of sovereignty, which the state cannot give up (Ingram and Oggins, 1992). The government can assert its rights over water by controlling the water allocation directly through government agencies, or by acting as a licensing or leasing agent for granting water rights (Paul, 2003). In Zimbabwe for example, the water reform in the 1990s declared all the water to be the property of the state.

People can get water rights by acquiring water permits, which give them legal licence to use but not own water. Water permits are issued in consideration of the needs of the applicant and the expected benefits of the proposed water use (Latham, 2000; Mtisi and Nicol, 2003). In Mozambique, the Water Act of 1991 regards water as a public good. People cannot have private ownership of water sources but can obtain rights to use water by acquiring a water licence (Vaz and Pereira, 2000). Water licences are granted for a period of 5 years and are renewable. The use of water for primary needs like small irrigation, domestic use and watering of livestock is free.

Common water rights refer to communal water rights where water can be used by people in ways that are specified by some community. For true common property, some form of community or user group should have rights to allocate water at some level, e.g. in specifying who may or may not use the water and in what ways, as seen in many farmer-managed irrigation systems in Asia (Tang, 1992). In most African customary water laws, water from natural resources is considered as a community property and private ownership of such water is not recognized (WFP, 2001).

Private property rights are rights held by an individual or legal individuals like corporations (Bruns and Meinzen-Dick, 2005). With regard to water, it is generally only use rights that are recognized for individuals, particularly permits or licences that give an individual a right to use water in certain ways (Paul, 2003). In Botswana, for example, people do not need to acquire water rights if they are using the water

for domestic purposes or for watering livestock. However, they are required to obtain water rights if they are using the water for irrigation or commercial purposes. In some cases, private rights go beyond just use rights, to include the rights to allocate the water, as in Chile's tradable water rights systems, in which a right-holder can transfer that water to others through sale or lease.

Although there are individual use rights in Africa, private water allocation rights are not widespread. There are some sources such as shallow wells or small dams that are considered private, in which the right-holder has the right to allocate water from that source. In the case of a private water source like a well, an individual is required to obtain land rights to be able to construct a well on a particular piece of land. After the well has been constructed, an individual holds the rights to both the land and water (Carlsson, 2003). Private water rights are also widely observed for groundwater in Asia, and farmers under farmer-managed irrigation systems in Nepal and Indonesia may have private rights to a share of the water in those systems.

In most treatments of property rights, these types of rights are contrasted with open-access situations in which anyone has unrestricted use of the resource. There are no specific rights assigned to anyone and no one can be excluded from using the resource. It is the lack of rules in open-access situations that is seen as contributing to the 'tragedy of the commons', wherein resources degrade because of lack of control over their use or lack of incentives for investment in their provision (Bromley, 1992).

Thus 'open access' has taken on a very negative connotation in much of the resource management literature. However, many discussions of African water rights use the term 'open access' with a positive connotation, which others might associate with the notion of human rights to water (e.g. Gleick, 1999). In African countries, the notion of free access is also applied to some rangelands, rivers and streams (FAO, 2002). Many of these notions were developed under conditions of low population densities, and may not stand up to increasing scarcity and competition. In practice, there are often some forms of restrictions on the use of the resource. It is important to address the questions of who will manage the resource, how well and why, whether they cannot exclude others and what consequences these have for the state of the land and water as they come under pressure. However, it is also important to recognize the value placed upon 'open access' to water for all, and to seek ways of accommodating this for growing populations.

Although these different property rights regimes can be distinguished analytically, in practice they often overlap. The state may claim ultimate ownership of the resource, but recognize communal rights over water in a stream and open-access primary use rights for outsiders. When that same water percolates into the water table and is accessed through a well, it may be considered the private right of the person who built the well.

South Africa provides an illustration of these overlapping property rights regimes, and how they change over time. During the apartheid era, state water law was based on the UK common law principle, which gave use and control rights over water to those who owned the overlying land. Thus, groundwater, springs and even small dams on a farm were effectively private property. However, the customary law of most black communities held that there is no private control of water, but the community leader such as the village chief had the right to control and determine the use of water resources for the benefit of the whole community (Tewari, 2002).

The new government reformed water rights through the National Water Act (Act 36 of 1998). This Act declared that the state is the guardian of all water resources in South Africa, but it also incorporated the African customary view on water rights by declaring water to be a public resource belonging to the whole nation and requiring to be available for common use by all South African citizens. All water required for basic human needs like drinking is guaranteed as a right (RSA, 1998; Perret, 2002). Under this Act, people cannot own water but can be granted water use rights through a licensing system, which requires users to pay for it. The money generated from water use charges is used for water service and management costs (Tewari, 2002; Farolfi, 2004). Individual water users are authorized to have

water use rights without any payment, registration or licensing if the water is taken for reasonable use for domestic purposes, small gardening and for animal watering. If the water is used for commercial purposes, then individuals are required to obtain a legal entitlement or licence to use water. Through the licensing system, an individual is granted water use right for a maximum of 40 years subject to renewal (Perret, 2002).

Regulations to public water rights are meant to control water use and to resolve problems that might occur as a result of overuse, and to resolve conflicts as results of competing uses. There are thus public rights to regulate the resource, collective rights of all to use water for basic needs and private individual use rights under licences.

The Relationship Between Land and Water Rights

Much of the current attention to water rights reform now is directed at investigating ways of making water rights separable from rights over land. This particularly applies to well-publicized cases in western USA, Chile and Australia, where growing demand for water for non-agricultural uses in cities and industries creates pressure to transfer water away from agriculture. However, from the point of view of many European statutory laws, water rights have been a subsidiary component of land rights (Hodgson, 2004). In much of Africa and Asia it is hard to identify the water rights because they are intrinsically linked to land. African customary land rights, in turn, depend on social relations – membership in communities or relations with land-allocating chiefs, for example. Indeed, in Ramazzotti's (1996) review of the ethnographic literature on customary water law, most of the information about water rights came from discussions of land law or the institutions of chieftaincies, demonstrating how water rights are embedded in both land tenure and social relations.

Two very different environmental conditions – wetlands and semi-arid rangelands – illustrate the linkages between land and water rights. In wetlands, control over land also gives control over water. Here, land is scarcer than water, and hence it makes sense to concentrate on the allocation of land. By contrast, in dry areas, water rights are the key to the control and use of land for pastures. Access to water points opens up the possibility of using large areas of grazing land for migratory pastoralists.[3] Enclosing a water point can make pastoral production – and even the lives of the pastoralists – unviable.

Keeping animals often overlaps with other land (and water) uses. At the more humid end of the spectrum, animals may be raised in agricultural areas, either by the farmers themselves or by pastoralist households. While there can be complementarity in resource use by letting animals graze on fallow fields, thus providing manure in exchange, there is also potential for conflict, especially where cattle must pass by or through growing fields to get to water. In the Kirindi Oya irrigation system in Sri Lanka, the irrigation development displaced pastoralists from land and did not provide enough alternative watering points for the cattle. Although the cattle farmers' association was included in irrigation Project Management Committee meetings to address cattle damages to crops as they walked through the system to get water, they were not included in the decision making about water allocation to ensure that their needs were met (Meinzen-Dick and Bakker, 2001).

At the drier end of the spectrum, there are important overlapping uses between pastoralists and wildlife that are particularly important in Africa. The interactions between humans, livestock and wildlife have often been studied in terms of land, particularly where parks or reserves are created for wildlife, excluding the people and their animals, but the interactions and even conflicts are often over water, particularly where tourism is developed entailing consumption of large amounts of water, or where fences are used to prevent people from accessing water points, thus denying basic needs.

The principles of interconnected land and water rights are important in understanding both wetlands and drylands, which are particularly important resources in Africa. But even in irrigation systems, land rights are key to obtaining water. There are clearly demarcated areas of land entitled to receive irrigation water. In South India, for example, land is even classified according to whether it is supposed to receive one season of irrigation per year or two, and land values and taxation rates differ accordingly. The development of many irrigation projects has also disrupted land tenure arrangements by expropriating the land to be irrigated, and then reassigning plots in the new system.

This is illustrated in van Koppen's (2000) study of the development of irrigation systems on *bas fonds* (wetlands) in Burkina Faso: women had held relatively strong use, decision making and even full ownership rights over the bas fond, where they cultivated rice. However, the project initially ignored the fact that women were the landholders, and assigned 'household' plots to the male heads of households, thereby weakening women's rights – an example of project law and customary law clashing. The result was a fall in productivity despite the 'improvement' of the technical infrastructure, because the underlying institutions – including not only property rights but also intra-household relations – were disrupted. Later sites under the project corrected this by involving the women in the land allocation.

In other cases of irrigation development, the state has expropriated all land in the area to be irrigated, and then reassigned (often smaller) plots within the irrigation system, as in Kenya, Malawi and Zimbabwe, for example. The result may be stronger water rights, but weaker land tenure security, as the farmers cultivating irrigated plots often shift from holding relatively strong customary use rights to their land to being 'tenants' on government land, and subject to the threat of eviction for failure to cultivate in prescribed ways, which often include growing specified crops.

Thus, farmers not only lose many decision-making rights over their land but also face uncertainty about the duration of their rights. And, because they often cannot transfer or sell their land in the irrigation scheme, they do not benefit from any improvements. This contrasts with the situation in much of Asia, where farmers generally have ownership rights to land within irrigation schemes, which provide for much greater security of tenure and a long-term view of irrigated production.

Even where land and water are not strongly connected for productive purposes (as they are for cultivation or herding), there are vital links between land and water rights. In Kenya, for example, there are strong norms specifying that everyone has rights to use water. However, much of the land has been privatized (Onyango et al., Chapter 11, this volume). In the Nyando basin, land-buying companies bought land from large-scale white farmers, subdivided and sold all of the land to smallholders, without regard for the slope or location of the plots relative to water. While no one should be denied water, it was not incumbent upon landowners to allow people and their animals to cross their private land to access the water. The result was that many people had no access to the springs or rivers, and hence could not get water, even for basic domestic needs. The few public access points, such as bridges, became overused. Moreover, communities faced considerable obstacles to developing water sources if they could not control the land, as well.

In the Kiptegan site referred to above, the spring development that benefited the whole community was only possible when, after discussions with World Agroforestry Center and government staff, several men with land surrounding a spring decided to devote that land to the spring protection, planting indigenous trees above it and setting aside an area in which people and cattle could (separately) access the water (Leah Onyango and Brent Swallow, Nairobi, 2004, personal communication).

This spring protection offers a positive example of how the way in which land is used has a major impact on both the quality and the quantity of water resources, and thus on water rights. Unfortunately, negative examples come to mind more readily: (i) cattle tracks or cultivation of hillsides contributing to soil erosion and hence lower water quality and silting up of reservoirs; (ii) pesticide use on farms polluting the streams and groundwater; or (iii) deforestation or reforestation affecting the run-off rates. This linkage between land and water in hydrological units lies at the heart of watershed management programmes. Swallow et al. (2001) point out that these relations are complex, and not all land is equally influential in this: there are particular types of land uses, including wetlands, riverine vegetation and paddy fields, that play critical roles as sinks or filters for water, sediment and other flows. Unfortunately, the property rights to riverine

vegetation and wetlands are often not clearly defined, nor are they under the effective control of a management entity that seeks to protect or enhance their watershed functions.

Alongside the burgeoning number of watershed management projects supported by governments and NGOs, land and water rights are being increasingly separated. This is in part fuelled by government structures: land and water are specified in different statutes and administered by different government agencies. Even international and donor organizations recommending policies for land tenure often neglect to mention water, and vice versa. There are also fundamental differences in the conceptualization of land and water rights, with state law often treating land rights in the abstract, without regard for location or topography (as exemplified by the land-buying companies in Kenya). Water rights, by contrast, are usually defined in terms of location, time and use. In reviewing both the functional linkages between land and water, and these divergences, Hodgson (2004) finds that: 'Few formal mechanisms exist in law to ensure a coordinated approach to the allocation and administration of land tenure rights and water rights.'

The growing trend toward integrated water resources management (IWRM) tries to link land and water management in overcoming the divide created by assigning authority over land and water to different government agencies. There are hopeful signs. Kenya's current land tenure and water rights reforms are taking place in parallel, but officials involved in the two processes are at least consulting one another. However, for the integration of land and water rights, state law and institutions may not be the best starting point. Rather, it is useful to look at the ways in which land and water rights and management have been linked in a range of customary institutions, and seek to identify principles upon which appropriate land and water rights linkages can be built.

Water, Rights and Conflict

Based on property rights theory and experiences with land, it would seem that clearly defined property rights – which, by definition, create shared expectations – would help reduce conflict over water, particularly as it becomes scarcer. This notion provides part of the impetus for water rights reforms and formalization (Rogers and Hall, 2003). However logical it may be, it is not necessarily true. When a fixed expectation comes up against a fluctuating resource, that in itself can be a source of conflict. This may explain why customary water rights are so often ambiguous.

In a collection of studies of water conflict in Nepal and India (Benda-Beckmann *et al.*, 1997), a recurring theme is that local norms, which form the basis for claiming water rights, are principles rather than precise rules, subject to recurring negotiation. Indeed, in many of these cases it was the attempts to formalize rights that often triggered conflict, rather than the use of the water itself. The same was found along the Tana river in Kenya, where a government land adjudication programme triggered violence between Pokomo farmers and Orma pastoralists, who had historically shared the resource under more flexible tenure arrangements (Weiss, 2004).

That ambiguous or flexible rules are particularly adapted to situations where the resource is very variable is seen in a study from Marsabit, a dry pastoral area in Northern Kenya. Although there has been recurrent violence and raiding between the different ethnic groups in the area, and both claimed rights to the water points based on different customary principles, Witsenburg and Adano (2003, p. 11) found that conflicts actually *decreased*, rather than increased, during drought because:

> Both ethnic groups claim ownership of the well site, but they both said that the other group had a legitimate claim as well, which they consider in crisis times of drought. Samburu/Rendille herdsmen said that the Boran have a rightful claim, because they have invested time, money and labour to develop the wells, whereas the Boran admit that the Samburu/Rendille have a rightful claim based on their history, having used this water site long before the Boran migrated from Ethiopia in the 1920s … many [violent] incidents take place at well sites, though not because they want to capture the well or to fight for access to the well. If they would really like to use the well, they would approach the other group peacefully. Instead, they fight at well sites because these are profitable places to raid when there is a concentration of people and animals …

situations of drought and hunger, as in 2000, are different from other situations: they now have a common enemy to fight.

Thus, recognition of the two groups' interrelationships and common need for water mitigates conflict over this vital resource.

Studies from Zimbabwe (Cleaver, 1998; Chikozho and Latham, 2005) have similarly found that customary water rights place a high value on conciliation and conflict avoidance. Although there may be rules governing the use of water, there is a reluctance to punish rulebreakers. 'Approximate compliance' is accepted, taking into consideration hardship circumstances of the rule-breakers. This is similar to *adat* (customary law) in Indonesia, which considers the intention behind an action as being as important as the act itself when meting out sanctions (Ambler, 1998). Meinzen-Dick and Bakker (2001) also found in Sri Lanka that communities allowed people to use water in ways that were against official government regulations when 'they need it and there is no other source'.

Aaron Wolf (2000) suggests that localized principles used to manage water and mitigate conflict could also provide valuable lessons for those dealing with water at the international level. Based on a study of the Berbers in Morocco and Bedouin in Israel, he suggests that principles such as prioritizing uses and protecting downstream and minority rights can be applied to international waters as well. From our examination of these cases we can suggest an additional principle to draw upon: the value placed on mutual survival, because people recognize that misfortunes that befall others today may affect themselves tomorrow. This leads to a sense that, especially in times of drought, there is a common enemy that competing users should cooperate to overcome.

Implications for Water Rights Reform Processes

Many countries in Africa have been, or still are, engaged in a variety of land tenure reform processes. Now, due to a range of internal and external pressures, many in Africa and Asia are also embarking on water rights reforms. Comparing the impetus between land and water rights reforms, Hodgson (2004, p. 30) finds: 'The concerns of water rights reform, scarcity and sustainability, are quite absent from the land reform debate.' But, on the other hand: 'Generally speaking, water rights reforms have had fewer re-distributive or socio-economic objectives than reforms to land tenure rights. An exception is South Africa whose recently enacted Water Act seeks to implement the two key principles of the 1997 National Water Policy, "sustainability" and "equity"' (Hodgson, 2004, p. 28).

Many land tenure reform programmes (e.g. Kenya's Swynnerton Plan (Swynnerton, 1954)) have imposed Western-style private property with cadastres and title. However, experience has shown problems with this approach in terms of the high costs and potential to exclude many people. Research on customary tenure (particularly in Africa) has also found that customary systems do not necessarily create tenure insecurity that limits investment (Bruce and Migot-Adholla, 1994). Consequently, new donor and government plans take more nuanced approaches, starting with more attention to existing land tenure (e.g. EU, 2004).

Even de Soto, a well-known advocate of land titling and privatization programmes, argues that it is essential to understand the customary rules and social contracts ('people's law') that are already in place before implementing any major reforms:

> Outside the West, extralegal social contracts prevail for a good reason: they have managed much better than formal law to build on the actual consensus between people about how their assets ought to be governed. Any attempt to create a unified property system that does not take into account the collective contracts that underpin existing property arrangements will crash into the very roots of the rights most people rely on for holding onto their assets.
>
> (de Soto, 2000, p. 171)

If that applies to land rights, it is even truer for water.

Yet, water-reform processes are often dominated by (statutory) legal scholars and/or hydrologists, and have not always started with a thorough understanding of existing water rights and governance systems. Programmes of formalizing, registering and individualizing water rights run the risk of creating 'cadastre disasters', unless they learn from the experience of land tenure reforms and take into account the range of existing water rights. In the remainder of this section, we examine how an improved understanding of the complexity of existing (pluralistic) water rights could contribute to effective reforms, and how the experiences from land and water rights reforms might inform one another.

It may yet be that the property rights school will be proved right, and rising demands and competition for limited water resources will prompt formalization of water rights in Asia and Africa. These changes are taking place in land rights, both through state and external intervention and endogenously through changes in the customary law itself (Otsuka and Place, 2001). With rising populations and growing per capita water consumption – for domestic uses, intensification of agriculture and industrialization – water uses and users are becoming even more interconnected, not just at the local level where face-to-face negotiations are possible, but over large distances, from rural areas to cities, and even across national boundaries. For example, in the Mara-Serengeti basin of Kenya and Tanzania, agricultural development in the upstream areas is affecting the quantity and quality of water available for the pastoralists and wildlife further down, a factor compounded by increasing tourism, which also creates high water demands. Some form of new institutional arrangements is called for to regulate or reconcile these competing demands.

Existing customary institutions are likely to be inadequate where the competing users are from different ethnic or religious backgrounds and where they do not share the same norms and customs. Thus, the emerging water law is likely to be based on state institutions. When the competing users do not even share the same government, then creating some form of international institution is often suggested. But, as these decision-making and regulatory bodies move away from the institutions, based on social relations in which much customary water law is currently embedded, the users affected are likely to have less direct say in the decision making. Just as importantly, they are likely to identify less with the other water users with whom they share the resource, or to understand and respect each other's needs. The lower influence on the rules and lower sense of identity with other users are likely to reduce compliance with the rules, unless there is strong enforcement, which is often beyond the capacity of those new formal institutions.

The question is whether the emerging (national or international) governance systems that set and enforce water rights at these higher levels can build on the principles of social relations and personal contact, by including mechanisms for members of different user groups to meet and understand each other's needs. Such 'multi-stakeholder platforms' may take longer in developing the rules, and may seem more costly than to just have 'experts' do the work, but in the long term it may pay off through increased legitimacy, and hence higher compliance at lower enforcement costs.

At the same time, we should not romanticize customary systems. There is ample evidence that customary law frequently reflects unequal power relationships in local communities. Such relationships greatly affect the ways in which land and water are distributed and managed. State law may seek to confer more rights on the less advantaged members of a given community, on paper at least. Formalization of water rights may also be called for to protect the livelihoods of existing users against new uses and users. This is especially relevant as water use increases, bringing local users into competition with other users.

However, there is ample evidence that groups like women or the poor often lose out in processes of formalization, particularly in land titling programmes (Lastarria-Cornhiel, 1997). One reason they lose out is that they often lack the resources (knowledge, time, travel and money) required to acquire security of tenure through the state. But, as the 'force field' of state law increases, the customary security of tenure through social relations can weaken. To understand the barriers that marginalized groups face in getting formally recognized rights, it would be useful for those who develop any water rights registration programmes to literally walk through the whole process with a poor rural woman. Seeing exactly what it would take for such a woman to acquire recognized rights through the state could provide both insights and motivation to modify the system to remove as many obstacles as possible for

people like her.

Another reason that the poor lose out is that formal state systems often accord less recognition to the overlapping rights to the resource, on which many poor people rely (Hodgson, 2004). We have seen that both land and water rights have multiple uses and users. These multiple users often have some shared understandings on who, how, when and how much of the resource can be used, the interlinkages between them and perhaps even quality issues. These are often lost in tenure reforms, particularly privatization, because such conditionality is seen to increase transaction costs and hinder the efficient redistribution of property rights. Even when the state declares itself the owner of all resources, as the custodian for all the people, Hodgson (2004) finds that the effect is the denial of customary rights as well as the erosion of local management authority over the resource.

Codification of rights often does not allow for considerations of special circumstances, such as basic livelihood needs, that are typically given substantial weight in customary systems. This is partly due to limitations of state capacity to interpret individual circumstances, but it also derives from the current emphasis on the 'rule of law', which implies that everyone should be treated equally, without special considerations. Reforms of both land and water tenure often have the objective of 'regularizing' all uses of water under the authority of a state agency (Hodgson, 2004) or of 'integrating all forms of property into a unified system' (de Soto, 2000, p. 162). Legal anthropologists who study the multiple types of 'law' that abound in any society would suggest that this is not possible – that pluralism will always persist, in some form. But, even if it were possible to fit all customary law within the ambit of state law, it may not be desirable, because the pluralism in water rights and basis for claims allow for dynamism, for adaptation to varying local circumstances (Berry, 1993; Meinzen-Dick and Pradhan, 2002).

One option that is increasingly used in land tenure reforms is for the state to recognize local authorities, who can set and administer rights within their areas. This builds on both local custom and uses the institutions to back those rights, instead of relying heavily on state apparatus, which is often costly or ineffective, especially in rural areas. Tanner (2002) discusses some of the challenges that this approach faced in Mozambique, particularly difficulties in codifying many different customary systems, protecting the rights of women (who are strongly disadvantaged under customary land law) and guarding against unscrupulous chiefs.

To this list of challenges should be added variation in the capacity of local leaders and of communities to manage the resource. Effective management of the resource itself is required to make water rights effective and, if the state does not deliver this, then local leadership and collective action are critical. But such local institutions do not function well in every community; hence, devolution of authority over water rights will not work well in all locations, and due attention should be given to local capacity building.

Whatever institutional reforms are chosen, the state cannot simply wave a magic legislative wand or issue an administrative order and expect to automatically change water rights on the ground. Effective changes – from de jure to de facto – require more than changes in the law itself: they need to become widely known, discussed and even debated. South Africa's water rights reforms exemplify this.

There was a prolonged process of public discussion over the Water Act, which not only served to refine the legislation itself but to ensure that it was widely known, so that people could appeal to the new laws to claim their rights and to see that the provisions of the law were implemented. In contrast to other countries in which reforms in water rights legislation have been passed in response to donor requests, but never discussed, the public is aware of South Africa's reforms, which makes implementation much more likely. The next step is to build the capacity of implementing institutions, which may require considerable investment of time, training and other resources, particularly if multi-stakeholder institutions are to be developed (Seetal and Quibell, 2005).

However, it is not only statutory water rights that can be changed. Customary and even religious law also evolve over time in response to changing environmental conditions, livelihoods and even changes in other types of law. Thus, a change in state law can stimulate changes in customary law.

Because of the fundamental importance of water, water rights reforms need to give particular attention to the question of how such changes in state or local law will affect the poor. State law can make special provisions for disadvantaged groups, to which they can appeal. But for this to have any effect requires legal literacy campaigns, so that even illiterate rural women will know of any new rights that they are supposed to be accorded.

Before rushing to formalize water rights, which have often involved either nationalization or privatization, it is important to consider the full range of options, including looking for new forms of property rights that build upon strong customary principles, especially the widespread norms that specify rights to water for basic needs. Here, the international discourse and customary law come together in emphasizing water as a basic human right. However, because water rights are meaningless without an institution to back them, serious questions of how much water can be used will need to be addressed, as well as what incentives there will be for anyone to supply it.

'Open access' to water may be desired (as indicated in many of the local laws) but not feasible. Yet, water rights reforms should strive to ensure that the basic principle is met: that water for basic livelihood needs will be available for all. Both restraint on use and investment in provision are required. Achieving this may require going beyond conventional measures of regulation or economic incentives, to also appeal to norms and values of sharing and caring for others, as well as for the earth. As Mahatma Gandhi reminded us, over 50 years ago: 'Earth provides enough to satisfy every man's need, but not every man's greed.'

Acknowledgements

This chapter draws upon work done in collaboration with Bryan Bruns, Rajendra Pradhan, Brent Swallow and Leah Onyango, particularly Bruns and Meinzen-Dick (2000, 2005), Meinzen-Dick and Pradhan (2002) and Onyango et al., Chapter 11, this volume. Esther Mwangi and Stephan Dohrn also gave valuable comments.

The intellectual input of all these colleagues is gratefully acknowledged.

Endnotes

[1] Although there is considerable talk of 'water wars', in fact there is little evidence of international violent conflict over water. Violence over water is more likely at the local level (Ravnborg, 2004).

[2] An exception in customary law is where someone has dug a well or developed a source that is considered private, and can bequeath that source to heirs, e.g. under Maasai tradition (Potkanski, 1994, cited in Juma and Maganga, 2005).

[3] In west Asia and North Africa, herders with large flocks increasingly bring water to their animals, rather than the reverse, but the higher costs of fuel and transport, as well as high poverty rates, make this less of an option in most of sub-Saharan Africa.

References

Alchian, A.A. and Demsetz, H. (1973) The property rights paradigm. *Journal of Economic History* 16 (1), 16–27.

Ambler, J.S. (1998) *Customary Law (Adat) in Indonesia: Perspectives on Colonialism, Legal Pluralism, and Change*. Background paper prepared for the visit of the delegation from the Institute of Folk Culture, National Centre for Social Sciences and Humanities, Hanoi, to Indonesia, 23 May–1 June 1998.

Arnold, C.A. (2002) The reconstitution of property: property as a web of interests. *Harvard Environmental Law Review* 26 (2), 281.

Benda-Beckmann, F.v., Benda-Beckmann, K.v. and Spiertz, J. (1997) Local law and customary practices in the study of water rights. In: Pradhan, R., Benda-Beckmann, F.v., Benda-Beckmann, K.v., Spiertz, H.L.J., Khadka, S.S. and Azharul Haq, K. (eds) *Water Rights, Conflict and Policy*. International Irrigation Management Institute, Colombo, Sri Lanka, pp. 221–242.

Berry, S. (1993) *No Condition Is Permanent: the Social Dynamics of Agrarian Change in Sub-Saharan Africa*. University of Wisconsin Press, Madison, Wisconsin.

Besley, T. (1995) Property rights and investment incentives: theory and evidence from Ghana. *The Journal of Political Economy* 103 (5), 903–937.

Bromley, D. (1992) The commons, property, and common-property regimes. In: Bromley, D. (ed.) *Making the Commons Work*. Institute for Self-Governance Press, San Francisco, California, pp. 3–16.

Bruce, J.W. and Migot-Adholla, S. (eds) (1994) *Searching for Land Tenure Security in Africa*. World Bank, Washington, DC.

Bruns, B.R. and Meinzen-Dick, R. (eds) (2000) *Negotiating Water Rights*. Sage, New Delhi, India.

Bruns, B.R. and Meinzen-Dick, R. (2005) Frameworks for water rights: an overview of institutional options. In: Bruns, B.R., Ringler, C. and Meinzen-Dick, R. (eds) *Water Rights Reform: Lessons for Institutional Design*. International Food Policy Research Institute (IFPRI), Washington, DC, pp. 3–26.

Carlsson, E. (2003) *To Have and to Hold: Continuity and Change in Property Rights Institutions Governing Water Resources among the Meru of Tanzania and the BaKgatla in Botswana, 1925–2000*. Almqvist and Wiksell International, Stockholm.

Chikozho, C. and Latham, J. (2005) The relevance of customary law: Zimbabwean case studies of water resource management. Paper presented at the *International Workshop on African Water Laws: Plural Legislative Frameworks for Rural Water Management in Africa*, 26–28 January 2005, Gauteng, South Africa.

Cleaver, F. (1998) Choice, complexity, and change: gendered livelihoods and the management of water. *Agriculture and Human Values* 15 (4), 293–299.

de Soto, H. (2000) *The Mystery of Capital: Why Capitalism Triumphs in the West and Fails Everywhere Else*. Basic Books, New York.

Easter, K.W., Rosegrant, M.W. and Dinar, A. (eds) (1998) *Markets for Water: Potential and Performance*. Kluwer Academic Publishers, Boston, Massachusetts.

EU (European Union) Task Force on Land Tenure (2004) *EU Land Policy Guidelines*. European Union, Brussels.

FAO (Food and Agriculture Organization of the United Nations) (2002) *Gender and Access to Land*. Land Tenure Studies 4, Food and Agriculture Organization, Rome.

Farolfi, S. (2004) *Action Research for the Development of a Negotiation Support Tool Towards Decentralised Water Management in South Africa*. Working Paper 2004-01, University of Pretoria, South Africa.

Furubotn, E.G. and Pejovich, S. (1972) Property rights and economic theory: a survey of recent literature. *Journal of Economic Literature* 10 (4), 1137–1162.

Gleick, P.H. (1999) The human right to water. *Water Policy* (1), 487–503.

Hodgson, S. (2004) Land and water – the rights interface. FAO legal papers, Online 36, Food and Agriculture Organization, Rome. Available at http://www.fao.org/legal/prs-ol/lpo36.pdf (accessed 3 January 2004).

Ingram, H. and Oggins, C.R. (1992) The public trust doctrine and community values in water. *Natural Resources Journal* 32, 515–537.

Juma, I.H. and Maganga, F.P. (2005) Current reforms and their implications for rural water management in Tanzania. Paper presented at the *International Workshop on African Water Laws: Plural Legislative Frameworks for Rural Water Management in Africa*, 26–28 January 2005, Gauteng, South Africa.

Lastarria-Cornhiel, S. (1997) Impact of privatization on gender and property rights in Africa. *World Development* 28 (8), 1317–1334.

Latham, J. (2000) Towards an understanding of local level adaptive management: matching the biosphere with the sociosphere. Paper presented at the *1st WARFSA/WaterNet Symposium: Sustainable Use of Water Resources*, 1–2 November 2000, Maputo, Mozambique.

McCarthy, N., Dutilly-Diane, C., Drabo, B., Kamara, A. and Vanderlinden, J.-P. (2004) *Managing Resources in Erratic Environments: an Analysis of Pastoral Systems in Ethiopia, Niger, and Burkina Faso.* Research Report 135, IFPRI, Washington, DC.

Meinzen-Dick, R.S. and Bakker, M. (2001) Water rights and multiple water uses: issues and examples from Kirindi Oya, Sri Lanka. *Irrigation and Drainage Systems* 15 (2), 129–148.

Meinzen-Dick, R.S. and Pradhan, R. (2002) *Legal Pluralism and Dynamic Property Rights.* CGIAR System-Wide Program on Property Rights and Collective Action Working Paper 22, IFPRI, Washington, DC. Available at http://www.capri.cgiar.org/pdf/capriwp22.pdf (accessed 3 January 2005).

Mtisi, S. and Nicol, A. (2003) *Caught in the Act: New Stakeholders, Decentralisation and Water Management Processes in Zimbabwe. Sustainable Livelihood in Southern Africa.* Research Paper No. 14, Institute of Development Studies, Brighton, UK.

Ngaido, T. (1999) Can pastoral institutions perform without access options? In: McCarthy, N., Swallow, B., Kirk, M. and Hazell, P. (eds) *Property Rights, Risk, and Livestock Development in Africa.* IFPRI/International Livestock Research Institute, Washington, DC, pp. 299–325.

Otsuka, K. and Place, F. (eds) (2001) *Land Tenure and Natural Resource Management: a Comparative Study of Agrarian Communities in Asia and Africa.* Johns Hopkins University Press, Baltimore, Maryland.

Paul, M.O. (2003) Defining property rights for water marketing. Paper presented at the *World Water Congress, International Water Resources Association*, Madrid.

Perret, S.R. (2002) Water policies and smallholding irrigation schemes in South Africa: a history and new institutional challenges. *Water Policy* 4 (3), 283–300.

Potkanski, T. (1994) *Property Concepts, Herding Patterns and Management of Natural Resources among the Ngorongoro and Salei Maasai of Tanzania.* Pastoral Land Tenure Series No. 6, IIED Drylands Programme, International Institute for Environment and Development, London.

Pradhan, R., von Benda-Beckmann, F., von Benda-Beckmann, K., Spiertz, H.L.J., Khadka, S.S. and Azharul Haq, K. (eds) (1997) *Water Rights, Conflict and Policy.* International Irrigation Management Institute, Colombo, Sri Lanka.

Ramazzotti, M. (1996) *Readings in African Customary Water Law.* FAO Legislative Study No. 58, Food and Agriculture Organization, Rome.

Ravnborg, H.M. (2004) *Water and Conflict – Lessons Learned and Options Available on Conflict Prevention and Resolution in Water Governance.* DIIS Brief, Danish Institute for International Studies, Copenhagen.

Rogers, P. and Hall, A.W. (2003) *Effective Water Governance.* TEC Background Papers No. 7, Global Water Partnership/SIDA, Stockholm.

RSA (Republic of South Africa) (1998) *The National Water Act.* Act Number 36 of 1998.

Schlager, E. and Ostrom, E. (1992) Property-rights regimes and natural resources: a conceptual analysis. *Land Economics* 68 (2), 249–262.

Seetal, A. and Quibell, G. (2005) Water rights reform in South Africa. In: Bruns, B.R., Ringler, C. and Meinzen-Dick, R. (eds) *Water Rights Reform: Lessons for Institutional Design.* IFPRI, Washington, DC, pp. 153–166.

Swallow, B.M., Garrity, D.P. and van Noordwijk, M. (2001) *The Effects of Scales, Flows and Filters on Property Rights and Collective Action in Watershed Management.* CGIAR System-Wide Program on Property Rights and Collective Action, Working Paper 16, IFPRI, Washington, DC. Available online at http://www.capri.cgiar.org/pdf/capriwp16.pdf (accessed 3 January 2005).

Swynnerton, R.M.M. (1954) *A Plan to Intensify the Development of African Agriculture in Kenya.* Government Printer, Nairobi.

Tang, S.Y. (1992) *Institutions and Collective Action: Self-Governance in Irrigation.* ICS Press, San Francisco, California.

Tanner, C. (2002) *Law-making in an African Context: the 1997 Mozambican Land Law.* FAO legal papers, Online No. 26, FAO, Rome. Available at http://www.fao.org/Legal/prs-ol/lpo26.pdf (accessed 3 January 2005).

Tewari, D.D. (2002) *An Analysis of Evolution of Water Rights in South African Society: an Account of Three Hundred Years.* Working Paper, University of Natal, Durban, South Africa.

van Koppen, B. (2000) Gendered water and land rights in rice valley improvement, Burkina Faso. In: Bruns, B. and Meinzen-Dick, R. (eds) *Negotiating Water Rights.* Sage, New Delhi, India, pp. 83–111.

Vaz, Á.C. and Pereira, A.L. (2000) The Incomati and Limpopo international river basins: a view from downstream. *Water Policy* 2 (1–2), 99–112.

Weiss, T. (2004) *Guns in the Borderlands, Reducing the Demand for Small Arms.* Monograph No. 95, Institute for Security Studies, Pretoria, South Africa.

WFP (World Food Programme) (2001) *Water Tenure, Natural Resource Management and Sustainable Livelihood.* Natural Resource Management No. 5, International Land Coalition Resource Centre, Rome.

Witsenburg, K. and Adano, W.R. (2003) The use and management of water sources in Kenya's drylands: is there a link between scarcity and violent conflicts? AGIDS/UvA, Amsterdam. Available online at http://www2.fmg.uva.nl/agids/publications/2003/documents/witsenburg_use.pdf (accessed 3 January 2005).

Wolf, A.T. (2000) Indigenous approaches to water conflict negotiations and implications for international waters. *International Negotiation* 5 (2), 357–373.

3 Community Priorities for Water Rights: Some Conjectures on Assumptions, Principles and Programmes

Bryan Bruns

Santa Rosa Beach, Florida, USA; e-mail: bryanbruns@bryanbruns.com

Abstract

Increasing policy support for community participation in natural resources management has been challenged by questions about the feasibility, risks and results of such approaches. The application of participatory approaches for improving basin-scale water governance should be considered in light of critical analysis of community-based natural resources management and institutional design principles for common-property resources management. Problems of conflicting interests and contextual contingency (politics and history) illustrate the need for revising assumptions and expectations. A community perspective on principles for institutional design leads to distinct priorities for improving basin water allocation. Measures to support community involvement in basin water governance – such as legislative reform, legal empowerment, networking, advocacy, participatory planning, technical advice and facilitation – should be formulated to fit community priorities for negotiating rights to water.

Keywords: water rights, water allocation institutions, river basin, governance, integrated water resources management, community-based natural resources management, institutional design principles.

Introduction

As governments and other organizations seek to improve the management of natural resources, participatory and community-based approaches have promised valuable advantages, and so they have received increasing support in the policies of national and international agencies. However, evidence and analysis indicate that the application of such approaches also faces serious challenges and constraints; see, for example, Agrawal and Clark (2001); Knox and Meinzen-Dick (2001); Ribot (2002); Young (2002); Agrawal (2003); Cleaver and Franks (2003); Mosse (2003); Sengupta (2004); Mansuri and Rao (2005);

Shah, Chapter 5, this volume. This chapter looks at the relevance of community-based approaches to the negotiation of water rights within basin water governance, considered in the light of critical analysis of community-based natural resources management and of institutional design principles for common-property resources management. It applies a community perspective to identify practical implications for revising assumptions about community participation, customizing application of institutional design principles and formulating more effective programme interventions.

Rights to water may be negotiated in many contexts (Bruns and Meinzen-Dick, 2000, 2001), not only within communities[1] but also between communities sharing rivers, aquifers and other common-pool water resources. Government assistance in developing irrigation and water supply systems may require agreements limiting how much water will be abstracted, as well as allocating access to enhanced supplies. As competition for water rises along rivers, water users may take part in deciding how scarce water will be shared between users within a sub-basin or basin. If government agencies seek to formalize water rights, then quantities and conditions in permits and plans may be negotiated. One source of water to supply the demands of growing cities may come through voluntary agreements that compensate irrigators for transfers. These situations not only offer important opportunities for government intervention in basin-scale water allocation but also present opportunities and challenges for communities.

From the perspective of rural communities, negotiating agreements about rights to water may be a necessary condition for aid in improving water supplies to farms and homes. More likely, though, is the need to defend access to water against threats from competing users. Drought intensifies conflicts, stimulating short- and long-term efforts to modify rules and procedures regulating rights to water. New projects for urban water supply or irrigation may take water away from existing users. Bureaucratic programmes, such as basin planning or registration of water rights, pose risks where rights will be impaired or lost unless water users act effectively to protect themselves. Legislative changes may imperil customary community-based water rights, denying them legal status or forcing fragmentation of rights.

Communities may respond by employing multiple strategies in various arenas, such as: (i) acting directly to acquire more water or block others' access; (ii) participating in planning and other formal administrative procedures; (iii) litigating in courts; (iv) lobbying to advocate their case to the public and politicians; and (v) pursuing agreements with other water users and with water management agencies.[2] Negotiation frequently plays an important part in such strategies, whether agreements are sought immediately or worked out later to settle disputes initially fought in other arenas. More broadly, government interventions in water allocation and community efforts to defend access to water create situations where water rights are negotiated.

Table 3.1 summarizes key contexts for negotiation of water rights, highlighting differences between situations that governments may see as opportunities for intervention to serve societal goals compared to what communities may see as threats to their rights to water. Such situations, as perceived and prioritized by communities, then provide contexts for reconsidering assumptions underlying community participation, principles for institutional design and formulation of programme interventions.

Participatory and community-based approaches to natural resources management promise important advantages in the development institutions for water allocation at sub-basin- and basin-scale but, as outlined in the next section of this chapter, limitations of politics and history should be expected to constrain and complicate their implementation. While general institutional design principles for the management of common-property resources have been proposed, the following section shows how a community perspective on the application of such principles to basin governance identifies priorities distinctly different from generic recommendations. Similarly, measures intended to support communities may fail to achieve intended results, unless adjusted to fit local circumstances and priorities, as discussed in the section on aiding community negotiation of this chapter. The final section of the chapter summarizes conjectures about community dynamics and priorities in securing access to water.

Table 3.1. Key contexts from two perspectives on negotiating water rights in river basins.

Government perspective (Opportunities for intervention)	Community perspective (Defence against threats)
Assistance–improvement projects	Competition for water
Basin allocation	Drought
Reallocation	Expropriation
Formalization of rights	Denial or fragmentation of rights

Advantages and Limitations of Community-based Approaches

Top-down approaches, emphasizing centralized government authority and control, have dominated most government efforts to manage water and other natural resources. There is now increased interest in, and support for, participatory and community-based approaches to natural resources management and conservation that may help address some of the limitations, disappointments and problems associated with top-down approaches (Knox and Meinzen-Dick, 2001; Ribot, 2002; Mosse, 2003). Participation may cover a range of interactions between decision makers and stakeholders, ranging from minimal dissemination of information, through consultation that listens to stakeholder inputs, involvement in dialogue, collaborative development of alternatives, joint decision making in co-management and delegation of specific authority and empowerment of communities to make autonomous or independent decisions.[3]

The rhetoric of community-based resources management often suggests strong devolution of authority, empowering communities to make decisions on their own, perhaps with some technical guidance and support from outside.[4]

However, in practice what often occurs are more limited forms of participation, for example where government approval for detailed management plans is required, or where mutual consensus for co-management, or agreements, or even narrower forms of stakeholder involvement with final authority for decisions remaining fully with government agencies, are necessary. Participation is used in this chapter as a general term for a variety of institutional arrangements that involve stakeholders in decisions, while community-based refers to arrangements that provide primary decision-making power to communities, either local governments or specialized organizations of resource users such as water user associations.

Participatory and community-based approaches may be valued for their own sake, as ways to support local cooperation and self-governance. Such approaches may also be pursued for practical reasons, for example as instruments to increase equity or raise water productivity, as ways to reduce transactions costs or simply as a means to shift costs away from government. Such approaches can utilize local knowledge in crafting management measures to match local conditions. Many of the advantages of these approaches potentially apply not just within communities but also in the situations that are the primary concern of this chapter, where water rights may be negotiated between communities as part of basin water management:

- Water users possess detailed local knowledge about how they use water, their needs and the possible consequences of changes. Community-based approaches cultivate channels through which this information can be considered in making decisions.
- Collective action to manage water weaves water users together in webs of relationships. These relationships can build social capital of trust and shared understanding that facilitates cooperation, at both local and larger scales.
- As part of their daily activities, it is often easy for water users to observe whether neighbours are fulfilling their commitments and obligations in using water. They can monitor and detect nearby violations with relatively little time and effort.
- Communities can selectively apply sanctions unavailable through formal institutions. The threat of being shamed or of losing one's reputation as respected and trustworthy may compel compliance. Water users possess strong incentives and willingness to struggle for their access to water.
- Community-based approaches may be able to resolve many conflicts at a local level, by those most concerned, with less cost or complication. Such subsidiarity, customized to local circumstances, reduces the transaction costs of coordinating resource use and implementing agreements.
- Involving communities in decisions builds legitimacy and support, reducing risks of rejection and resistance. Participation realizes principles of democracy and empowerment.
- Water management may become more effective when it utilizes the capabilities of users, not only as individuals but also as communities linked by ongoing relationships, with shared views and common interests that facilitate cooperation.

However, participatory and community-based approaches have been the subject of growing critical scrutiny.[5] Community-based approaches have frequently been applied with unrealistic assumptions and expectations. They have sometimes been advocated and applied with inadequate attention to the variety of people involved in using and managing resources in local areas, and the intricate arrangements through which they compete and cooperate. Simplistic stereotypes of isolated, small, stable and homogeneous groups sharing the same interests and traditional norms for preserving local resources often fit poorly with the complexity of how diverse local and external actors struggle to make and break rules about exploiting and replenishing resources that may be mobile and interconnect broad areas (Agrawal and Gibson, 2001). The conditions and limitations of community-based approaches need to be considered along with their advantages, within particular contexts. Critiques of community-based natural resources management concentrate on core themes of conflict, differences between actors, incentives and contextual contingency.

Conflict

The concept of community itself is problematic, presuming local solidarity and cooperation that may be absent or achieved only through exceptional effort. Romanticism and ideological aspirations risk obscuring recognition of the tensions, strife and flaws that characterize collective action, past and present. Thus, for example, accounts portraying Balinese *subaks* and other irrigation communities as highly cohesive encourage exaggerated assumptions about what exists or may be feasible for water user associations.[6] Access to water and other resources is politically contested, so 'management' is not purely a neutral technical exercise in optimizing water productivity but also a process of continuing struggle among competing claimants. Incentives to take part in collective action depend on, among other things, the distribution of anticipated benefits and costs, and conflicts about the distribution of gains and losses can obstruct agreement about defining rights and arranging cooperation.

Heterogeneity

Assumptions of homogeneous actors are invalid, with gender, age, wealth and other distinctions differentiating communities internally. Within an irrigation system, head-enders have different interests and options than tail-enders. Similarly, communities differ from one another in resources, livelihoods, organizational capacity and other characteristics. Theoretical and empirical analyses indicate that heterogeneity may impede or facilitate collective action (Olson, 1971; Mansuri and Rao, 2005), but diverse situations of different actors inevitably shape perceptions and actions. Collective action is not simply a matter of aggregating identical interests but one of forging coalitions among diverse participants.

Asymmetry

Differences in knowledge, wealth, power and other characteristics matter not only within communities but also in wider interactions. Such asymmetries often (but not always) place communities at a disadvantage in negotiating with outside water users. Communities, especially rural communities, may have little room for manoeuvre beyond compliance or muted resistance. If an opportunity exists to negotiate, they may have few alternatives for maintaining or improving their access to resources, leaving them in a weak bargaining position.[7]

Inequity

Aid that may help people who are generally poor by national standards does not necessarily do much for those who are the poorest. Biased decisions may reinforce and worsen inequities in access to resources (Mansuri and Rao, 2005). Poor people, women, ethnic minorities, youths and elderly, and others who are not part of local elites may be left out and their views and concerns neglected unless special outreach efforts are arranged. However, a degree of control by local elites, although not necessarily 'capture,' seems almost inevitable.[8]

Within communities, it may not be realistic to expect community-based approaches to reduce inequalities, unless specific conditions and measures direct change in pro-poor directions. Specific targeting measures may help to provide more benefits for those who are poorer. In general, rather than idealistically assuming that community-based approaches will automatically or necessarily favour equality or yield 'pro-poor' results, a more realistic assumption may be that community-based approaches are likely to reproduce existing inequalities, and may even worsen them, unless offset by countervailing conditions and measures.

Local incentives

Participation imposes substantial transaction costs, particularly for the poor, and may not be worthwhile for participants. This is due not only to problems in organizing collective action but also to the risks of manipulated and meaningless participation, and policies that transfer responsibility without authority. Furthermore, the incentives of both leaders and ordinary resource users are not necessarily consistent with conservation and sustainable use. In practice, transfer to local control may be almost as prone to biased access and neglect of longer-term sustainability as state control of resources, unless adequately offset by local and external regulation to promote broader societal interests, such as legal equality, social equity and environmental conservation (Ribot, 2002). Rather than simplistic state withdrawal for full local control, the need may be to find an institutional mix that better combines community, market and state action, as in forms of co-production, co-management or regulated autonomy.[9]

Context

The complexity of local resource characteristics, social relationships, external linkages and other circumstances conditions the impact of interventions, making them prone to fail unless carefully customized to context (Mansuri and Rao, 2005). Communities have been, and will continue to be, strongly shaped by external linkages including trade, migration, politics and culture (Wolf, 1983). Potential pathways for change are shaped by existing conditions.

Uniform implementation and outcomes are unlikely. Attempts to impose solutions from outside often founder because they fit poorly with local resource characteristics and institutions[10] and are resisted as inappropriate and illegitimate. Existing institutional arrangements shape perceptions and the potential for modifying or replacing rules, so that paths for change depend on past and present perceptions and practices that are not easily altered. Institutional rearrangements that occur under exceptional circumstances, such as outstanding local leaders, strongly integrated communities, abundant funding and skilled advice are hard to replicate, and prone to revert when the unusual circumstances disappear (Bruns, 1992; Shah, Chapter 5, this volume).

In simple terms, politics and history condition what is possible. These factors influence the applicability of community-based approaches to natural resources management in general and to water allocation in particular. Community-based approaches are not a panacea: they do not offer a way to escape politics, bypass elites or safely shortcut to social justice. However, the thrust of most critiques is not to say that community-based management is impossible, but rather to challenge invalid assumptions, oversimplified implementation and unrealistic expectations. Revised assumptions, as summarized in Table 3.2, and further discussed below, may provide a more realistic foundation for community-based approaches.

Applying Institutional Design Principles

One important source of ideas about community-based natural resources management comes from research on common-pool resources, such as forests, fisheries, rangeland and irrigation systems. Principles of institutional design, as summarized in the first column of Table 3.3, have synthesized findings from analysis of long-enduring institutions managing common-property resources (Ostrom, 1990). The principles identify means to overcome the 'tragedy of the [unmanaged] commons', where individual self-seeking behaviour would degrade shared resources, unless regulated through suitable institutional arrangements.[11] Research on such arrangements has documented the potential for successful self-governance (Hardin, 1968; 1988).

Resource users, acting as insiders, design institutions through various conscious and unconscious processes including deliberate rule

Table 3.2. Revising assumptions for community-based natural resources management (CBNRM).

Conventional	Critical
Community solidarity	Conflicting interests, coalitions
Homogeneity	Heterogeneity
Equality	Asymmetry
Technical optimization	Political contestation
Equitable outcomes	Reproduction of inequities unless countervailed
Independence	Regulated autonomy
Self-sufficiency	Interlinkages
Replicable intervention	Contextual (path dependent, ergodic, bricolage, improvised, episodic, adaptive, experimental)

making, imitation, trial and error learning and improvisation. While detailed local rules for resource use vary widely, the design principles summarize general patterns. Many studies of common property have focused on small communities, apparently managing resources through relatively autonomous self-governance, often analytically treated as relatively homogeneous and isolated from external political and economic forces. Such relatively simple conditions ease theoretical analysis. The principles emphasize 'long-enduring' institutions, able to recover from shocks and adapt to changing conditions, especially since there may be no stable ecological equilibrium and no 'one best way' to manage a resource (Ostrom, 1999; Anderies *et al.*, 2003).

Further analysis has challenged simplistic interpretations of institutional design principles. Research has highlighted differentiation within communities, interactions with external social and economic forces and implications of resources and livelihood strategies that extend beyond small localities (Agrawal, 2003). The capacity of government intervention to disrupt local institutions for managing common property resources has been extensively documented, but less has been learned about ways that states can support and sustain local management (Sengupta, 2004).

Attempts to apply the principles of institutional design to prescriptively determine how institutions for river basin water allocation *must* be designed may fit badly with the complexity of local history and politics (Cleaver and Franks, 2003; Ravnborg, 2004 (cited with permission)). Institutional change may be less a process of careful and deliberate craftsmanship,

and more a messy process of institutional *bricolage*, an improvised recombination of available arrangements.[12] Thus, application of institutional design principles needs to take into account the influence of politics, history and the improvisational and contested ways in which institutions are modified, as well as incomplete information and uncertainty about outcomes of modifying complex systems.

Nevertheless, within an appropriately contextualized approach, institutional design principles usefully outline key challenges facing stakeholders concerned with governing shared water resources. While institutional design principles are insufficient by themselves to devise solutions, they provide a framework for analysing some of the challenges facing communities seeking to negotiate rights to water in contexts of competition with other communities and significant state influence on water governance.[13] Based on experience and analysis of common property resources management in general, and water allocation in particular, some preliminary ideas can be proposed about priorities for communities negotiating rights to water.

Clearly defined boundaries

Watersheds delimit catchments within which water flows into streams that merge to form rivers, delineating sub-basins and basins that appear to clearly define boundaries for water management. As water becomes scarcer in a basin, the scope of interaction and competition between users increases, increasing the need for, and potential benefits from, coordination among those sharing a common resource.

Table 3.3. Institutional design principles, issues and conjectures on community priorities (The first column repeats 'design principles derived from studies of long-enduring institutions for governing sustainable resources', as presented in Anderies *et al.*, 2003, which was based on Ostrom, 1990, p. 90. For column two, see this chapter, and also Cleaver and Franks, 2003 and Ravnborg, 2004).

Principle	Issues	Community priorities
Clearly defined boundaries The boundaries of the resource system (e.g. irrigation system or fishery) and the individuals or households with rights to harvest resource units are clearly defined	Basins offer clear boundaries, but: shortages are uncertain and concentrated in particular times and places; administrative boundaries, livelihood activities and other linkages cross-cut basins	Coalitions for problemsheds
Proportional equivalence between benefits and costs Rules specifying the amount of resource products that a user is allocated are related to local conditions and to rules requiring labour, materials and/or money inputs.	Volumetric allocation difficult and expensive; infrastructural subsidies distort linkages between receiving water and paying costs	Local water allocation practices accommodated, e.g. shares and time-based allocation
Collective choice arrangements Most individuals affected by harvesting and protection rules are included in the group who can modify these rules.	Scale makes representation necessary; platforms may be biased, manipulated or lack authority	Representation in decisions, in multiple forums, especially during crises
Accountable monitoring Monitors, who actively audit biophysical conditions and user behaviour, are at least partially accountable to the users and/or so are the users themselves.	Agency accountability weak; complex factors affect basin water availability; information technologies make more information available, but threaten information overload	Local and scientific expertise to demystify information
Graduated sanctions Users who violate rules-in-use are likely to receive graduated sanctions (depending on the seriousness and context of the offence) from other users, from officials accountable to these users or from both.	Lack of relationships between distant users impedes trust and informal sanctions; formal sanctions hard to enforce	Remedies if rights infringed
Low-cost conflict-resolution mechanisms Users and their officials have rapid access to low-cost, local arenas to resolve conflicts among users or between users and officials	Courts problematic for resolving water conflicts	Efficient mediation, backed by government authority
Minimal recognition of rights to organize The rights of users to devise their own institutions are not challenged by external governmental authorities, and users have long-term tenure rights to the resource	National legal frameworks ignore or disrupt customary water rights and organizations; insecure tenure	Customary water rights recognized, including local processes for dispute resolution
Nested enterprises (for resources that are parts of larger systems) Appropriation, provision, monitoring, enforcement, conflict resolution and governance activities are organized in multiple layers of nested enterprises	Participation is costly; multiple government units and agencies	Community autonomy; strategic alliances

However, other factors blur the seeming clarity of basins as management units (see, for example, Cleaver and Franks, 2003). Administrative jurisdictions, such as districts and provinces, cross-cut basins. Resource users engage in activities inside and outside of basins. Within basins, conditions are not uniform: shortages become severe at particular times and places, meaning that specific sub-areas will be much more concerned about particular problems. Regulation of land-use changes that affect water flows engages different sets of people and agencies. Health agencies hold responsibilities for water quality, while environmental agencies and organizations pursue agendas for conservation. Groundwater basins overlap surface basins. Irrigators steer water around hillsides, moving water between different sub-basins and basins, as do cities reaching out to expand their water supplies. Physical linkages within a basin offer a foundation for management, but social and economic linkages follow different patterns, raising the transaction costs of coordination.

Conceptual frameworks for integrated water resources management (IWRM: Agarwal *et al.*, 2000; Rogers and Hall, 2000) offer the appealing prospect of coordinating solutions to many of these complexities, but may presume or be interpreted to require ambitious projects for design and implementation of elaborate new institutional arrangements. From a community perspective, if negotiation is costly it may be most important to engage those most affected by, and able to contribute to, in solving an immediate problem and crafting coalitions within and between communities. Thus, the most relevant scope may cover a *problemshed* (Halaele and Knesse, 1973) rather than necessarily including an entire river basin or comprehensively integrating water resources management. Rather than clearly defined boundaries and complete membership, the immediate challenge from a community perspective may be to form an ad hoc coalition among a fuzzy set (Kosko, 1994) of people with widely differing stakes in a problemshed.

Proportionality between costs and benefits

Within communities, access to shared water infrastructure for household or irrigation use is usually linked with obligations to contribute to investment, or at least maintenance. However, government subsidies for water infrastructure often encourage expectations of receiving benefits without paying costs. From an economic perspective, raising water prices may appear to be a logical way to link costs and benefits. However, users are likely to oppose formalization of water rights if it is seen as primarily a means to impose new charges.

Few governments have enough political power to establish themselves as water lords, extracting marginal cost prices for water, although recovery of some operation and maintenance costs may be feasible. Alternatively, tradable water rights could open a politically more feasible pathway to voluntary win-win exchanges. However, establishing tradable rights requires working through a variety of complex issues, resolving conflicts and clarifying rights, as well as developing institutions for more precise water accounting and protection of third parties.

Shifting to volumetric water allocation of surface water offers theoretical benefits, and practical problems in measurement and control that grow larger as the volumes involved become smaller. From a community perspective, arrangements that accommodate existing local practices – such as proportional sharing of shortages and measuring water based on time rather than volume – are likely to be much more feasible and acceptable than drastic changes in how water is measured and priced.

Collective-choice arrangements

The scale of basins prevents direct participation of all stakeholders, but representation risks reinforcing biases (Wester *et al.*, 2003). The danger that participatory platforms (Boelens *et al.*, 1998; Steins and Edwards, 1998) are co-opted, manipulated and lack meaningful decision-making power makes it wiser to take a selective and strategic approach to participation and coalition-building, carefully considering whether or how to 'come to the table', and retaining options to employ a mix of strategies in multiple forums.[14] Representation is most crucial during crises, such as droughts, when modifications in water allocation rules receive urgent attention.

Accountable monitoring

Communities lack information about conditions elsewhere in a basin. Agencies with monopoly control over infrastructure may escape accountability, and tend to develop information systems primarily to serve their internal purposes. Advances in information technology promise abundant information, accompanying problems of information overload and difficulty in understanding the complex impact of land-use changes, return flows and other factors on water availability. Local and outside experts can help demystify knowledge, improving the capacity of communities to make and monitor agreements.

Graduated sanctions and conflict-resolution mechanisms

Rights mean little unless there are ways to enforce them when they are infringed and, as the legal saying goes, 'there is no right without a remedy'. The asymmetry of water flowing downhill lets upstream users act without consideration of the consequences for those downstream, a lack of reciprocity that impedes the emergence of self-enforcing cooperation. Lack of social ties between distant communities further limits the potential influence of sanctions based on reputation and repeated interaction. A framework of government authority can enable strangers to contract credible commitments (North, 1990), and this can include agreements about government-recognized water rights. However, legal proceedings that are prolonged, costly, hard to enforce or construed in ways that fit poorly with the practical needs of water management often make courts problematic for resolving conflicts, although they sometimes offer useful bargaining leverage (see, for example, Sengupta, 2000).

If effective conflict resolution mechanisms and sanctions are absent, then problems such as unchecked upstream abstraction and mining of aquifers may be inevitable (Shah *et al.*, 2001; Shah, Chapter 5, this volume). Conditions in many basins mean that having any form of effective recourse is a higher priority for communities than minimizing transaction costs or precisely calibrating sanctions. In the absence of effective alternatives, mediation by government authorities typically plays a central role in dealing with disputes over water, and mediation processes can be further improved.

Rights to organize

Formalization of water user associations in government-driven projects sometimes does more to disrupt than to sustain local collective action in irrigation (Bruns, 1992; Mosse, 2003; Shah, 2004). Constitutional and legislative provisions asserting government sovereignty over natural resources, including water, are often construed to ignore or deny community rights rather than recognizing customary rights and the pluralism of different forms of rights to water. However, advocates can develop other legal interpretations that support community-based property rights (including both common and individual rights derived from community rights) (Lynch, 1998).

Various legal mechanisms are available by which customary rights can be recognized, with legal standing, without requiring formal registration. As one example, Japan's River Law includes transitional clauses stating that existing users are 'deemed' to have permission, and must be so treated, without requiring a formal permit (IDI, 1997; Sanbongi, 2001).[15] Such legal frameworks establish a default situation where community rights are recognized. The burden of proof would then lie with those who would seek to challenge such rights, or processes that may seek to balance their claims against others. For communities, finding ways to assert customary community rights may well be more important than establishing a government-prescribed organization or formal registration of water rights.

Nested organizations

The logical structure of basins, sub-basins and localities invites multiple layers of organization, but makes no guarantee that such a hierarchy will be effective, worthwhile or even feasible (Ravnborg, 2004). Water rights systems may be more successful and reduce transaction costs by avoiding government micro-management of water allocation within communities (Guillet, 1998). Legal frameworks can enable the formation of special districts, with the necessary authority to manage water and mobilize funds, while leaving it up to water users to initiate polycentric organizations on scales that fit their needs and capabilities (Blomquist, 1992).[16] Even if local government jurisdictions mismatch hydraulic boundaries, some support from local authorities will probably be essential to put new or modified rules into practice. From a community perspective, local autonomy and external alliances are likely to be more important than establishing elaborately nested organizations.

If principles for institutional design are interpreted as necessary conditions for coordinating water use within basins, then the limiting and complicating conditions reviewed above might be used to conclude that participatory governance will be impossible. Even if institutional design principles are interpreted more modestly, as desirable conditions that favour good management, they still highlight the many challenges facing basin water management and the need to customize and prioritize how principles are applied. In most cases, especially in the short term, it is unlikely that all or even many of the principles will be completely fulfilled. The question then becomes not one of prescriptively designing an ideal institution, but one of what communities, agencies and other actors in water governance, improvising institutional design as insiders, might accomplish under the conditions that actually prevail.

Aiding Community Negotiation

Water users who want to negotiate water rights may choose various means to pursue their interests. They may study relevant statutes and regulations and gather other information on their own about water problems and potential solutions. They may organize themselves, working through existing local organizations or forming new organizations and coalitions. They may share experiences and coordinate with other groups, through informal contacts and more structured activities such as conferences or workshops. They may participate in planning activities related to water allocation. They may advocate their interests through the media or by directly lobbying politicians and agency officials. They may establish forums covering broader areas such as a basin or sub-basin and develop such organizations to provide effective platforms for negotiation.

Complementing the means available to water users are various measures available to improve community participation in basin governance. Table 3.4 summarizes potential programme interventions from a government perspective and potential community priorities.

Legislative reform

Legal frameworks can empower existing user communities if their rights are recognized and backed by legal recourse if rights are harmed. Legal reforms that provide formal water rights and legal status for user organizations may be useful in providing legal standing to sue in courts or to participate in administrative procedures, strengthening strategic options for litigation and participation. Stronger rights to resources may be very valuable over the long term, not just for encouraging investment but more directly by empowering people to protect and improve their livelihoods (de Soto, 2000). More generally, transparency, accountability and other characteristics of the rule of law in good governance provide conditions that enable stakeholders to act more effectively to protect their interests.

However, from a community perspective, one major problem is that legislative reforms take a long time. Passage of new legislation requires the construction of political coalitions: institutional bargaining that is often contingent on propitious circumstances may be more a matter of luck than of planning. Political conditions and coalitions shape the space available for institutional changes. If reforms are enacted, they may

Table 3.4. Priorities for support programmes.

Government perspective	Community priorities
Reform laws and policies	Rights, recourse procedures
Legal education	Paralegals, legal aid
Protection for subsistence	Meaningful livelihoods
Technical analysis	Community experts
Facilitating organizations	Networks, coalitions
Participatory platforms	Authority, strategic allies

make a big difference, or not. Even after legislation is passed, implementing regulations are often needed. Government agencies may or may not be active about applying what has been put into law. If ambiguities or conflicts exist with other legislation, then legal rulings or amendments may be needed. Nevertheless, even with carefully drafted legislation, if courts are unable or unwilling to enforce legislation then regulation of social and environmental externalities is difficult (Bauer, 2004). For communities, minor modifications of existing regulations on the one hand, and long-term rights to resources on the other, may be more important than the medium-term policy reforms that attract much attention from researchers and reformers.

Legal empowerment

Legal aid, legal education and related approaches, sometimes referred to as legal literacy or legal empowerment, cover a range of activities for improving the capacity of people to understand and use legal systems (Lynch, 1998; Harwell and Lynch, 2002). This includes opportunities for creative use and reinterpretation of existing national and international law. Even if legal protection for local rights is weak, ambiguous or uncertain, litigation may still play a useful role in combination with other strategies for defending community access to water.

While conventional 'rule of law' efforts to develop good governance tend to focus on courts, lawyers and government officials, legal empowerment approaches emphasize improving the capacity of communities to know and use the law (Golub, 2003). Local people who develop some expertise can play crucial roles as paralegals. Legal aid may be provided by non-governmental organizations, law schools and government programmes (see, for example, NNMLS, 2000). Habits, concepts and prejudices sometimes lead disputants to behave in ways that may not be conducive to reaching agreement. Specific techniques, such as interest-based negotiation and assistance from facilitators or mediators, may play a valuable role.

For communities whose water rights are under immediate threat, legal empowerment measures offer some of the most promising opportunities. A first challenge is to enable communities to link with sources of assistance. Media publicity and networking, for example through civic organizations, may play a key role. The second challenge, and probably the main constraint, is the availability of resources, such as funds and skilled lawyers. Usually, governments are not particularly enthusiastic about providing resources to those who want to challenge agency actions. Legal empowerment requires detailed work on the ground, much less exciting and much more prone to failure than advocacy.

In practice, it requires lots of compromise, deciding which struggles to prioritize, which goals seem achievable, working with government officials and seeing what can be done within the constraints of an existing system. What may be most relevant for communities is to have knowledgeable local people and outside counsellors who know the existing legal framework, and what bases it may offer communities for securing water rights.

Advocacy

Advocacy draws attention to community concerns, concepts and roles in the management of water. Outside groups may provide links with reporters, document problems, convene forums to discuss issues and strengthen capability to prepare and deliver messages. Advocacy can open access to additional forums for defining community rights to water. If links can be obtained to media or decision makers, then advocates may be able to mobilize allies and reframe issues in ways that favour community concerns. Advocates may play influential roles in policy debates at the national and international level.

Non-governmental organizations (NGOs), local communities and others concerned about adverse impacts on communities have played major roles in blocking the passage of new water laws in countries such as Thailand, Sri Lanka, Peru and Ecuador (Gunatilake and Gopalakrishnan, 2002; Trawick, 2003; Bauer, 2004, p. 146). In the case of Indonesia's recent water law, key provisions regarding water rights were revised with intentions of better protecting poor farmers' access to water, in response to concerns of NGOs, academics and some parliamentarians.

However, communities themselves cannot earn a living from advocacy, and rather than endless ideological struggle they are likely to prefer pragmatic engagement that expands meaningful opportunities. Governments and NGOs may focus on protecting subsistence, while communities also want to gain the benefits of new technologies and markets. Time scales for local advocacy may differ from those of organizations that would like to aid them. On the one hand, communities want pragmatic solutions to immediate problems, and so may have less interest in medium-term struggle for policy reform and intricate basin planning. On the other hand, communities may pursue their efforts over decades or even centuries outliving opponents, overcoming temporary setbacks and applying patient persistence to achieve their local objectives (Maass and Anderson, 1978).

Technical advice

Lack of technical information is often a key constraint. For example, technical analysis can help to clarify how much water is available and how it is being used. This may help correct misconceptions and focus attention more precisely on feasible solutions. Participatory rural appraisal, participatory geographic information systems, scenario models and related methods offer a variety of techniques for blending local and outside knowledge in ways that can be relatively fruitful and efficient in terms of local people's time. Information technologies such as remote sensing, databases, modelling, e-mail and web sites are reducing the costs of monitoring, but they still face constraints including limited funding for acquiring data, scientific uncertainty and information overload for those who want to use such data.

Information may be useless if it seems irrelevant, incomprehensible or confusing. A few people within a community may be interested and able to learn deeply about an issue, but most people are busy with their lives and are not interested in becoming technical experts. Specific studies focused on problems perceived as important and framed in ways that reflect community concerns are much more likely to be worthwhile than more academic and general research. For a community, an attractive option may be to have their own experts, both local and external, to counter – at least partially – the weight of expertise that government agencies can mobilize.

Networking

Establishing and strengthening of local organizations can be facilitated by outside assistance. However, in dealing with basin- and sub-basin-scale issues, strengthening of external links may be more crucial. Networking between communities cross-fertilizes experiences and enables coordinated efforts. As discussed earlier, one of the main challenges for water management is the scale of conflicts that can extend across broad areas. Local people may be able to make use of existing linkages with other areas, through relatives and friends living elsewhere, formal organizations and political and other contacts. Outsiders may be in a good position to foster linkages between distant groups with few existing connections, creating 'bridging' social capital (Putnam, 1993). An outside organization may be able to convene a workshop, seminar or other activity that brings people together across a basin or sub-basin. This may facilitate constructing coalitions for coordinated efforts to pursue shared or complementary interests.

However, networking for its own sake risks dissipating time and energy on prolonged discussion. Reforms that offer a voice in consultation processes but not genuine power, e.g. representation on advisory basin committees, may be useful, or may consume effort out of proportion to outcomes, especially if they require high costs in time and money to congregate dispersed networks of participants. Networks might be most useful when engaged for specific objectives, such as sharing solutions, lobbying government agencies and legislatures, or coordinating responses to a crisis.

Participatory planning

Opportunities can be opened for communities to take part in preventing and resolving problems, increasing input from stakeholders, promoting dialogue, facilitating joint problem solving and structuring processes through which decisions can be made jointly with user representatives. For water rights, this may apply across a range of activities from managing a particular crisis, seasonal planning for water allocation during periods of scarcity to long-term basin planning. A 'participation audit' could assist an agency to assess the ways it allows and supports participation, and to determine whether stronger, more empowering participation may provide greater incentives for stakeholders. Stakeholders may not know about opportunities for participation and, even when they do, they may be sceptical about what potential there is for genuine influence. Signs of credibility, such as participation of senior agency staff and honesty about how final decisions will be made, may provide important signals.

Methods for reducing the transaction cost of participation, particularly the time required, can make a difference, for example by providing information, accepting input and engaging in dialogue through multiple forms, rather than restricting interaction to a single stylized approach such as conventional public hearings. Many efforts labelled as participation or decentralization fail to convey genuine power, while others that do transfer power, money and other resources fail to consider the risks of local abuse, inequities, overexploitation of resources and other problems. A key question is: 'who decides?' Empowerment is far more meaningful if both sides must agree, or when decisions are delegated, authority transferred or local institutions enabled to make decisions on their own, while governments and civil society act to provide appropriate regulatory checks and balances.

Platforms

Availability of particular forums or platforms (Steins and Edwards, 1998; Boelens and Hoogendam, 2002) can make negotiation possible, providing focused arenas within which problems can be discussed, alternatives considered and agreements formulated. This may occur as part of other activities, as discussed earlier in terms of participatory planning, or through establishment of special-purpose organizations, such as alliances of concerned groups, basin committees or water councils. Groups can be brought together to discuss issues and consider establish-

ing arrangements for cooperation. Facilitators may help to convene stakeholders and strengthen organizations.

However, ostensibly neutral processes convening stakeholders to create consensus, based on shared information and improved communication, risk perpetuating and worsening existing differentials in power, wealth and status (Edmunds and Wollenberg, 2001). Rather than using a pure strategy of relying on a single forum, communities may want to employ a mixed strategy of working through multiple forums and asserting multiple bases for their claims to water. Outsiders intending to preferentially aid particular groups, e.g. poor people, women, ethnic minorities or other disadvantaged groups, may want to take a careful and strategic approach to the development of platforms, as may communities themselves. A strategic approach to platforms may involve selective alliances, controlling release of information about community conditions and objectives, waging struggles in multiple forums, opportunistically improvising responses to particular events and accepting pragmatic compromises conceived of as only temporary concessions during continuing contests over rights and resources.

Concluding Conjectures

Critical analysis of community-based natural resources management and institutional design principles provides a basis for proposing some working hypotheses[17] about how communities may be expected to act to secure rights to water. These may help to understand how communities may act to defend customary rights to water, and to manoeuvre within a plural framework of national and local laws and other normative orders regulating access to water, and the potential results of changes in institutional arrangements. Conjectures about priorities, principles and programmes need to be customized to specific contexts where communities are involved in basin water governance, but they may offer some practical starting points for discussion, research and practical application. Realistic expectations about community priorities may reduce the risks of waste, disruption and disappointment due to inappropriate interventions.

Critiques of community-based natural resources management and of institutional design principles clarify some of the challenges and constraints to interventions intended to change water allocation institutions. The scale of competition over water makes negotiation of credible agreements (commitments) contingent on the availability of government enforcement. Coalitions and compromises to forge cooperation among heterogeneous users may reflect and amplify differences due to wealth, power, gender, ethnicity and other characteristics, unless there are particular countervailing conditions and arrangements that promote equity. Political contests over claims to water, budgets and related resources often impel participatory reforms more towards allowing a voice in agency decisions than towards partnership (where both sides would have a veto) or fuller empowerment of communities.

Principles for institutional design can be made more applicable by suitable adaptation to the context of community perceptions and practical priorities. Communities may be more concerned about: (i) problemsheds than hydrologic catchments; (ii) protecting local practices more than precise proportionality of rights, costs and benefits; (iii) representation during crises more than participation in deliberative platforms; (iv) effective recourse to remedy harm to rights more than carefully calibrated sanctions; (v) administrative mediation more than consensual forums or courts; (vi) recognition of customary rights more than formal registration; and (vii) local autonomy and strategic coalitions with local governments and other allies more than elaboration of nested hydraulic enterprises.

A community perspective on water governance suggests that the dynamics of community collective action to secure water rights are likely to be:

- Primarily *defensive*, concerned with protecting against threats to existing claims.
- Constructed of heterogeneous *coalitions*, within and between communities.
- Employing *mixed strategies* using multiple claims and forums.
- Opportunistically *improvised in response to particular crises*.

Therefore, interventions aimed at optimizing and reallocating water use, assuming shared interests, attempting to monopolize water allocation decisions in a single forum, and pursuing comprehensive, anticipatory planning, such as ambitious projects for basin master planning, and IWRM, may fit poorly with the dynamics of community collective action, and so they may be prone to being ignored, resisted and rejected. Modest institutional modifications that fit the dynamics of community collective action and help secure rights and resolve urgent crises may meet with greater success.

Interventions in basin governance intended to support community-based natural resources management and strengthen local organizations may have better prospects if carefully fitted to the contours of institutional landscapes and oriented towards promising pathways for institutional transformation. From a community perspective, short-term *regulatory adjustments* that solve immediate problems and long-term *rights to resources* may be more important than medium-term reforms to build basin management organizations. Targeted training for local *paralegals* and access to *legal aid* may do more to make laws effective than extensive broadcasts, brochures and lectures.

Facilitating *strategic links* to outside groups and agencies may do much more for community capacity than intensive internal organizational development. Lobbying in opposition to changes that threaten to further disadvantage people may be helpful, but advocacy that pragmatically expands meaningful opportunities for people to sustainably improve their lives may accomplish even more. Participatory planning that honestly promises influence over decisions creates credibility, but *empowerment* that establishes partnerships, delegates decisions, transfers authority or enables autonomy (within appropriate regulatory checks and balances) may do even more to improve basin resource governance.

Information technologies are expanding availability of information but, to make abundant information useful, communities need *local and external expertise* to apply knowledge to serve their objectives. Platforms may facilitate formation of acceptable agreements but they may be only part of developing a *portfolio of community strategies* to negotiate rights to water.

Endnotes

1 Communities as used here include villages, irrigators' organizations and other groups of people acting collectively, and may include not only small face-to-face groups where all members know each other, but also larger groupings, for example based on shared ethnic identity and social relationships. For an example of such larger communities in the Andes, see Boelens *et al.*, Chapter 6, this volume.

2 For a discussion of negotiation of American Indian water rights, see Checchio and Colby (1993); McCool (1993, 2002); Colby *et al.* (2005). Danver (2004) compares three strategies: litigation, participation in project planning and negotiation, used by three American Indian groups in New Mexico, noting the tendency of different strategies to converge into similar processes.

3 For a discussion of various levels and scales of participation, following Arnstein's original (1969) 'ladder of participation', see Bruns (2003).

4 It is also important to distinguish clearly decentralization that deconcentrates power to local branches of central government from devolution that actually shifts authority (including authority over funding and conflict resolution) to local bodies.

5 See, among others, Agrawal and Gibson (2001), for a review of community-based conservation, Agrawal (2003), for a recent review of research on common-pool resources management and Mansuri and Rao (2005), for a recent synthesis of peer-reviewed studies of community-based and community-driven development and related approaches.

6 For a nuanced empirical and theoretical discussion of conflicts in Balinese subaks see Spiertz (2000).

7 Fisher *et al.* (1991) define power in negotiation operationally in terms of the 'best alternative to a negotiated agreement' (BATNA) – in other words, the 'fallback position', the outcome that a party could obtain if agreement is not reached.

8 Mansuri and Rao (2005) note that elite control may be almost inevitable. As apparent in most of the literature, they use the notion of 'capture' in a rather unexamined way. Capture is assumed to be undesirable and detrimental, and not clearly distinguished from other forms of local political support or 'buy-in'. The literature does not seem concerned about the finding from the study of regulated industries that capture by regulated interests may be a less important phenomenon than the tendency of new institutions to pursue their bureaucratic interests in expanding budget, staff and authority ('turf'). Similarly, analysis of

'elite' roles might benefit from more attention to competition within and between elites, and the 'circulation of elites'.

9 For co-production, see Lam (1997); Ostrom (1997). Berkes (1994) discusses co-management. For regulated autonomy, see Ribot (2002); Bruns (2003).

10 For institutional 'fit' see Young (2002).

11 Hardin's (1998) commentary belatedly corrected his earlier (1968) article to clarify that the tragedy is a problem for 'unmanaged' commons, those without effective institutional arrangements (state or community) to regulate access.

12 Cleaver and Franks (2003); see also Lévi-Strauss' original discussion of bricolage (1966 [1962]) (available at http://varenne.tc.columbia.edu/bib/info/levstcld066savamind.html).

13 It should be clear that the emphasis in this chapter is on community priorities and institutional arrangements that may be effective in meeting their priorities. This need not necessarily mean that these are the arrangements that would be the most economically efficient, socially equitable, ecologically sustainable or institutionally robust, or the ones that would best serve the interests of other stakeholders or the entire society. The intention here is to highlight relevant institutional options from a community perspective, within a larger landscape of social contestation concerning resource management.

14 Edmunds and Wollenberg (2001) critique the neutrality and inclusiveness of forums. For challenges in transferring meaningful authority over irrigation management, see Bruns (2003) and Vermillion (2005). 'Shopping' among forums need not require choosing only one forum: instead, a disputant may employ a portfolio (or basket) of forums, i.e. a mixed strategy.

15 It should be noted that this provision provides a way for such recognition of existing use to occur within a civil law system, which does not offer the same means for recognizing past practices as would be available within a common law system.

16 For polycentric governance, see Ostrom (1997). Applications to water resources include Ostrom (1990, 1992); Blomquist (1992); Tankimyong *et al.* (2005).

17 The ideas developed here are offered as conjectures, suggesting what might be the most likely (initial or 'prior') expectations based on currently available knowledge, subject to customization, testing and refutation or revision based on additional information. Thus, for example, rather than naively expecting an equal (or even pro-poor) per capita distribution of benefits, it seems more likely (i.e. an appropriate working assumption or null hypothesis) to start from an expectation that

outcomes will probably reproduce existing distributions of power and benefits (or skew them even more) unless countervailing measures are employed. From a practical perspective, particularly interesting questions then concern the extent to which outcomes may be affected by specific measures such as targeting, empowerment and advocacy.

References

Agarwal, A., delos Angeles, M.S., Bhatia, R., Chéret, I., Dávila-Poblete, S., Falkenmark, M., Fernando, G.V., Jønch-Clausen, T., Aqt Kadi, M., Kindler, J., Rees, J., Roberts, P., Rogers, P., Solanes, M. and Wright, A. (2000) *Integrated Water Management.* Global Water Partnership, Denmark.

Agrawal, A. (2003) Sustainable governance of common-pool resources: context, methods and politics. *Annual Review of Anthropology* 32, 243–262.

Agrawal, A. and Gibson, C.C. (2001) The role of community in natural resource conservation. In: Agrawal, A. and Gibson, C.C. (eds) *Communities and the Environment: Ethnicity, Gender, and the State in Community-Based Conservation.* Rutgers University Press, New Brunswick, New Jersey.

Anderies, J.M., Janssen, M.A. and Ostrom, E. (2003) Design principles for robustness of institutions in social-ecological systems. Presented at *Joining the Northern Commons: Lessons for the World, Lessons from the World,* Anchorage, Alaska.

Bauer, C.J. (2004) *Siren Song: Chilean Water Law as a Model for International Reform.* Resources for the Future, Washington, DC.

Berkes, F. (1994) Co-management: bridging the two solitudes. *Northern Perspectives* 22, 2–3.

Blomquist, W. (1992) *Dividing the Waters: Governing Groundwater in Southern California.* Institute for Contemporary Studies, San Francisco, California.

Boelens, R. and Hoogendam, P. (eds) (2002) *Water Rights and Empowerment.* Van Gorcum, Assen, Netherlands.

Boelens, R., Dourojeanni, A., Duran, A. and Hoogendam, P. (1998) Water rights and watersheds: managing multiple water uses and strengthening stakeholder platforms. In: Boelens, R. and Dávila, G. (eds) *Searching for Equity: Conceptions of Justice and Equity in Peasant Irrigation.* Van Gorcum, Assen, Netherlands.

Bruns, B. (1992) Just enough organization: water users' associations and episodic mobilization. *Visi: Irigasi* 1992, 33–41.

Bruns, B. (1993) Promoting participation in irrigation: reflections on experience in Southeast Asia. *World Development* 21, 1837–1849.

Bruns, B. (2003) Water tenure reform: developing an extended ladder of participation. Presented at *Politics of the Commons: Articulating Development and Strengthening Local Practices,* Chiang Mai, Thailand.

Bruns, B. and Meinzen-Dick, R. (eds) (2000) *Negotiating Water Rights.* Vistaar, New Delhi, India.

Bruns, B. and Meinzen-Dick, R. (2001) Water rights and legal pluralism: four contexts for negotiation. *Natural Resources Forum* 25, 1–10.

Checchio, E. and Colby, B. (1993) *Indian Water Rights: Negotiating the Future.* Water Resources Research Center, University of Arizona, Tucson, Arizona.

Cleaver, F. and Franks, T. (2003) How institutions elude design: river basin management and sustainable livelihoods. Presented at *The Alternative Water Forum,* Bradford, UK.

Colby, B.G., Thorson, J.E. and Britton, S. (2005) *Negotiating Tribal Water Rights: Fulfilling Promises in the Arid West.* University of Arizona Press, Tucson, Arizona.

Danver, S.L. (2004) Liquid assets: a history of tribal water rights strategies in the American Southwest. PhD thesis, University of Utah, Salt Lake City, Utah.

de Soto, H. (2000) *The Mystery of Capital: Why Capitalism Triumphs in the West and Fails Everywhere Else.* Bantam Press, London.

Edmunds, D. and Wollenberg, E. (2001) A strategic approach to multi-stakeholder negotiations. *Development and Change* 32, 231–253.

Fisher, R., Ury, W. and Patton, B. (1991) *Getting to Yes: Negotiating Agreement without Giving in.* Penguin, New York.

Golub, S. (2003) *Beyond Rule of Law Orthodoxy: the Legal Empowerment Alternative.* Carnegie Endowment, New York.

Guillet, D. (1998) Rethinking legal pluralism: local law and state law in the evolution of water property rights in northwestern Spain. *Comparative Studies in Society and History* 40, 42–70.

Gunatilake, H.M. and Gopalakrishnan, C. (2002) Proposed water policy for Sri Lanka: the policy *versus* the policy process. *Water Resources Development* 18, 545–562.

Haefele, E.T. and Kneese, A.V. (1973) Residuals management, metropolitan governance, and the optimal jurisdiction. In: Haefele, E.T. (ed.) *Representative Government and Environmental Management.* Johns Hopkins University Press, Baltimore, Maryland.

Hardin, G. (1968) The tragedy of the commons. *Science* 162, 1243–1248.

Hardin, G. (1998) Extensions of 'the tragedy of the commons'. *Science* 280, 682–683.

Harwell, E.E. and Lynch, O.J. (2002) *Whose Resources? Whose Common Good? Towards a New Paradigm of Environmental Justice and the National Interest in Indonesia.* Center for International Environmental Law, Washington, DC.

IDI (Infrastructure Development Institute, R.B., Ministry of Construction, Japan) (1997) *The River Law* (as Amended in July 1997, English Translation). Infrastructure Development Institute, Tokyo.

Knox, A. and Meinzen-Dick, R. (2001) *Collective Action, Property Rights and Devolution of Natural Resources Management: Exchange of Knowledge and Implications for Policy.* A workshop summary paper, International Food Policy Research Institute, Washington, DC.

Kosko, B. (1994) *Fuzzy Thinking: the New Science of Fuzzy Logic.* Hyperion, New York.

Lam, W.F. (1997) Institutional design of public agencies and coproduction: a study of irrigation associations in Taiwan. In: Evans, P. (ed.) *State–Society Synergy: Government and Social Capital in Development.* University of California Press, Berkeley, California, pp. 11–47.

Lévi-Strauss, C. (1966) *The Savage Mind.* University of Chicago Press. Chicago, Illinois.

Lynch, O.J. (1998) Law, pluralism and the promotion of sustainable community-based forest management. *Unasylva* 49, 194–207.

Maass, A. and Anderson, R.L. (1978) *… and the Desert Shall Rejoice: Conflict, Growth and Justice in Arid Environments.* MIT Press, Cambridge, Massachusetts.

Mansuri, G. and Rao, V. (2005) Community-based and -driven development: a critical review. *The World Bank Research Observer* 19, 1–39.

McCool, D. (1993) Indian water settlements: the prerequisites of successful negotiation. *Policy Studies Journal* 21, 227–242.

McCool, D. (2002) *Native Waters: Contemporary Indian Water Settlements and the Second Treaty Era.* University of Arizona Press, Tucson, Arizona.

Mosse, D. (2003) *The Rule of Water: Statecraft, Ecology, and Collective Action in South India.* Oxford University Press, New Delhi, India.

NNMLS (Northern New Mexico Legal Services) (2000) Acequias and water rights adjudications in northern New Mexico. In: Bruns, B. and Meinzen-Dick, R. (eds) *Negotiating Water Rights.* Vistaar, New Delhi, India.

North, D.C. (1990) *Institutions, Institutional Change and Economic Performance.* Cambridge University Press, New York.

Olson, M. (1971) *The Logic of Collective Action: Public Goods and the Theory of Groups.* Harvard University Press, Cambridge, Massachusetts.

Ostrom, E. (1990) *Governing the Commons: the Evolution of Institutions for Collective Action.* Cambridge University Press, Cambridge, UK.

Ostrom, E. (1992) *Crafting Institutions for Self-Governing Irrigation Systems.* Institute for Contemporary Studies Press, San Francisco, California.

Ostrom, E. (1997) Crossing the great divide: coproduction, synergy, and development. In: Evans, P. (ed.) *State–Society Synergy: Government and Social Capital in Development.* University of California Press, Berkeley, California, pp. 85–118.

Ostrom, E. (1999) Coping with tragedies of the commons. *Annual Review of Political Science* 2, 493–535.

Putnam, R.D. (1993) *Making Democracy Work: Civic Traditions in Modern Italy.* Princeton University Press, Princeton, New Jersey.

Ravnborg, H.M. (2004) Water management and the poor – organizing to (re)gain access to water in the Nicaraguan hillsides. Presented at the *Tenth Biennial Conference of the International Association for the Study of Common Property*, Oaxaca, Mexico.

Ribot, J.C. (2002) *Democratic Decentralization of Natural Resources: Institutionalizing Popular Participation.* World Resources Institute, Washington, DC.

Rogers, P. and Hall, A.W. (2000) *Effective Water Governance.* Global Water Partnership, Denmark.

Sanbongi, K. (2001) Formation of case law and principles in watershed management. In: Bogdanovic, S. (ed.) *Proceedings of the Regional Conference on the Legal Aspects of Sustainable Water Resources Management.* International Association for Water Law, Novi Sad, Serbia, pp. 121–127.

Sengupta, N. (2000) Negotiation with an under-informed bureaucracy: water rights on system tanks in Bihar. In: Bruns, B. and Meinzen-Dick, R. (eds) *Negotiating Water Rights.* Vistaar, New Delhi, India.

Sengupta, N. (2004) Common mistakes about common property. Paper prepared for the *Tenth Biennial Conference of the International Association for the Study of Common Property*, Oaxaca, Mexico.

Shah, T., Makin, I. and Sakthivadivel, R. (2001) Limits to leapfrogging: issues in transposing successful river basin management institutions in the developing world. *Intersectoral Management of River Basins, Proceedings of an International Workshop on 'Integrated Water Management in Water-Stressed River Basins in Developing Countries: Strategies for Poverty Alleviation and Agricultural Growth'*, Loskop Dam, South Africa, 16–21 October 2000. International Water Management Institute (IWMI) and German Foundation for International Development (DSE), Colombo, Sri Lanka.

Spiertz, H.L.J. (2000) Water rights and legal pluralism: some basics of a legal anthropological approach. In: Bruns, B. and Meinzen-Dick, R. (eds) *Negotiating Water Rights.* Vistaar, New Delhi, India.

Steins, N.A. and Edwards, V.M. (1998) Platforms for collective action in multiple-use CPRs. Presented at *Crossing Boundaries, the Seventh Annual Conference of the International Association for the Study of Common Property*, Vancouver, British Columbia, Canada.

Tankimyong, U., Bruns, P.C. and Bruns, B.R. (2005) The emergence of polycentric water governance in northern Thailand. In: Shivakoti, G., Vermillion, D., Lam, W.F., Ostrom, E., Pradhan, U. and Yoder, R. (eds) *Asian Irrigation in Transition: Responding to Challenges.* Sage, New Delhi, India.

Trawick, P. (2003) Against the privatization of water: an indigenous model for improving existing laws and successfully governing the commons. *World Development* 31, 977–996.

Vermillion, D.L. (2005) Irrigation sector reform in Asia: from 'patronage with participation' to 'empowerment with accountability'. In: Shivakoti, G., Vermillion, D., Lam, W.F., Ostrom, E., Pradhan, U. and Yoder, R. (eds) *Asian Irrigation in Transition: Responding to Challenges.* Sage, New Delhi, India.

Wester, P., Merrey, D.J. and Lange, M.D. (2003) Boundaries of consent: stakeholder representation in river basin management in Mexico and South Africa. *World Development* 31, 797–812.

Wolf, E. (1983) *Europe and the People without History.* University of California Press, Berkeley, California.

Young, O.R. (2002) *The Institutional Dimensions of Environmental Change.* MIT Press, Cambridge, Massachusetts.

4 Dispossession at the Interface of Community-based Water Law and Permit Systems

Barbara van Koppen

International Water Management Institute, Southern Africa Regional Program, Pretoria, South Africa; e-mail: b.vankoppen@cgiar.org

---

Abstract

This chapter challenges the assumption that permit systems are the best legal device to address the challenges of water scarcity in the 21st century, as widely held in the global trend of water law revisions. It analyses the origins of permit systems and their dual obligations and entitlement dimensions in Roman water law. It then highlights their differential development paths in high-income countries compared with middle- and low-income countries. As argued, permits may work in high-income countries as a hook for governments to impose obligations, like registration, taxation or waste discharge charges. In exceptionally arid closing basins, like Australia and the western USA, the century-old permit systems may facilitate water sharing, including trade. However, in middle- and low-income countries in Latin America and sub-Saharan Africa, permit systems were introduced by the colonial powers with the primary goal of dispossessing indigenous water users of their prior claims to water. Evidence from Chile and elsewhere shows how 'modern' water law revision risks reinforcing this colonial legacy for the large majority of informal water users. Permits as individual water rights based on an administrative act, first, ignore the intrinsically different nature of communal indigenous water rights regimes; secondly, favour the administration-proficient; thirdly, may entail explicit discriminatory conditions; and fourthly, discriminate against poor women even more than poor men. The chapter concludes with recommendations for formal legal tools that strengthen water entitlements of informal small-scale water users.

Keywords: water law, formal water rights, permits, customary water rights, Roman water law, water trade, Chile, sub-Saharan Africa, informal, poverty, gender.

Rationale, Aim and Structure of the Chapter

Background and rationale

The present chapter focuses on the highly problematic interface between community-based water law on the one hand and permit systems (also called administrative formal water rights, licences, concessions, royalties or leases) on the other. It is well known that both community and permit systems coexist, though not necessarily smoothly (see Meinzen-Dick and Nkonya, Chapter 2, this volume). In particular, as Boelens *et al.* (Chapter 6, this volume) highlight, after the colonization of the Andes, when permit systems were imposed over existing community-based systems, they created conflict and divested indigenous peoples of their claims to water and its use and management. At the same time, this imposition impacts the ways in which communities and their allies can engage with the contemporary state in the future design and negotiation of alternative water management arrangements.

A closer look at this interface is critical because permit-based formal water rights are now also rapidly gaining popularity in the region with the largest proportion of indigenous and informal water users: sub-Saharan Africa (see Meinzen-Dick and Nkonya, Chapter 2, this volume for the rationale of the proponents). There, neither the risk of dispossession by superimposition of permits over the widely prevailing indigenous water governance arrangements nor the alternatives emerging in Latin America have received much attention.

Fully fledged permits are written certificates that state: 'such matters as the approximate location of the land to be supplied, the purpose(s) for which water is sought, the source from which it is to be drawn, the proposed point of diversion, the volume to be diverted, the nature of existing and proposed hydraulic structures, and drainage and treatment' (Caponera, 1992). Permits entail the 'agreement to abide by conditions imposed in the permit' (Hodgson, 2004), usually for a fixed duration after which a review is performed. Permits are the legally binding contracts between the state and individual or organized water users.

Permit systems are now being promoted as the single most effective legal device to address the water management problems of the 21st century. They are increasingly perceived as a standard ingredient of Integrated Water Resources Management (IWRM). Virtually all water law reforms of the past few decades have introduced or strengthened this legal device: in high-income countries such as the UK in 1963 and in France in 1964; in middle-income countries in Latin America, e.g. Chile (Water Code of 1981) and Mexico (National Waters Law of 1992); and in low- and middle-income countries in sub-Saharan Africa, including Mozambique (Ley de Agua 1991), Uganda (Water Statute 1995), Ghana (Water Resources Commission Act 1996), Tanzania (1997 and 2002 Amendments to Water Ordinance [Control and Regulation] Act No. 42 of 1974, and currently redrafting the law), Zimbabwe (Water Act No 31/1998), South Africa (National Water Act 1998), Burkina Faso (Loi d'orientation relative a

la gestion de l'eau 2001) Kenya, (The Water Act 2002) and Swaziland (Water Act 2002).

As an intrinsic part of permit systems, these new water laws invariably confirm and strengthen the role of the state as trustee, owner or custodian of the nation's water resources. They typically increase the scope of water resources declared as being public and so under state control, for example, including groundwater as part of public water. Finally, they tend to expand the uses of water under state control requiring state authorization through permits including, for example, waste discharges.

Obligations, entitlements and dispossession

There are two dimensions to a permit (or administrative water right): (i) an obligation dimension (as the name 'permit' conveys); and (ii) an entitlement dimension (as expressed by the name 'right'). Users' obligations are conditions attached to the permits. Permits serve as a 'hook' for the state to impose such obligations. Not surprisingly, many government water managers tend to be most interested in this obligation dimension as they expect permits to be vehicles allowing more effective regulation of water resources. Global debates on permit systems often refer to this role as a hook to impose obligations. An exception is found in Latin America, where the focus is on the entitlement dimension, a point returned to later. For example, in high-income countries, 'the polluter pays' principle is increasingly implemented through waste discharge permits. Underresourced governments in sub-Saharan Africa, advised and financed by international organizations like the World Bank and donors, are often attracted to these systems, in part because they can provide financing for the basin organizations, as these international organizations often prescribe to establish as a conditionality of aid. The obligatory registration of water users also provides indispensable information for water managers about the use of the resource that is to be managed, certainly in low- and middleincome countries where such information is largely lacking.

Legally, though, permits are only one way of imposing obligations. States have fiscal, administrative and policing tools that can achieve the same purpose. As found in Mexico, Tanzania, South Africa and elsewhere, these other methods may even perform considerably better in enforcing obligations, at least if well targeted at specific water users (van Koppen, 2007, unpublished).

It is beyond the scope of this chapter to discuss the obligation dimension of permit systems in further depth, except in the sense that the effectiveness of permits as a hook to impose obligations fully depends upon the way in which the other dimension, the entitlement dimension, works out. This chapter focuses on the latter.

As argued here, for the Andean region as well as for sub-Saharan Africa, permit systems boil down to the formal dispossession of rural informal water users who manage their water under community-based arrangements. What is at stake becomes clear in the case of Ghana, where the legal power of the traditional author-ities, or 'stools', and the customary links between land and water rights are still strong enough to provide a voice that rural communi-ties elsewhere often lack. There Sarpong (undated), an expert in water law, made the following comments on The Water Resources Commission Act of 1996 and its establishment of permits:

By a stroke of the legislative pen and policy intervention, proprietary and managerial rights which had been held from time immemorial by families, stools, and communities have been taken away from a people some of who probably had no prior knowledge of the matter.
Significantly, water in view of its appurtenance to land, has all along been regarded as part of land. The Constitution of 1992 recognizes customary landholdings and bars state intervention and/or appropriation of lands except under stringent conditions laid down under Article 20. Indeed, the 1992 Constitution puts behind us the era of unbridled acquisition of land without payment of compensation. The issue is whether the Water Resources Commission Act can unilaterally hive off water from land and provide a separate institutional and legislative framework to address its use. If the Constitution provides the regime of land tenure ought to be in conformity with customary law, then any attempt by the state to fashion out a separate regime for water that runs counter to this constitutional edict will offend the letter, if not the spirit, of the Constitution. This is

an issue that deserves to be examined having regard to the massive nature of the assault of the legislation on customary proprietary water rights. [...] If the law on appropriation of land by the state is to be used as a guide on the matter, then it may be surmised that the Water Resources Commission, in spite of its far sweeping powers with regard to water appropriation, would have to yield to the constitutional requirement of providing prompt, adequate, and effective compensation in accordance with Article 20 of the Constitution for the compulsory acquisition of customary water rights as obtains in the case of compulsory land acquisition by the state.

It is remarkable indeed that this dispossession of indigenous water rights has received so little attention up till now in sub-Saharan Africa. One explanation may be that the colonial water laws which, on paper, entailed dispossession at a large scale, were only partially implemented. Water administrations focused instead on gradu-ally formalizing water sectors of settlers involved in large-scale irrigation, mining, urbanization, hydropower and upcoming industries. It was only recently that, under the banner of IWRM, water laws were revised to include permits more explicitly and nationwide. Now nationwide laws incorporating permit systems are also imple-mented with more force. The good news is that the limited implementation in sub-Saharan Africa still allows timely adaptation of the paper laws in accordance with the lessons that have been learned by now.

Structure of the chapter

For a better understanding of the rationale for and double-sided nature of permit systems with obligations and entitlements, a closer look at their historic origins is revealing: the second section (Roman Water Law) and the third section (The Transformation of Roman Water Law in High-income Countries) highlight these origins.

This history highlights how dispossession through the powers of the ruling aristocracy has been contested since the early 1800s and that it was only recently that permit systems became more popular again. The older permit systems in arid areas in former colonies, in particular in high-income Australia and the western USA are exceptional, although much cited, and beyond the scope of this chapter[1] (van Koppen, 2007, unpublished). The point is that both the recent and older permit systems work in the very specific context of highly sophisticated and formalized water economies in fully industrialized societies. This specific context tends to be ignored when the international donor community finances their replication in low- and middle-income countries of the south with entirely different settings.

In the south, the vast majority of water users are informal. As primary water takers, they develop their own water resources. The lack of state-sponsored infrastructure and water management institutions means that self-initiative and climate determine water availability. Yet, public water and permit systems, without many obligations attached, have already existed since the colonial era. One reason for this explored in further depth in the fourth section (The Colonial Legacy of Water Law in Latin America and Sub-Saharan Africa) is the still omnipresent legacy of the colonial water laws in Latin America and sub-Saharan Africa.

The fifth section (Permits as Property Rights in Low- and Middle-income Countries Today) develops a more abstract analysis of the resulting water rights systems in low- and middle-income countries in these two continents today. It exposes the essence of permits as formal entitlements to a public and shared resource that are basically vested by a mere administrative act. While administration as a basis for rights to water may be meaningful in the highly controlled conditions of high-income settings, such a legal system is ludicrous in societies with deep divides between the administratively knowledgeable, who can easily obtain such paper rights, and the large majority of informal users who cannot, or, if they can, can often only do it too late.

The sixth section (Resource Grab by Design: Evidence from Chile and Elsewhere) confirms this essence by tracing the real-life implications of permit systems for the earliest and best-documented case in the developing world: Chile. The seventh section (Discrimination by Water Administration) focuses in depth on two generic sets of discriminatory processes at stake when permit systems with their seemingly 'neutral' and 'orderly' administrative measures are imposed over informal rights systems in societies with deep administrative divides, not only in Chile, with its minority of informal water users, but even more so where the informal sector is larger. Conclusions and recommendations, particularly for sub-Saharan Africa, are given in the last section (Recommendations: challenging the colonial legacy of dispossession).

Roman Water Law

The dispossession dimension of permit systems has existed ever since the Romans invented the famous notions of public as opposed to private water and the requirement to obtain permits for the use of public water. From the outset, permits served the double purpose of providing the hook for the state to impose obligations *and* dispossessing conquered tribes from their existing claims to land and water resources. Caponera's (1992) fascinating classic analysis of historical and contemporary water law provides the following information, (pp. 29–48) although he, as most other water lawyers, has never explicitly mentioned the element of dispossession.

Throughout the 1500 years of Roman expansion, from 1000 BC till about AD 500, the core principle of Roman water law was that collectivities classified water resources into public waters[2] subject to regulation by the collectivity, for example for navigation (*res populi*, and later *res publica*) on the one hand and private waters, where the private title-holders (and his neighbouring private title-holders) all had rights to use and abuse surface water and groundwater as they liked (*ius utendi et abutendi*), on the other. This included the right to sell water. The underpinning 'statement of principle' that running water, like air, was a thing common to everyone (*res comunis omnium*) to which no one could claim ownership because of its nature, remained throughout, although with limited practical implication other than the classification mentioned.

While these core principles stayed, important changes took place with regard to 'collectivity', 'public' and their hierarchies. By 500 BC, 'collectivity' was still confined to the three agricultural communities founding Rome and the Republic in Latium immediately surrounding

Rome. By AD 500, history's early and aggres-
sive conquest of neighbouring tribes and their
land and water resources had led to an empire
stretching from continental western Europe to
Byzantium in the Near East. From the very
outset, the classification of land and water
resources as public versus private was linked to
this military conquest. Initially, the legal status
of water entirely followed that of land: springs
and artesian wells were appurtenances of land,
so if land was declared as being public, all water
running, springing, lying or gathering thereon
was deemed public. All water that fell within
private land (rain, groundwater and minor
water bodies) was deemed private.

Thus, in these early days of Roman
conquest, the 'lawful' way to appropriate territories conquered and, hence, their water
resources, was typically by ranking it as 'public'
land. Public land also included all mountain
land and such strips of land marking the
borders between existing colonies or, within a
colony, between allotted plots of land. These
borders often corresponded to a perennial river
and, sometimes, to streams – typically reliable
borders for delimiting land. As a consequence,
all rivers and some streams, the springs feeding
urban aqueducts, mountain lakes and such
rainwater as was collected by natural mountain
pools or artificial tanks, were also declared as
being public. In later phases, the Romans even
further expanded their definition of waters that
were seen as public. All perennial rivers and
some non-perennial watercourses became
'public rivers' (flumen publicum). In the last two
centuries of the Roman Empire, more non-
perennial rivers were included in this category.

While more waters became 'public,' the
'public' itself that owned the water resources
narrowed to reflect the evolving Roman centralizing hierarchies into, ultimately, the Emperor.
Initially, some autonomy was left for conquered
tribes. In the Republican period until 27 BC,
when the legal regime of water ownership
was extended from Italy to the provinces
conquered, water administration fell under the
responsibility of the Roman governors in territories entirely subject to direct Roman rule.
However, in territories governed by a treaty, a
large degree of autonomy was left to the
local authorities, also in the field of water
administration.

After the Republican period, power gradually
shifted from the 'people' to the Emperor and the
Senate. From the third century AD onwards, this
diarchy further evolved toward absolute monar-
chy, as all powers were ultimately concentrated
in the hands of the Emperor alone. The sovereignty of 'the people' was transferred to the
Emperor, also in water administration. In Italy
and the provinces, water administration responsibilities passed entirely to the Emperor's vicars,
parallel with the gradual suppression of the
surviving local autonomies. Res publica came to
mean only 'people's right of use' (res in publico
uso). Moreover, throughout the Roman world,
the Senate of Rome had supreme control of
state finances, both with regard to public expen-
ditures (including public works) and to revenues
(including water rates). In sum, by declaring
land and increasingly more water resources as
being public, more existing customary water
rights regimes were superseded by the more
authoritarian Roman water laws, controlled by
the more centralizing Roman administration.

The new 'right' to use the expropriated
'public' waters was through the administrative
permit or concession. In some situations – well
discussed in the literature ever after – permits
kept serving as a hook to impose obligations in
return to clear water service delivery by the
administration. In the city of Rome, for example,
a specialized technical water service and administration governed water use. The administration also kept registers both on water sources
and availability, and on distribution, with one on
modifications of water rights, water users and
water distribution. As soon as a concession
came to an end, this was recorded and the water
returned to the administration for reallocation to
a new concessionaire.

In most cases, nevertheless – although quite
ignored in the literature – the requirement of
'administrative' authorization of public water
use through permits had very little to do with
delivering any water service. Its main purpose
was to allow the rulers to 'lawfully' appropriate
resources from conquered tribes, at least on
paper. Gradually, administrative concessions
became the only legitimate mode of acquisition
of a right to divert water from public watercourses for irrigation and/or industrial purposes.
Also, it was generally prohibited to divert water
from navigable watercourses. The only two

ways to recognize existing water rights were, first, through the legal provision that some long-lasting use or *usus vetus* could evolve into a mode of acquisition of a right to use public waters and, second, through what Hodgson (2004) calls *de minimis* uses. The latter are micro-scale uses for domestic purposes, homestead gardening, small-scale livestock watering and sometimes a bit of irrigation.

To summarize the old Roman pattern that has remained so very intact ever since: by declaring land and waters as 'public', the authorities representing 'the public' could impose their ownership and rules. The declaration of land and waters as being public formally nullified prior resource claims of conquered tribes. The only way to regain 'lawful' access to their former water resources was to recognize the authority of the powers that were trying to establish their rule by asking *them* permission, thus negating own rules and surrendering to the new owner of the water resources. The new authority then 'granted' administrative authorization in the form of permits.

The Transformation of Roman Water Law in High-income Countries

In Europe itself, Roman water law was profoundly transformed and only revived very recently under entirely different conditions. As Caponera's (1992) study highlights, after the fall of the Roman Empire in the sixth century AD, Roman water law blended with customary laws. Yet, the emperors, kings, dukes and higher feudal lords kept their claims of ownership over land and water, which they vested, from the top down, in their lower-ranking vassals. In the feudal system there was no concept of private ownership of water, and the feudal lords had full control over land and water within their jurisdiction, including the authority to charge levies.

It took more than 1000 years before this changed. In the civil law countries, i.e. France and most of continental Europe, the aristocratic powers ended, among others, with the French Revolution. A bourgeoisie emerged as the new social class with new economic interests. In civil law countries Roman law was revived, but this time to strengthen the private rights of the

emancipating users against state interference. The Napoleonic Code of 1804 classified water into private waters (located below, along or on privately owned land) and public waters (which were confined to 'navigable' or 'floatable' waters only) requiring a permit for rights of use (with related water rates).

Around the same time, users also exerted their claims in the UK. Common law was adopted, which held that water could not be owned, neither by the Crown nor by individuals, but would be owned by all (*res comunis omnium*). Through a (riparian) use right, the riparian doctrine that evolved out of this new UK common law allowed riparian landowners the free utilization without the need for administrative intervention. The riparian doctrine gave equal status among riparians and strong rights to the riparians vis-à-vis newcomers beyond the riparian strips, who had to negotiate hard for their entrance. The many laws, ordinances, regulations or other legal enactments for administering or regulating specific subjects related to water were bottom-up. They all sprang from needs arising from local conditions. A similar system developed in the eastern USA.

During the following 150 years profound economic, social and political changes took place in Europe. Extensive state investments were made in public infrastructure to catalyse the evolving water economies. In France, the definition of public waters slightly expanded after 1910, to include waters that the state needed to acquire for the purpose of public works. Water economies developed in which public agencies, parastatals, public-private partnerships, hydropower plants, municipal and industrial water service providers and private companies established effective technical and institutional control over the nation's water resources. Gradually, almost all former primary water takers became secondary users as clients of these water service providers or as members of irrigation groups and water user associations. Extensive institutionalization took place, which assured that virtually all water users were known to the relevant authorities, registered and were paying their subsidized bills. Pollution issues became more important. The 'environment' emerged as a new water user in its own right. Stronger state control was needed and accepted for such regulatory roles. Further development and harmonization towards permit systems was *only by then* increasingly seen as a legitimate 'hook' to impose such obligations in a legitimate public interest.

For example, in France, it was only in 1964 that more waters were included in the public domain, such as that necessary for domestic water supply, navigation and agricultural and industrial production. And the law no longer spoke of private waters but of non-domanial waters – but they still required compensation in case the state revoked. A new criterion of 'public interest' was also introduced at this time, which further limited the sector of privately owned waters (Caponera, 1992, p. 77).

In the UK the common law riparian system also changed with the Water Resources Law of 1963, when licensing for the abstraction of water was imposed generally by statute. An authority became responsible for authorizing water abstractions above certain thresholds. Nevertheless, many features of riparianism were preserved. The common law notion of water ownership as being vested in the whole community (*res comunis omnium*) was also preserved: common law countries avoid the expression of 'water ownership' in legislative texts. Instead, the texts generally declare that the state has the power to control water utilizations (Caponera, 1992, p. 114).

Significantly, the expansion of public waters requiring permits in high-income countries was accompanied by the full recognition that there are plural legal regimes to govern water. In common law countries, a large part of formal entitlements still remains attached to customary law under the name of common law. In other countries, customary arrangements are also recognized, if not preserved. For example, in the Netherlands, the centuries-old customary water boards are well respected and their merging into the state apparatus has been gradual and negotiated. Similarly, the Water Tribunal of Valencia, Spain, which has held customary rules since time immemorial, is respected and enforced (Caponera, 1992; Hodgson, 2004).

High-income countries outside Europe also respect other existing water rights regimes. For Japan, Bruns has noted (2005):

> Acceptance of traditional water rights, even when these have not been formally registered, has been a key principle underlying river management in Japan (Sanbongi, 2001). The law established the principle that existing users have legal standing to protect their interests when necessary. The River Laws of 1896 and 1964 provided a formal basis in state law, through which agencies and courts could take account of existing rights. The principle of being 'deemed to have obtained permission' reduces conflicts between state and local law without forcing local rules to explicitly conform to the criteria and formulations of state law.
>
> (Bruns, 2005).

Full respect for non-permit systems, strong users' entitlements and more centralized authority only after nationwide, inclusive and highly sophisticated formal water economies were developed are in sharp contrast to the origins and development of water laws in Europe's former colonies.

The Colonial Legacy of Water Law in Latin America and Sub-Saharan Africa

Latin America

According to Roman military tradition, water laws in Europe's colonies in Latin America and sub-Saharan Africa were primarily designed to overrule prior claims and customary arrangements. Water laws 'lawfully' vested ownership to most, if not all, of the conquered areas' water resources in the colonial minority rulers. Often, permits were imposed as the only formal way to render existing and new water use 'lawful'. This enabled settlers to obtain rights that were declared formal and hence first-class compared with other water rights regimes. If indigenous inhabitants were allowed at all to apply for permits, they were forced to recognize the legitimate authority of the invaders as the 'lawful' new owners of waters that were already theirs. It introduced a divide-and-rule mode of obtaining water rights which only settlers and, at best, a small portion of indigenous people could obtain. It relegated all prior water rights regimes to a second-class status, and also in the cases in which the 'free' use of small quantities, so *de minimis* rights, was granted.

The water laws that the Spanish conquerors of Latin America vested were based on the Papal Bull in 1493, by which Pope Alexander VI

gave the catholic kings all newly discovered lands, including waters. Water use became the object of special king's permits (Mercedes) granted by the Spanish government authorities for certain purposes, such as domestic drinking needs and irrigation. Such permits could be revoked. […] The violation of permit requirements could be punished with a fine.
(Caponera, 1992, p. 49)

A few decades later, the Spanish phrased their encroachment upon prior appropriation claims in a subtler way, aligning with the community-based arrangements that they found alive among the Incas, Aztecs, Mayas and other indigenous water users. The Leyes de Indias, promulgated in Spain in 1550, declared for her American colonies that:

Rivers, ports, and public ways belong to all men jointly, so that any person coming from a foreign land may use them in the same way as those living in their vicinity. These common goods were attributed to the Crown and their ownership vested in the Prince as the representative of the community […] These principles were combined in the Laws of the Indies together with the existing local customs which were not contrary to them. In the indigenous agricultural practice the collective use of land by the clan necessarily implied a collective use of water. Thus the Laws of the Indies accepted the concept that water is a common good which must be distributed within the community for the benefit of its members, but vested its ownership in the Crown, and entrusted its administration to the Spanish authority, considered as the representative of the community.
(Caponera, 1992, p. 49)

By the 19th century, the privatization tendencies of the 1804 Civil Code of France found their way to Latin America. While some countries strengthened private waters, others states kept their declaration of most water resources as being public, but codified such use rights into such strong private rights that the prerogatives assigned were the same or almost the same as those associated with ownership. Caponera indicates how this has 'promoted expansion in water use, in that it offered the user certainty before the law and a freedom of action […]. Such a system called for only a very simple administrative organization for the application' (Caponera, 1992, p. 110). Thus, the colonial settlers kept carving out strong formal first-class rights to shared water resources, while

'lawfully' depriving indigenous communities of their prior customary water rights.

As listed above, recent water laws have revived and reinforced this colonial legacy. Chile's Water Code of 1981 is the world's most extreme example in which refurbished concessions offer *certain* users 'certainty before the law and a freedom of action […], while calling for only a very simple administrative organization for the application', as elaborated below.

In Mexico, the concept of concession was introduced with the Spanish conquest in 1512, which stipulated that ownership of water resources was vested in the Spanish king and that a royal grant was required to use it. However, the factual 'granting' of concessions remained dormant up till 1992. By then only 2000 concessions had been granted (Garduno, 2001). From then onwards, however, the system of concessions was revived nationwide, partly inspired by the Chilean experience (van Koppen, 2007, unpublished).

As documented by Boelens *et al.* (Chapter 6, this volume), indigenous peoples in the Andean region have increasingly contested the revival of this colonial legacy.

Sub-Saharan Africa and the revival of colonial law in Tanzania

When France (and Belgium) colonized Africa, water was originally classified, as in France, as public or private, public waters being those that were 'navigable or floatable' – and vested in the colonial governors. Caponera (1992, p. 99) describes the mindsets of the conquerors in more detail:

Later, due to climatic circumstances, i.e. of the fact that most African streams are seasonal and therefore non-navigable during certain periods of the year with the consequence that very little is left to the public domain, the distinction between navigable and non-navigable waters disappeared and, generally, all waters were placed in the public domain. Under this regime, every use of public water is subject to the obtention of an administrative authorization, permit or concession. In addition, specialized institutions, government, private or mixed, have been set up to deal with particular water development activities such as domestic and municipal water supplies, power generation and distribution, irrigation and others.

Countries under former British administration

have adopted the British system according to
which water is *res comunis onmium* (common to
all), of which the riparian landowners can make
use, unless it has been brought under government
control through legislation or judicial decisions.
Crown land did not generally include water
resources, with the result that every specific use of
water had to be the object of special legislation.
This has produced a large number of legal
enactments concerning specific water utilizations.

(Caponera, 1992, p. 100)

However, in various colonies the British
minority was quick to introduce permit systems,
as in Zimbabwe (Derman *et al.*, Chapter 15,
this volume) or under certain conditions as in
Ghana (Sarpong, undated) and Kenya
(Mumma, Chapter 10, this volume). In South
Africa, the British land title deed system had
vested strong paper titles for whites only on
91% of the territory. By adopting riparian rights
throughout the Union of South Africa, most of
the water resources were appropriated with a
stroke of the pen.

The case of Tanzania illustrates both the
history of dispossession and the revival of colonial law under the banner of IWRM. In line with
the German colonial tradition before German
East Africa was ceded to Britain as Tanganiyka
in 1919, the Water Ordinance of 1923 required
registration to vest water rights. This was open
to white settlers only. The Water Ordinance of
1948, Chapter 257, stipulated: 'The entire
property in water within the Territory is hereby
vested in the Governor, in trust for His Majesty
as Administering Authority for Tanganyika.'
Under this Ordinance, water uses 'under native
law and custom' were recognized but native
users could only participate in decision making
through 'duly authorized representatives' or
'natives in addition to the District Commissioner'. Customary law was tolerated, but
only where it did not conflict with the interests
of the colonial state.

In the Water Ordinance of 1959, urban
water supply and water use for mining operations were regulated separately. For other uses,
obtaining a water licence, permit or right from
the colonial water authority was emphasized.
The option of registration was extended to
native water users, but the status of those who
did not comply was left somewhat undeter-

mined. After independence in 1961, the new
government under Julius Nyerere shifted
ownership to the new state, declaring that 'All
water in Tanganyika is vested in the United
Republic' under the Water Utilization (Control
and Regulation) Act 1974, Section 8. In section
14, registration was rendered obligatory for all
who 'divert, dam, store, abstract and use'
water. From then onwards, throughout the
nation, only registered water use was considered to be lawful. However, the water law
remained rather dormant up till the early 1990s
(van Koppen *et al.*, 2004).

The dormant laws were revived when a
Rapid Water Resources Assessment was carried
out by the World Bank and DANIDA
(URT/MOW, 1995). This project identified a
need for stronger state regulation to better
divide what was seen as an inevitably limited
pie – ignoring Tanzania's abundant water
resources but lack of means to develop them.
Especially the 'user pays' principle was
promoted, both to create the awareness that
was expected to lead to wiser water use and to
finance new basin Water Offices. The Staff
Appraisal Report of the World Bank report that
formulated a River Basin Management project
to implement the reform discovered that, ever
since the Water Ordinance of 1948, the respective governments had ascribed to themselves
the authority to 'prescribe the fees payable in
respect of any application or other proceeding
under this Ordinance'. But this had never been
operationalized. Underlining, in essence, the
suitability of a slightly altered colonial water law
for modern water management, the report says:

The conceptual framework for integrated river
basin management is already laid out in the 1974
Act, as amended in 1981. However, the
legislation has never been effectively
implemented. The Government has submitted a
letter of Water Resources Management Policy
outlining measures to be taken to update the
legislation and improve management of this
resource.

(World Bank, 1996, section 2.13)

Thus, in 1997 and 2002, the government
promulgated amendments to the law of 1974 in
order to raise water tariffs considerably (URT,
1997, 2002). Besides a once-off registration fee
of $40,[3] an annual 'economic water use fee'
was introduced. The rate was proportionate to

the annual volume allocated and dependent upon the water use sector, with lower rates for the smaller users. As Tanzania has no exemptions for small users, at least in the current version of the law, the minimum flat rate for an individual or, more often, a group of users was set at $35, irrespective of the actual flow or volume used. This is more than a monthly income for over half of Tanzanians. According to the World Bank, in Tanzania, 58% of the population live on less than $1/day (World Bank, 2000).

The water rights registers of the Rufiji basin, with several million water users, illustrate how permits had factually been implemented, primarily by formal and foreign users (Sokile, 2005). By mid-2003 the Rufiji Basin Water Office's database contained 990 water rights. Of these, 14% had been issued between 1955 and 1960 (just before independence) and 29% administered after the establishment of the Rufiji Basin Office in 1993, although these are still largely in the stage of application or have only a provisional status.

Of these rights, 40% were held by governmental agencies, 12% by Brooke Bond Tea Company and 8% by various Catholic dioceses. The remaining 40% of registered users included private irrigation schemes, such as those belonging to Baluchistani and other Asian immigrants who were brought by the British colonialists. As many as 47% of the registered rights, especially the older rights, were 'not operated' anymore, which may reflect the outflow of Germans, Baluchis and Greeks after independence in 1961 and the Arusha Declaration in 1967, which announced further nationalization (for the study of the implementation of the revived water rights system in the Upper Ruaha catchment among customary water users, see van Koppen *et al.*, 2004; Mehari *et al.*, 2006; van Koppen 2007, unpublished; Chapter 14, this volume).

Tanzania and Ghana, mentioned above, are no isolated cases. Other chapters in this volume touch upon similar revival and revision of colonial water laws that have focused on dispossession, towards more widespread application and implementation of permit systems (globally: Meinzen-Dick and Nkonya, Chapter 2, this volume; the Andean regions: Boelens *et al.*, Chapter 6, this volume; other authors on

Kenya, Malawi and Zimbabwe); for Uganda, see Garduno, 2001.

Although further study of the legal revisions and their implications in these and other countries is clearly warranted, some general characteristics of today's permit systems in low- and middle-income countries emerge, and are discussed in the remainder of this chapter. The fifth section discusses the notion of 'property' rights to water in the more abstract sense. The sixth section presents the empirical consequences of such notion of 'property' for the case of Chile. The seventh section builds upon the Chilean case, and also upon evidence from elsewhere, to identify the two key processes that render administration-based permit systems highly discriminatory entitlement systems for informal water users. Again, the obligations dimensions of permits are not discussed here. See van Koppen (2007, unpublished) for the argument that permits can only be vehicles for registration, taxation and waste discharge charges in low- and middle-income countries, if well targeted at the few formalized users and disconnected from entitlement dimensions.

Permits as Property Rights in Low- and Middle-income Countries Today

This section takes a closer look at the nature of this peculiar form of property rights in countries with deep divides between the few administratively knowledgeable large-scale users and many less administratively knowledgeable informal water users. In the light of contemporary notions of justice and fairness, it is odd that a formal property right can be vested primarily through an administrative act. Indeed, a formal property right boils down to the formal legal backing of a user's claim to such a resource, and sometimes to compensation if taken away (Bromley, undated). That is also the core of administrative water rights. However, unlike other objects of property rights, like land, the contents of the rights to water are difficult to define. The physical nature of water as a fugitive, highly variable and unpredictable resource renders any quantification and verification highly problematic. This is certainly the case for under-resourced water departments without measuring devices and with underdeveloped water control infrastructure. It should be remembered that the inability to quantify was of little concern to the Roman and colonial conquerors, whose primary interest was to establish *whose* water it was in order to establish who could authorize its use, not how much precisely.

For modern states, such formal legal backing remains the primary role. However, even in high-income settings where water is fully controlled physically and institutionally, water laws typically include clauses which stipulate that, in no way, can water rights holders hold the state accountable to make the waters available as stipulated in the rights. Water laws in middle- and low-income countries contain similar clauses. There, any quantification is even more unreliable and inaccurate because of the weak monitoring capacity of water departments and the even greater unpredictability of available water resources in the absence of infrastructure. Average annual volumes stipulated in permits may give some indication of water use and may work as some basis for taxation, but have little to do with factual water quantities and even less with low flows far below any average, when entitlements count most. This renders vesting formal rights to water resources primarily an administrative act in low- and middle-income countries. A permit with formal state backing of the entitlement is a first-class right compared with any claim without such formal backing, which automatically becomes second class when it regards competition for the same resource.

The exemption of domestic and micro-scale water uses, or *de minimis* water uses, from the obligation to register and apply for permits only confirms the inadequacy of a property rights system which defines an administrative act, without much reliability of the contents, as the primary basis for vesting rights. Or, in the words of Hodgson (2004) commenting on *de minimis* rights as a 'curious type of residuary right':

> There is no great theoretical justification for exempting such uses from formal water rights regimes. Instead a value judgement is made by the legislature that takes account of the increased administrative and financial burden of including such uses within the formal framework, their relative value to individual users and their overall impact on the water resources balance. [...] While they may be economically important to those who

rely on them, it is hard to see how they provide much in the way of security. [...] The problem is that a person who seeks to benefit from such an entitlement cannot lawfully prevent anyone else from also using the resource even if that use affects his own prior use/entitlement. Indeed the question arises as to whether or not they really amount to legal rights at all.

The second-class status of *de minimis* rights is also manifest in the fact that no state has any compensation measure if water for micro-scale uses is taken away. In low-income countries, exemptions for *de minimis* uses relegate the majority of citizens, including the poorest who depend on micro-scale domestic and productive water uses for basic livelihood needs, to having only second-class rights. They are given a status of being negligible and invisible by design for the mere reason – not their own fault – of not being administrable.

As described earlier, this administrative property system is now increasingly imposed to replace indigenous water rights, and reforms are further reaching out into rural areas. Especially in many countries of sub-Saharan Africa, prior water claims are declared as illegal until they undergo an administrative process and are 'converted' or 'regularized' into registrations and permits. Usually, the high costs for registration are with the water user and the period for registration is extremely short. The invariably needed extensions are called 'grace' periods. South Africa is an exception, as it recognizes existing lawful use as continuing to be lawful under the 1998 National Water Act.

Thus, although often unintentionally, contemporary dispossession of prior indigenous and informal water claims, as under colonization, occurs essentially by forcing users to recognize administrative water rights as the first-class titles and denouncing the status and nature of their own, earlier rights. A burden of proof of centuries-old claims is suddenly imposed, assuming that the old claims can be expressed in terms of permits at all. The revised laws and their re-energized implementers seek to finish the unfinished business of colonial dispossession.

Idealistically, it is assumed that everybody will be equally subsumed under the new system and that administrative systems are equitable and fair because 'everybody can apply for a permit'. This ideal of reaching everybody equally further increases the pressures to register quickly and in an encompassing way. Yet, this ideal is totally unrealistic in low- and middle-income countries with strong differences in administrative adeptness between the few formal users and a majority of informal small-scale users who have hardly any contact with the state, local governments or water departments. Equal treatment is as unlikely today as it has been in the past. Evidence from Chile (in the sixth section) and elsewhere (seventh section) corroborates this and debunks the myth that administration-based permit systems foster justice by treating all citizens, in principle, equally in low- and middle-income countries.

Resource Grab by Design: Evidence from Chile and Elsewhere

The following analysis of Chile is the author's interpretation of the findings of Bauer's in-depth studies on the Chilean Water Code (Bauer, 1997, 1998, 2004), unless indicated otherwise.

The Chilean experience, which has now lasted over 20 years, gives insights in the essence of administrative water entitlements of permit systems. The Water Code of 1981 is an extreme case because it cancelled all earlier restrictions and obligations for users, even the obligation to use the water. In line with the general colonial practice in Latin America sketched above, Chile's Civil Code of 1855 codified that 'administrative concessions' could be obtained to water defined as 'the national property for public use'. Besides these use rights to a public resource, some categories of water use were recognized as private.

With Chile's first Water Code of 1951, some formalization of administrative procedure for granting use-rights started. The law also began encouraging registration of those rights in the local Real Estate Title Offices. The users' rights remained subject to various legal conditions. Rights were tied to landownership and their owners were required to actually use the water within 5 years. The state could revoke without compensation and had well-defined regulatory authority. In the 1960s, state power over water was further enhanced when the socialist government started implementing distributive

land reform and also needed to redistribute water. A new Water Code of 1967 was adopted. This Code reallocated water rights according to new principles, such as plot-size-based crop water requirements of the smaller-sized plots of the 'parceleros' benefiting from the land reform. A new agency, the General Water Directorate, was created to implement this package (Bauer, 1997, 1998, 2004).

Pinochet's military coup of 1973 halted the land reform and introduced Chile's extreme neo-liberal economy, with absolutely minimal state interference. A new constitution was formulated in 1980, which defined water use rights unambiguously as 'private property'. The right encompasses 'the right to alienate the water owned through sale, donation, transfer, inheritance, or to constitute different rights on the same, whatever their nature, at the discretion of the owner'. Not being an 'administrative concession' any more, the state was now also obliged to pay compensation if water was taken away.

Water rights were, for the first time in history, legally separate from landownership and could be freely bought, sold, mortgaged, inherited and transferred like any other real estate. There were no requirements to prevent pollution attached to a water right. Originally, owners had not even the legal obligation to actually use their water rights, and they faced no penalty or cancellation for lack of use. Lawyers realized the peculiarity, unique for water rights, that an individual can have absolute private ownership rights to a public resource. The Chilean legal 'solution' for this contradiction intrinsic to the entitlement dimension of permit systems is that an individual can own a water right but not the water itself, 'since it is only the former that he is free to sell' (Bauer, 2004, p. 141).

These sophisticated definitions of what a 'right' entails in Chile may suggest that the substance is sophisticated as well. This is not the case: water rights are mostly not even registered. Formal property rights to existing uses were based on actual water use in 1981, which somehow has to be proved. The Water Code of 1981 addressed the potential uncertainty of existing claims by declaring a presumption of ownership in favour of those who were using water rights de facto at that moment. The high courts confirmed that unregistered rights had full constitutional protection as property, insisting that they are not lost through failure to be registered. Hence, the large majority of water rights in Chile are not formally registered as they pre-dated the 1981 Water Code, but at least they can be established by proving factual water use.

Rather than being sophisticated and well defined, administrative procedures and registration of water rights opened up a resource grab, both for existing and new water uses. Registration of existing uses without clear measurement and checking by the government as a third party implied that basically any claim held (unless verified by other bodies such as Water User Associations). Not surprisingly, a recent study in the Valley of Codpa showed that individual water rights ranged from 200 to 10,000 m³/ha (Hendriks, 1998).

The possibility of vesting new claims merely through application and registration with the centralized General Water Directorate proved a very easy way to lawfully gain access to water by the expanding foreign mining and irrigated export fruit cultivation under the neo-liberal economy.

Moreover, and most heavily criticized, the option even to claim water without obligation to use led to the hoarding and speculation by a minority of administratively knowledgeable vested powers. The large hydropower companies especially laid massive claims on still uncommitted water resources – anticipating reducing gas supplies from Argentina. After 1990, the newly elected government and the National Water Directorate agreed that this was socially unjust as well as economically undesirable – letting private parties profit from public resources without fulfilling a useful social function in return, and holding back economic development by disallowing others to use the water for productive activities (Bauer, 1997, 1998, 2004).

The Water Code Reform of 2005 introduced licence fees for unused water rights and the limitation of water use rights requests to genuine needs as a deterrent against speculation and hoarding (GWP, 2006). However, as in other low- and middle-income countries, the government lacks the implementation and monitoring capacity to factually check how much water requested in new applications is to be used beneficially.

As widely recognized now within and outside Chile, the expected water market did not come about. Although rights had become saleable, there were hardly any transfers of water from (registered) willing sellers to (registered) willing buyers. The main transfer that took place was between the government that gave water away for free and speculators who now lawfully demand payment from both government and new users wanting a new water right.

Not even informed about the laws and also otherwise structurally disadvantaged to make use of the laws, informal and indigenous small-scale water users have been most injured by the vesting of water rights through administration. As they were too late to claim their share of the nation's available water resources, access to new water resources has been severely hampered, stifling further water development by them. Moreover, even their existing rights are increasingly under attack. Boelens et al. (Chapter 6, this volume) cite the Mapuche leader, bitterly complaining about water originating and used in his areas that has been appropriated by vested powers downstream.

Also, in many other cases, settlements that previously included natural access to water were by now given restricted and irregular access. By the time peasants and their organizations learned of the new procedures, they found that rights to available water had already been granted by the General Water Directorate or regularized by those more legally adept: large farmers, agro-industries and mining and logging companies (Bauer, 2004). Even in a number of government-created programmes to promote small-scale irrigation, subsidies have been denied because of inability to get legal title to unused waters (Maffei and Molina, 1992, cited in Bauer, 1997).

In spite of government support from the 1990s onwards to 'regularize' local rights in the formal property owners' records, the gap has remained. Legal advice and financial support, including considerable expense to repurchase rights on behalf of indigenous groups, still left most of the indigenous claims unanswered. Even specific legislation for minorities' rights was of little avail in the encounters with the powerful Water and Mining Codes (Boelens et al., Chapter 6, this volume).

At the same time existing indigenous water rights, which were formally protected as factual water use in 1981, are increasingly challenged. The business sector keeps promoting registration by indigenous and peasant communities. According to them, the water rights market and investment in water resources cannot operate if there are local and customary rights that are not registered but do entail a certain legal protection (Boelens *et al.*, 2005). Registration would 'provide a broad catalogue of legal certainties for outside investors in rural areas and indigenous territories'. However, registration for outsiders' 'certainties' imposes heavy and costly burdens of proof, if possible to prove at all, on indigenous users. It traps them further in the recognition of an administrative system that is designed to overrule and erode other legal water rights systems and, as elaborated below, is intrinsically discriminatory vis-à-vis informal small-scale users.

The discriminatory processes at stake are not limited to Chile, but intrinsic to water administration in low- and middle-income countries in Latin America and sub-Saharan Africa in general. Ever since colonization, they have deepened structural inequalities and favoured the powerful at the expense of the less powerful, including informal water users. Today's liberal language that 'everyone can apply for a permit' hides and entrenches these structural inequalities even further. Below, we summarize two sets of generic discriminatory processes when administration is the basis of vesting rights.

Discrimination by Water Administration

Forcing the informal into the formal

The first form of discrimination is, obviously, that permit systems are declared as the superior system and as the norm to which other existing arrangements have somehow to adapt. It is simplistically assumed that customary water rights systems, which are very different legal systems, can be formulated in terms of an administrative right without violating the essence of customary water rights systems. Yet, the differences are substantive. For example, in indigenous water rights regimes, ownership is usually defined as a communal right in contrast to permit systems that vest ownership in the state and permits in individuals and formal entities. Caponera (1992) has also advocated,

fully respecting these essential features in high-, middle- and low-income countries alike: 'In the countries where customary rules exist regarding the ownership of water, such ownership, generally deemed to be community ownership, should be recognized in the legislation' (Caponera, 1992, p. 139).

However, 'recognition' of one legal system in terms of the other system is not easy. Boelens *et al.* (Chapter 6, this volume) discuss the complexities of the politics of recognition in the Andean region. A common option, also adopted in Chile, is vesting permits in collectives. However, this still creates new problems rather than solving existing ones. Typical issues include the definition of 'the community' and the risk of male elite capture that further polarizes internal gender, class and ethnicity hierarchies.

For sub-Saharan Africa, where the proportions of informal rural users are largest, the issue at stake may be as fundamental as changing the norm of which legal system should be the first law. In this regard, the water sector can learn much from the indigenous land tenure debates, where it was found out in the hard and costly way that one cannot simply replace one legal system by another. Ever since independence, governments, development organizations and academics have deployed huge efforts to 'formalize' indigenous land tenure through centralized formal land titling. They have all failed up to the point that now, after five decades, it is recognized in mainstream debates that indigenous land tenure should be recognized as the first and superior law (McAuslan, 2005). The 'received' colonial and statutory formal land laws have a modest role only, which can only take shape gradually and in a problem-based and bottom-up way. While land tenure policies and debates have abandoned centralized titling, the water sector seems to want to reinvent the same wheel all over again by promoting centralized titling through permit systems for a much more complicated natural resource.

For both land and water tenure, 'regularizing' communal systems into individual saleable ownership rights can be highly destructive. These negative impacts should be fully considered when opting for a certain legal system. In Chile, the novel possibility for individuals to sell water rights, which were by now de-territorialized, to outsiders eroded the precious social capital of communities' collective water-sharing arrangements. The 'soaking off' of water rights from collective and community-controlled frameworks _created_ the 'tragedy of the commons' by encouraging individuals to pursue their own individual interest at the direct expense of others and the collective as a whole. Indeed, 'the individualization of formerly collective rights and management systems has created internal chaos' (Boelens _et al._, 2005). In sum, different legal systems are like apples and oranges: one cannot compare them, and it is even less possible to change the one into the other.

Discrimination by administration

Unequal access to information and communication

The second set of processes that lead to differential impacts of administrative water rights concern the working of administration in general. One main reason why the resource grab in Chile by the elite could happen was simply that only very few people were informed about the possibility of registering and obtaining water rights. After the promulgation of the 1981 Water Code, the Chilean government undertook no campaign of public information about the Code's new features, nor did it offer legal or technical advice about how to apply for new rights or regularize older ones. Even if publicity had been better, major gaps in access to 'public' information, or rather timely access in order to be the first to take the share, would have remained.

Those informed and submitting their claims just a couple of years later found that they were already too late. The unequal access to the main information channels and the structural differences in the ability and skills to communicate in the language of the powerful have been amply documented. They include inequalities in: (i) literacy; (ii) access to audio-visual media and written documents; (iii) personal means of communication, like mobiles, internet, post office or bank accounts; (iv) mobility and relative costs of transport; (v) experience with bureaucracy; (vi) distance to state offices; (vii) officials' acquaintances; and (viii) vulnerability to and adeptness for bribery.

Disproportionate costs

A less documented form of structural discrimination is a matter of scale. The transaction costs in applying for permits are disproportionately high for small-scale users compared with those for large-scale users. Both have to undergo largely the same procedures with the exorbitant high costs for the applicant, as in Chile. Costs include presentation of technical antecedents (geographical coordinates, flows, etc.), publication in the official gazette, public registration and lawyers' fees, travel and lodging etc. to arrange this paperwork (Hendriks, 1998). Yet, for small-scale users the profitability of water use is by definition much less than for large-scale users, for whom the application costs are just a tiny proportion of the profits made. Another example of increasing costs for permits that are disproportionate, if not unaffordable for small-scale users in the colonial past, is the obligation to install expensive measuring devices, as imposed by governors in Zimbabwe in the 1950s (Manzungu and Machiridza, 2005). Collective applications mitigate only partially for these disproportionate costs, as they require extensive internal transaction costs as well.

Explicit discriminatory conditions

On top of this implicit discrimination through administration, there may also be conditions attached to permits that discriminate explicitly against small-scale informal users. One common condition for formal permits tied to land is that they apply only to formally titled land. For example, the Kenyan Act of 2002 allocates permits only for titled land that only a small proportion of Kenyans possess (Mumma, Chapter 10, this volume). Such conditions formally exclude all other Kenyans from water titles.

Conflict management and law enforcement

Differential proficiency in conflict management and law enforcement are illustrated in the Chilean case. Even if small-scale informal users in Chile had been able to prove their existing water uses as formally protected by the 1981 Water Code, and even if they had obtained well-recognized and registered formal water rights, they would have no recourse if such rights were infringed upon. Even state legal advisors cannot do much if large-scale users violate smaller users' rights, for at least two reasons. First, the Code stipulates that decisions on water management are weighted according to actual possession of certain water rights. So rights holders with more water shares (volumetric right per time unity) have stronger decision-making power. This contrasts with indigenous management, where collective interests are negotiated according to the rule of 'one man, one vote'. This minority that possesses the majority of shares, many of whom, moreover, live in the city, has no interest whatsoever in using water more efficiently. They are legally allowed to continue depriving others, even if the latter try hard to increase the efficiency of water distribution and enhance water productivity (Hendriks, 1998).

A second reason for the weak bargaining position in the case of conflicts is that the Water Code reduced all state intervention possible and relegated all conflict management to the regular civil courts. Their judges are powerful, but rarely competent in technical aspects of water rights, and tended to hold a narrow and formalistic concept of law (Bauer, 2004). The costs of their specialist adjudication are high and unaffordable for peasant farmers and out of proportion compared with the limited profits they make with the low volumes of water. Even if small-scale users were to win such court cases, there would be no agency to ensure enforcement (Hendriks, 1998).

Gender

Women as a gender are most excluded. Their legal status in indigenous arrangements is often a second-class status of minor only; their individual resource rights are overruled by men claiming to be the head of the household and therefore deserving control over all household resources towards external parties; their literacy rates are lower and their other forms of access to information and communication are also less than for men; women can even less afford the costs of regularization, let alone formal adjudication for the relatively small quantities of water that they use which, nevertheless, are crucial for basic well-being. In virtually all formal property regimes in the world nowadays, women's

individual titling or joint titling by spouses is debated and gradually taken up in policies and legislation (Lastarria-Cornhiel, 1997). This gender issue is addressed to some extent in Latin America. However, it has been entirely ignored in any debate on permit systems in sub-Saharan Africa up till now.

Thus, for the widely assumed merit of formal water rights systems: 'When formal water rights are secure and tradable [...] they allow for orderly allocation of water resources' (Hodgson, 2004).

Recommendations: Challenging the Colonial Legacy of Dispossession

This chapter attempts to show that permit systems, the favourite in the discourse on IWRM, may function in high-income countries but risk repeating the divestment of rural informal water users from their prior claims to water in Latin America and sub-Saharan Africa. Reviving the strong but still largely ignored legacy of colonial water law, the entitlement dimensions of revised permit systems allow, again, the 'lawful' grab for water resources by the minority of administratively knowledgeable large-scale users. Although the experiences in Chile are exceptional in some senses, the underpinning design of administrative water rights and the processes of discrimination have general validity.

Administrative water rights systems are highly problematic in low- and middle-income countries, first because of the structural social differences between the administratively knowledgeable formal sectors, well acquainted with the state, and those who are not; and, second, because the state lacks the capacity to check and control. This implies that the administratively knowledgeable can lawfully obtain water resources by such measures as: (i) 'regularizing' their existing water uses and claiming higher volumes than actually used; (ii) submitting requests for claims to new water resources as they like, forcing the state without the factual information to allocate whatever is 'still available'; (iii) being legally empowered to treat any other existing water use governed by other regimes than permit systems as second class only, if not illegal; and (iv) intimidating other users with the volumes claimed and asking for the support of formal lawyers to corroborate their case.

The administratively knowledgeable are faster than others and the first to claim still uncommitted water resources. When the others catch up, they will probably be too late. While losing out to outsiders, communities also lose when administratively knowledgeable individuals within their own communities destroy social capital and create the tragedy of the commons.

To conclude, the following measures are recommended for policy and law in low- and middle-income countries. In countries that are still in the process of redrafting their laws these lessons will be timely.

1. Existing indigenous and informal water rights systems should be recognized and obtain at least equal formal legal status as other legal systems without any burden of proof. From there, adequate forms of written recognition are to be developed.

2. For providing a higher status of entitlements that formally empower informal users, innovative measures are required, e.g. reserved rights doctrine in western USA (Getches, 2005) or General Authorizations that have priority over permits, as currently discussed in South Africa (RSA, 2006).

3. The 'regularization' of existing non-permit systems into permits by the administratively knowledgeable users should be discouraged, as this opens up opportunities for abuse by these users to claim more water than actually used. If applied at all, this should be accompanied by accurate assessments of actual use.

4. Permit systems should, at best, be used as hooks to impose targeted obligations. They need to be well targeted, for example to newcomers only, or as vehicles to impose certain obligations to certain users. In both cases, other legal tools that can achieve the same goal, e.g. registration or taxation, need to be considered as well, as they may appear more effective, requiring considerably leaner administrations.

5. If permits are used as hooks to impose obligations, the entitlement dimensions of the permit need to be removed so that permits are not pursued as an easy way of claiming rights to more water.

Acknowledgement

The author gratefully acknowledges the critical and constructive review of earlier drafts by Mark Giordano.

Endnotes

[1] The specific context in which tradable water rights have evolved in high-income countries is illustrated by the arid and under-populated states of Australia. Here, strong state intervention with permit systems evolved over more than a century. Neither extended irrigation nor gold mining would have been achieved if the use of water had been limited to riparian land, as the earlier common law from the UK had envisaged. In New South Wales, for example, licences had already existed since 1884, numbering 130,000 today – a number that is manageable with Australia's modern institutions and information technologies. The step to tradability was small. Licences became transferable in the 1980s in response to droughts that made it impossible to put all water licences to productive use. In 1994, all federal states of Australia were committed to engagement in water reform, driven by a nationwide concern for salinization and other environmental problems. In 2000, New South Wales promulgated its Water Management Act. Even in the fully dammed rivers of arid New South Wales, annual precipitation is too variable for secure water delivery. So the security that the state was willing and able to offer as legal backing to its licence holders (and their buyers) was limited, and expressed in an annual volume with the long-term computed probability of availability in any one year. Computations are based on long-term data collection and sophisticated modelling. The more expensive high-security licences have a probability of 99%, while general security licences (for irrigation) are in the 35–70% range. This system is still being perfected, and is now also being extended into proportional rights. Trade is stimulated, among others, through the statewide, internet-based water exchange (http://www.waterexchange.com.au). However, permanent trade regarded only 4.5% of the total water rights in 1997–1998, largely because people did not like to leave the already made on-farm investments idle (Haisman, 2005).

[2] The word 'waters' is used here and in many other instances in this chapter following Caponera's (1992) usage.

[3] In this book, $ means US$.

References

Bauer, C. (1997) Bringing water markets down to earth: the political economy of water rights in Chile, 1976–1995. *World Development* 25 (5), 639–656.

Bauer, C. (1998) Slippery property rights: multiple water uses and the neoliberal model in Chile, 1981–1995. *Natural Resources Journal* 38 (1), 109–155.

Bauer, C. (2004) *Siren Song. Chilean Water Law as a Model for International Reform.* Resources for the Future, Washington, DC.

Boelens, R., Gentes, I., Guevara, A. and Arteaga, P. (2005) Special law: recognition and denial of diversity in Andean water control. In: Roth, D., Boelens, R. and Zwarteveen, M. (eds) *Liquid Relations. Contested Water Rights and Legal Complexity.* Rutgers University Press, New Brunswick, New Jersey.

Bromley, D. (undated) *The Empty Promises of Formal Titles: Creating Potempkin Villages in the Tropics.* University of Wisconsin, Madison, Wisconsin.

Bruns, B. (2005) Routes to water rights. In: Roth, D., Boelens, R. and Zwarteveen, M. (eds) *Liquid Relations. Contested Water Rights and Legal Complexity.* Rutgers University Press, New Brunswick, New Jersey.

Caponera, D.A. (1992) *Principles of Water Law and Administration. National and International.* Balkema, Rotterdam, 260 pp.

Garduno, V.H. (2001) *Water Rights Administration. Experiences, Issues, and Guidelines.* FAO Legislative Study 70, Food and Agricultural Organization of the United Nations, Rome.

Getches, D. (2005) Defending indigenous water rights with the laws of a dominant culture: the case of the USA. In: Roth, D., Boelens, R. and Zwarteveen, M. (eds) *Liquid Relations. Contested Water Rights and Legal Complexity.* Rutgers University Press, New Brunswick, New Jersey.

Global Water Parntership (GWP), Technical Committee (2006) *Water and Sustainable Development: Lessons from Chile.* Policy Brief 2, Global Water Partnership, Stockholm.

Haisman, B. (2005) Impacts of water rights reform in Australia. In: Bruns, B., Ringler, C. and Meinzen-Dick, R. (eds) *Water Rights Reform: Lessons for Institutional Design.* International Food Policy Research Institute, Washington, DC.

Hendriks, J. (1998) Water as private property. Notes on the case of Chile. In: Boelens, R. and Dávila, G. (eds) *Searching for Equity, Conceptions of Justice and Equity in Peasant Irrigation.* Van Gorcum and Comp, Assen, Netherlands, pp. 297–310.

Hodgson, S. (2004) *Land and Water – the Rights Interface.* FAO Legislative Study 84, Food and Agriculture Organization of the United Nations, Rome.

Lastarria-Cornhiel, S. (1997) Impact of privatization on gender and property rights in Africa. *World Development* 28 (8), 1317–1334.

Maffei, E. and Molina, J. (1992) Evaluación del Programa de Riego Campesino (Convenio FOSIS/INDAP). Photocopy FOSIS, Santiago.

Manzungu, E. and Machiridza, R. (2005) Economic–legal ideology and water management in Zimbabwe: implications for smallholder agriculture. In: van Koppen, B., Butterworth, J.A. and Juma, I. (eds) *African Water Laws: Plural Legislative Frameworks for Rural Water Management in Africa. Proceedings of a Workshop held in Johannesburg, South Africa. 26–28 January 2005.* International Water Management Institute, Pretoria, South Africa.

McAuslan, P. (2005) Legal pluralism as a policy option: is it desirable, is it doable? Paper presented at the *UNDP – International Land Coalition conference 'Land Rights for African Development: From Knowledge to Action',* Nairobi, 31 October–3 November 2005. Proceedings available at http://www.undp.org/drylands

Mehari, A., van Koppen, B., McCartney, M. and Lankford, B. (2006) Integrating formal and traditional water management in the Mkoji sub-catchment, Tanzania: is it working? Paper presented at the *7th WaterNet/WARFSA/GWP-SA Symposium,* Lilongwe, Malawi, 1–3 November 2006.

RSA (Republic of South Africa, Department of Water Affairs and Forestry) (2006) Assignment to develop and test methodologies for determining resource specific General Authorizations under the National Water Act. Prepared by Ninham Shand (Pty) Ltd. and Umvoto Africa and Synergistics Environmental Services; prepared for Director: Water Allocation, Department of Water Affairs and Forestry. November 2006. Department of Water Affairs and Forestry WFSP/WRM/CON6002, Pretoria, South Africa.

Sanbongi, K. (2001) Formation of case law and principles in watershed management. Paper read at the *Regional Conference on Water Law: Legal Aspects of Sustainable Water Resources Management,* Sarajevo. Bosnia. Cited in: Bruns, B. (2005) Routes to water rights. In: Roth, D., Boelens, R. and Zwarteveen, M. (eds) *Liquid Relations. Contested Water Rights and Legal Complexity.* Rutgers University Press, New Brunswick, New Jersey.

Sarpong, G.A. (undated) *Customary Water Law and Practices: Ghana.* http://www.iucn.org/themes/law/pdfdocuments/LN190805_Ghana.pdf

Sokile, C. (2005) Analysis of institutional frameworks for local water governance in the Upper Ruaha catchment. PhD thesis, University of Dar es Salaam, Dar es Salaam, Tanzania.

URT (United Republic of Tanzania) (1997) *Water Utilization (General) Regulations of 1997.* The United Republic of Tanzania, Dar es Salaam, Tanzania.

URT (2002) *Water Utilization (General) (Amendment) Regulations, 2002.* The United Republic of Tanzania, Dar es Salaam, Tanzania.

URT/MOW (United Republic of Tanzania, Ministry of Water) (1995) *Rapid Water Resources Assessment Volumes I and II,* January 1995. DANIDA/World Bank, Washington, DC.

van Koppen, B. (2007) Administrative water rights from a poverty and gender perspective: Discrimination and dispossession by design? IWMI Research Report, International Water Management Institute, Colombo, Sri Lanka.

van Koppen, B., Sokile, C., Hatibu, N., Lankford, B., Mahoo, H. and Yanda, P. (2004) *Formal Water Rights in Tanzania: Deepening a Dichotomy?* IWMI Working Paper 71, International Water Management Institute, Colombo, Sri Lanka.

World Bank (1996) Staff appraisal report. In: *River Basin Management and Smallholder Irrigation Improvement Project.* Report No. 15122-TA, Agriculture and Environment Operations, Eastern Africa Department, Washington, DC.

World Bank (2000) *Poverty Headcount at $1.00 a Day, Tanzania.* http://ddp-ext.worldbank.org/Export_Ext/557e019c-cb44–41a8–9a7e-fbf2ef439168.csv

5 Issues in Reforming Informal Water Economies of Low-income Countries: Examples from India and Elsewhere

Tushaar Shah

International Water Management Institute, South Asia Program, Anand, India;
e-mail: t.shah@cgiar.org

Abstract

The past decade has witnessed a growing sense of urgency in reforming water sectors in developing countries like India faced with acute water scarcity. India, like many other developing countries, is still focused on building water infrastructure and services, and making these sustainable in all senses of the term. The new wave of ideas is asking it to move from this supply-side orientation to proactive *demand management* by reforming water policy, water law and water administration, the so-called 'three pillars' of water institutions and policies. But making this transition is proving difficult in India and elsewhere in the developing world. Here, making water laws is easy – enforcing them is not. Renaming regional water departments as basin organizations is easy – but managing water resources at basin level is not. Declaring water an economic good is simple – but using the price mechanism to direct water to high-value uses is proving complex. This chapter explores why.

It distinguishes between Institutional Environment (IE) of a country's water economy, which comprises the 'three pillars', and the Institutional Arrangements (IAs), which refer to the humanly devised rules-in-use, which drive the working of numerous informal institutions that keep a vibrant economy well lubricated. The relative influence of IE and IAs varies in high- and low-income countries because the water economies of the former are highly *formalized*, while those in the latter are highly informal. In high-income countries' formalized water economies, IE has an all-powerful presence in the water economy; in contrast, in highly informal water economies of low-income countries, IAs have a large role with the IE struggling to influence the working of countless tiny players in informal water institutions. The emerging discussion exhorting governments to adopt demand-side management overestimates the developing-country IE's capacity to shape the working of their informal IAs through direct regulatory means, and underestimates the potential for demand management through indirect instruments.

Demand-management reforms through laws, pricing and rights reforms in informal water economies are ill advised, not because they are not badly needed but because they are unlikely to work. The real challenge of improving the working of poor-country water economies lies in four areas: (i) improving water infrastructure and services through better investment and management; (ii) promoting institutional innovations that reduce transaction costs and rationalize incentive structures; (iii) using indirect instruments to work towards public-policy goals in the informal sectors of the water economy; and (iv) undertaking vigorous *demand management* in formal segments of the water economy such as cities and industrial water users. Facilitating these requires that water resources managers adopt a broader view of policy and institutional interventions they can catalyse to achieve policy goals.

Keywords: informal water economies, water institutions, institutional environment, irrigation management transfer, groundwater markets, groundwater recharge, energy, fishery, fluoride, India, China, Mexico.

Institutions and Policies in Formal and Informal Water Economies

A recent review of institutional changes in the water sector in 11 countries by Saleth and Dinar (2000) deals with water law, water policy and water administration, as the three pillars of institutional analysis in national water economies. This focus on law, policy and organizations as central themes of institutional analysis has been the concern of many analysts and practitioners of water resources management (see, e.g. Bandaragoda and Firdousi, 1992; Merrey, 1996; Frederickson and Vissia, 1998; Holmes, 2000; Saleth, 2004). However, if institutional change is about how societies adapt to new demands, its study needs to go beyond what government bureaucracies, international agencies and legal/regulatory systems do. People, businesses, exchange institutions, civil society institutions, religions and social movements – all these too must be covered in the ambit of institutional analysis (see, e.g. Livingston, 1993; Mestre, 1997 cited in Merrey, 2000, p. 5).

The current chapter takes this broader view in attempting a preliminary analysis of water institutions in India and elsewhere (see Fig. 5.1). In doing so, it draws upon the vast emerging field of New Institutional Economics (NIE) whose goal is to 'explain what institutions are, how they arise, what purposes they serve, how they change and how – if at all – they should be reformed' (Klein, 2000). We begin by borrowing from North (1990) the notion of institutions as 'formal rules, informal constraints (norms of behaviour, conventions, and self-imposed codes of conduct) and the enforcement characteristics of both'; and also the notion that 'if institutions are the rules of the game, organizations are the players'. It is also useful to borrow the important distinction drawn in the NIE between *institutional environment* (IE) and *institutional arrangements* (IAs). IE refers to the background constraints or 'rules of the game' – formal and explicit (constitutions, laws, etc.) and informal and implicit (norms, customs). Thus aspects that Saleth and Dinar (2000) include in their 'institutional analysis' represent, mostly, IE. IAs, in contrast, 'are the structure

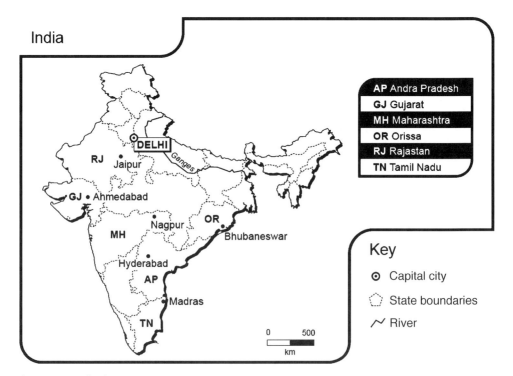

Fig. 5.1. Map of India.

that humans impose on their dealings with each other' (North, 1990).

In the Indian context, then, IE would include various government agencies at different levels that directly or indirectly deal in water, international agencies, governments' water policy and water-related laws and so on. And *institutions* or IAs – what Williamson (1985) calls 'governance structures' – refer to entities like groundwater markets, tube well cooperatives, water user associations (WUAs), Tarun Bharat Sangh's *johad* (small pond) movement in Alwar (Shah and Raju, 2001), groundwater recharge movement in Saurashtra (Shah, 2000), tank fishery contractors in Bundelkhand (Shah, 2002), emergence of defluoridation plants in the cottage sector in North Gujarat's towns (Indu, 2002), private lift irrigation provisioning on a large scale from Narmada canals in Gujarat (Talati and Shah, 2004) and from government reservoirs in the Upper Krishna basin in Maharashtra (Padhiari, 2005), and urban tanker water markets operating throughout cities in India and many other developing countries (Londhe *et al.*, 2004) and so on.

We begin with three propositions:

- Water institutions existing in a nation at any given point in time depend critically upon the level of *formalization* of its water economy; by formalization, we mean the proportion of the economy that comes under the ambit of direct regulatory influence of the IE.[1, 2]

- In this sense, water sectors are highly informal in poorly developed economies and become more formalized as national economies grow.

- The *pace* of water sector formalization in response to economic growth varies across countries and is influenced in a limited way by a host of factors but principally by the nature of the 'state'[3] (i.e. how hard or soft it is) (Myrdal, 1968). How much difference these other factors make is unclear; what is clear is that India or Tanzania cannot have Netherlands' level of formalization of its water sector at their present state of economic evolution.

The level of formalization of a country's water sector is best indicated by the low level of interface between its water IAs and its water IE –

or by what North (1990) calls the 'transaction sector'[4] of the water economy. Informal water economies, where the writ of 'the three pillars' does not run, are marked by heavy dependence of water users on self-provision (through private wells, streams, ponds), on informal, personalized exchange institutions or on community-managed water sources. In contrast, in highly formalized water economies – as in Europe and North America – self-provision disappears as a mode of securing water service; all or most users are served by service providers – private-corporate, municipal or others – who form the interface between users and the institutional environment. Volumetric supply and economic pricing are commonly used in highly formal water sectors for cost recovery as well as for resource allocation. Here, water emerges as an organized industry easily amenable to a host of policy and management interventions that become infeasible in informal water economies.

Just how informal the water economy of a developing country can be was explored by a large nationwide survey (NSSO, 1999b, p. 46) carried out in India during June–July 1998. Based on interviews with 78,990 rural households in 5110 villages throughout India, its purpose was to understand the extent to which they depended upon common property (and government) land and water resources for their consumptive and productive uses. The survey showed that only 10% of water infrastructural assets used by survey households were owned and managed by either a public or community organization. The rest were mostly owned and managed by private households or owned by the government/community but *not* managed by either.[5]

If receiving domestic water from a 'tap' is an indicator of getting connected to a formal water supply system, the same survey also showed that over 80% of rural households were not connected with *any* public or community water supply system: they self-supplied their domestic water needs. In urban households (sample = 31,323 households), the situation was the reverse: 75% were connected to a public water supply system.

A somewhat different 2002 survey (NSSO, 2003) showed that, of the 4646 villages covered, only 8.8% had a public/community water supply system. People living in the rest of the villages depended on wells or open water bodies for domestic water supply. A strong imprint of economic growth was evident too. The proportion of villages with a public water supply system increases rapidly as we move from a poor state to a relatively rich one. In Bihar, one of India's poorest states, none of the 364 villages covered had a public/community water supply. In the somewhat richer Haryana state, over half the villages surveyed had a public water supply system and, in still richer Goa, every village surveyed had a public water supply system.

The irrigation economy of India is equally informal. A 1998 survey of 48,419 cultivators around India showed that nearly 65% used irrigation for five major field crops cultivated by them. For nearly half of these, the source of irrigation was informal, fragmented pump irrigation markets (NSSO, 1999b, p. 42), which are totally outside the ambit of *direct* influence of the 'three pillars'. In a 2002 survey of 4646 villages around India (NSSO, 2003), 76% of the villages reported they irrigated some of the lands. However, only 17% had access to a *public* irrigation system: the rest depended

primarily on wells and tube wells, tanks and streams.

All these surveys suggest that rural India's water economy – both domestic and irrigation use – is predominantly *informal*, based as it largely is on self-supply and local, informal water institutions. It has little connection with public systems and formal organizations through which the 'three pillars' typically operate in industrialized countries.[6]

Figure 5.2 presents a clutch of empirically verifiable hypotheses – a set of 'iron laws of economic development'[7] – about how the economic organization of a country's water economy metamorphoses in response to economic growth and the transformation of society that comes in its wake. It is difficult to find a country in, say, sub-Saharan Africa with a modern water industry of the kind we find in a European country. South Africa is an exception: white South Africa – inhabiting its towns or operating large, commercial farms in the countryside – is served by what approximates a modern water sector. In the rural areas of the Olifants basin, for example, only 0.5% of this

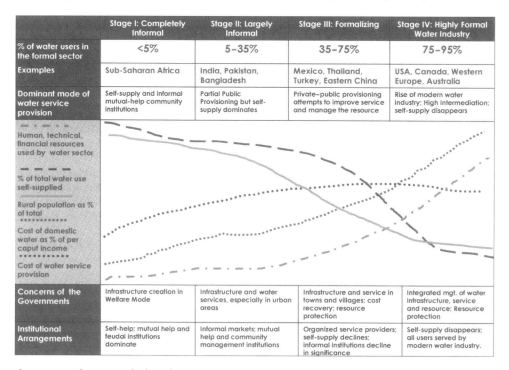

	Stage I: Completely Informal	Stage II: Largely Informal	Stage III: Formalizing	Stage IV: Highly Formal Water Industry
% of water users in the formal sector	<5%	5–35%	35–75%	75–95%
Examples	Sub-Saharan Africa	India, Pakistan, Bangladesh	Mexico, Thailand, Turkey, Eastern China	USA, Canada, Western Europe, Australia
Dominant mode of water service provision	Self-supply and informal mutual-help community institutions	Partial Public Provisioning but self-supply dominates	Private–public provisioning attempts to improve service and manage the resource	Rise of modern water industry; High intermediation; self-supply disappears
Human, technical, financial resources used by water sector / % of total water use self-supplied / Rural population as % of total / Cost of domestic water as % of per caput income / Cost of water service provision				
Concerns of the Governments	Infrastructure creation in Welfare Mode	Infrastructure and water services, especially in urban areas	Infrastructure and service in towns and villages; cost recovery; resource protection	Integrated mgt. of water infrastructure, service and resource; Resource protection
Institutional Arrangements	Self-help; mutual help and feudal institutions dominate	Informal markets; mutual help and community management institutions	Organized service providers; self-supply declines; informal institutions decline in significance	Self-supply disappears; all users served by modern water industry.

Fig. 5.2. Transformation of informal water economies in response to overall economic growth (from author).

formal sector – some 1600 registered users in a population of 2.5 million – uses 95% of the water resources (Cullis and van Koppen, 2007). The former homelands, where half of South Africans live, are served by a water economy even more informal than India's.

Water institutions that exist in a country or can be expected to be successfully catalysed by external actors depend upon, besides several other factors, the stage of formalization of its water economy which, in turn, depends upon the overall economic evolution of that country as outlined in Fig. 5.2. Water IAs we found in India, Pakistan and Bangladesh – such as, say, pump irrigation markets or urban tanker water markets – are unlikely to be found in Australia or Spain because they would serve nobody's purpose there. Likewise, water IAs that are standard in industrialized countries – multinationals managing a city's water supply system – would not begin to work until Dhaka has a water service market evolved, at least, to the level of Manila or Jakarta.[8]

The Process of Institutional Change

In understanding how societies adapt their institutions to changing demands, Oliver Williamson (1999) suggests the criticality of social analysis at four levels. At the highest level (say L1) of social embeddedness are customs, traditions, mores and religion, which change very slowly because of the spontaneous origin of these practices in which 'deliberative choice of a calculative kind is minimally implicated'. At the second level (L2), evolutionary processes play a big role; but opportunities for design present themselves through formal rules, constitutions, laws and property rights. The challenge here is getting the rules of the game right through better definition and enforcement of property rights and contract laws. Also critical is the understanding of how things actually work – 'warts and all' in some settings, but not in others. However, it is one thing to get the rules of the game (laws, policies, administrative reforms in the IE) right; it is quite another to get the play of the game (enforcement of contracts/property rights) right.

This leads to the third level (L3) of institutional analysis: transaction costs of enforcement of contracts and property rights, and the governance structures through which this is done.

Governance – through markets, hybrids (like public–private partnerships), firms and bureaus – is an effort to craft order, thereby mitigating conflict and realizing mutual gains. Good governance structures craft order by reshaping incentives, which leads to the fourth level (L4) of social analysis – getting the incentives right.

L1 and L2 offer possibilities for change only over the long term.[9] Sectoral interventions aiming to achieve at least L2 level changes[10] – property rights on water through a permit system or reorienting the bureaucracy – are not uncommon; but it is virtually impossible to enduringly[11] transform *only* the water bureaucracy while the rest of the bureaucracy stays the same. All things considered, L3 and L4 comprise the most relevant playing field for institutional reform in the short term.

An important question that New Institutional Economics (NIE) helps us explore is: 'Why do economies fail to undertake the appropriate activities *if they had a high pay-off?*' (North, 1990). The response to this question depends largely on L3 and L4 levels of institutional analysis. India's water sector is replete with situations where appropriate activities can potentially generate a high pay-off and yet fail to be undertaken; in contrast, much institutional reform being contemplated or attempted may not work, in the current context, because, among other things, high transaction costs make them inappropriate to undertake.

An institutional change creates a 'structure' of pay-offs with gains varying across different groups of agents and, therefore, inviting different 'intensities' of responses. A small group of agents each threatened with large loss may put up a stiff resistance to a change that is beneficial for the society as a whole, and vice versa. Likewise, different groups of agents in IAs as well as in IE may experience different levels of incidence of transaction costs attendant on a change. In NIE, transaction costs are seen to include: (i) costs of search and information; (ii) costs of negotiation, bargaining and contracting; and (iii) costs of policing and enforcement of contracts, property rights, rules and laws. Our key proposition in this chapter is: for a policy or institutional intervention, all these three increase *directly* with the number of agents involved as well as with the strength of their preference for or against the intervention.

All three costs come into play in determining the 'implementation efficacy' of an institutional intervention because each depends on the number of agents involved in a transaction, which in an informal water economy is large. Just take the case of groundwater regulation in a country like Mexico which, in some parts, faces problems of resource over-exploitation similar to those of India and the North China plains. Mexico's new Law of the Nation's Water provided for the registration of all groundwater diverters and issue of 'concessions' to each, with an entitlement to pump a permitted quota of water per year. Nearly a decade later, the 'implementation efficacy' of this policy regime has varied across different segments of groundwater diverters: municipal and industrial diverters – all large, visible entities in the formal sector – have been promptly and effectively brought within the ambit of the new Law because these large diverters are few in number. Household wells – far too numerous, and each diverting small quantities – were wisely kept out of the ambit of the law; the transaction cost of regulating them was not worth the gains in 'implementation efficacy'.[12]

The real problem was with over 96,000 agricultural tube wells, some of them abstracting up to 1 million m^3 of groundwater each per year. Having registered agricultural tube wells, Mexico's CNA (*Comisión Nacional del Agua*) found it impossible to police and enforce concessions with the staff and resources at its command. To reduce policing and enforcement costs, CNA created COTAS (Comités Técnicos de Aguas Subterráneas), assuming that farmers would police each other better. A slew of recent studies, however, have shown that Mexico's new Law of the Nation's Water, its national water policy as well as institutions like COTAS have had no perceptible impact on groundwater abstraction for agricultural use (Shah *et al.*, 2004b).

If Mexico is serious about groundwater regulation, it will need to either find effective ways to reduce policing and enforcement costs of tube well concessions or else allocate much larger resources to absorb the high costs of policing and enforcement of groundwater concessions on 96,000 tube well owners scattered over the countryside. And if India were to try a similar strategy, it would need to provide for policing and enforcement costs for some

20 million private tube well owners scattered over 600,000 villages.

One core NIE idea – especially, of the Transaction Cost Economics (TCE) branch – is that economizing on transaction costs is a key determinant of the nature of IAs that economic agents evolve. Our proposition is that players in IE of sectoral economies too are sensitive to transaction costs in designing, implementing or abandoning institutional interventions. This implies that the state too indulges in transaction cost-economizing behaviour. This is indicated by the fact that water regulations in most countries exclude small users from their ambit. Mexico's Law of the Nation's Water does not apply to anyone who stores less than 1030 m^3 of water. Australia's water law excludes users who irrigate less than 2 ha (MacDonald and Young, 2001). Water withdrawal permits instituted in South Africa and many African countries in recent years exclude domestic users, homestead gardening and stock watering (Shah and van Koppen, 2005).

One rationale for leaving these out is that these represent lifeline uses of water. But another equally important reason is that the inclusion of these would hugely increase search, information and policing and enforcement costs involved in implementing the new intervention. Under its new water law, China has instituted a system of water withdrawal permits to be obtained by each tube well owner. But, in reality, except in selected provinces such as Beijing, Hebei and Shandong where tube wells are deep and heavy duty, the permits are issued to the village as a whole. Doing this defeats the intent of the law but it reduces transaction costs (Shah *et al.*, 2004a). When transaction costs of implementing an institutional intervention become prohibitive, players in IE relinquish it rather than enforcing it *at any cost*.

Alternatively, IE players discover well-thought out approaches to drastically reduce transaction costs. Provincial and city water bureaus in eastern China have for long tried to regulate pumping of urban groundwater aquifers that are under great stress. An array of regulatory measures – imposition of a water withdrawal fee, increases in water price, sealing of urban tube wells, etc. – failed to control urban groundwater depletion. More recently, many cities have begun sourcing water from distant reservoirs and supplying it to urban water service providers. Alternative water supply assured, many cities have quickly brought urban groundwater diverters within the regulatory fold (Shah *et al.*, 2004a).

Another example of 'transaction cost economizing' behaviour of IE players is the Mexican government's decision of levying a penal charge for electricity use by tube wells withdrawing groundwater beyond the concessioned volume. Having failed to police and enforce groundwater abstraction concessions through COTAS, the CNA found the second best approach, whose key merit is that it imposed little 'incremental' transaction cost because metered electricity use already provided a good surrogate of volumes of abstraction (Scott and Shah, 2004).

In analysing the Indian institutional experience in the water sector, then, our key propositions are embodied in Fig. 5.3. It suggests that several kinds of institutional reform tried or suggested in the Indian water sector have tended to have entailed either high transaction costs (quadrant 2), low pay-offs (quadrant 4) or both (quadrant 3). In contrast, institutional changes that have quietly occurred because

pay-offs are high *and* transaction costs low (quadrant 1) are either ignored or thwarted or, at least, not built upon. In the following sections, we briefly analyse a sample of situations in each of these four quarters in Fig. 5.3 before drawing some general implications arising from this analysis.

Interventions with Poor Implementation Efficacy (Quadrants 3 and 4)

When policing and enforcement costs of an intervention are high, the tendency often is to design frivolous interventions without serious intention to implement them or to abandon an intervention even if designed with serious intent. International pressure has often led to a persistent demand for a modern legislative and policy framework for orderly and effective management of the water economy and sustainable husbanding of the resource. Conditionalities imposed by donors sometimes oblige developing-country governments to agree to interventions without a local buy-in. One possible reason they submit to such pressures is their dependence on them for financial

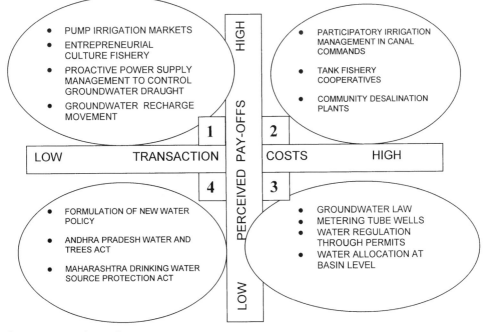

Fig. 5.3. Expected pay-offs versus transaction costs.

resources; however, it may also be that donors can pressurize governments to make laws but not to enforce them. Even if governments had a genuine intent to enforce, in a predominantly informal water economy such as India's, the transaction costs of enforcing a 'strong' water law effectively are so high that these attempts often remain cosmetic, essentially setting 'targets without teeth'. Indeed, laws and policies are often written to minimize transaction costs by progressively removing clauses that bite and are likely to be extensively violated, thereby reducing the *effective* regulatory powers of a law. When this is not done, decision makers responsible for enforcement shy away.

The Model Groundwater Law developed by the Government of India circa 1970 is a case in point; it has been tossed around for 35 years across state capitals but it has found no takers, not only because of the virtual impossibility of reasonable enforcement but also because of the invidious political economy of rent-seeking that it may create at the local levels. The Gujarat assembly passed the law but the Chief Minister decided, wisely, not to gazette the act in view of high transaction costs of enforcing it.[13]

The chief ministers of some other Indian states were, however, less transaction cost-savvy. So in 1993, Maharashtra made a law with a limited ambition of disabling irrigation wells within 500 m of a Public Water Source during droughts, with a view to protecting drinking water wells. Ten years after its enactment, the International Water Management Institute (IWMI) commissioned a study of the enforcement of this law (Phansalkar and Kher, 2003). The law provides for stern action against violation but has a 'naughty' clause requiring that the law be invoked only when a '*gram panchayat* (village council) files a written complaint' (which, at one stroke, reduces to a fraction the transaction costs as well as the potency of the law).

The study found numerous cases of violations of the 500 m norm, yet not a single case of legal action has resulted because gram panchayats have failed to file a written complaint. It concluded that: 'There is a near complete absence of social support for the legislation. The rural lay public as well as the office bearers of gram panchayats appear inhibited

and reluctant to seem to be "revengeful" towards those who are doing no worse than trying to earn incomes by using water for raising oranges.'

Instead of invoking the law, supply-side solutions in the form of upgraded drinking water facilities and water tankers during droughts are preferred by people, gram panchayats as well as *zilla parishads* (district councils). IWMI also did a quick assessment of the Andhra Pradesh Water and Trees Act (Narayana and Scott 2004),[14] and concluded on a similar pessimistic note. A similar exercise has been the formulation of the official Government of India Water Policy of 1987 and 2002. Both these pieces are an excellent example of bland, almost tongue-in-cheek, enunciations that are *not* designed to change anything in any manner.[15] As a result, they have low transaction costs, but also no pay-off.

Other widely espoused proposals entail high transaction costs and promise doubtful benefits – at least in the prevailing circumstances. A good example in India is the effort to introduce volumetric pricing of electricity supply to groundwater irrigators after having given up on it decades previously. It was the high transaction costs of metering over a million irrigation pump-sets – which involved installing and maintaining meters, reading them every month, billing based on metered consumption of power but, more importantly, controlling pilferage, tampering with meters with or without collusion with meter readers, etc. – that obliged State Electricity Boards (SEBs) to switch to a flat tariff during the 1970s (Shah, 1993).

A flat tariff, collected based on the size of the pump horsepower rather than on the metered consumption of electricity for pumping, succeeded in reducing transaction costs of serving a market where derived demand for electricity was confined to periods of peak irrigation requirements. It would have been a viable system if SEBs had learnt to ration power supply to agriculture and gradually raise the flat tariffs to break-even levels. However, neither happened; farmer lobbies have managed all along to prevent upward revision in the flat tariff while compelling the SEBs to maintain electricity supply to the farm sector. The invidious nexus between energy and irrigation – which has contributed to the bankruptcy of the

Indian power sector and rampant over-exploitation of groundwater – has been discussed by Shah *et al.* (2004c). We simply summarize its conclusion here.

In the thinking of SEBs and multilateral donors about ways out of this imbroglio, a return to metering power is critical, even if it means taking on farmer lobbies. Several chief ministers have tried to bite the bullet in the past few years. But farmers' opposition has been so strong, swift and strident that they have been either felled or obliged to retract. Some, as in Punjab and Tamilnadu, have done away with farm power tariff altogether. Recommending metering farm electricity in today's setting is asking politicians to do hara-kiri.

But even if a politician were to succeed in metering farm power supply, it would probably change little because, if anything, transaction costs of metered power supply are much higher today than they were in the 1970s. Most states have at least eight to ten times more irrigation tube wells today than they had during the 1970s; and farming livelihoods depend far more critically on electricity today than 30 years ago. If metering must work in the India of today, we must learn from the Chinese experiments, which always stuck with metering, and then focus on modifying the incentive structures to address many of the problems metering faces in India (see Shah *et al.*, 2004a).

Surprisingly, the electricity–irrigation nexus is not a subject of discussion in China at all. The Chinese electricity supply industry operates on two principles: (i) total cost recovery in generation, transmission and distribution at each level, with some minor cross-subsidization across user groups and areas; and (ii) each user pays in proportion to their metered use. Unlike in much of South Asia, rural electricity throughout China was charged at a higher rate than urban; and agriculture paid more than domestic and industrial use until a few years ago (Wang *et al.*, 2004).

Until 1997, the responsibility for operation and maintenance of the village electricity infrastructure and user charge recovery lay with the village committee. The standard arrangement in use was for the village committee and the township electricity bureau to appoint and train one or more local farmers as part-time village electricians with dual responsibility for: (i) main-taining the power supply infrastructure in the village; and (ii) collecting user charges for a transformer assigned to him/her based on metered individual consumption from all categories of users. The sum of power use recorded in the meters attached to all irrigation pumps had to tally with the power supply recorded at the transformer for any given period. The electrician was required to pay the township electricity bureau for power use recorded at the transformer level.

This arrangement did not always work easily. Where power supply infrastructure was old and worn out, line losses below the transformer made this difficult. To allow for normal line losses, a 10% allowance was given by the township electricity bureau to the electrician. However, even this must have made it difficult for the latter to tally the two; as a result, an electricity network reform programme was undertaken by the national government to modernize and rehabilitate rural power infrastructure.[16] Where this was done, line losses fell sharply,[17] and among a sample of ten villages I visited in 2003, none had a problem tallying power consumption recorded at the transformer level with the sum of consumption recorded by individual users, especially with the line loss allowance of 10%.

It is interesting that the village electrician in Henan and Hebei provinces in North China is able to deliver on a fairly modest reward of US$24–30/month plus an incentive bonus of around $24/month (Zhang, 2004), which is equivalent to the value of wheat produced on 1 mu (or 0.67 ha) of land. For this rather modest wage, China's village electrician undertakes to make good to the township electricity station the full amount on line and commercial losses in excess of 10% of the power consumption recorded on the transformers; if he can manage to keep losses to less than 10%, he can keep 40% of the value of power saved. This generates a powerful incentive for him to reduce line losses.

In the way that the Chinese collect metered electricity charges, it is well nigh impossible to make financial losses since these are firmly passed on downstream from one level to the next. Take, for example, the malpractice common in South Asia of end-users tampering with meters or bribing the meter reader to under-report actual consumption. In the Chinese system, it is very unlikely that such malpractices could occur on a large scale, since the village electrician is faced with serious personal loss if he fails to collect from the farmers electricity charges for at least 90% of power consumed as reported at the transformer meter. And since malpractice by a farmer directly hits other farmers in the village, there is likely to exist strong peer control over such practices.

In making metered power pricing work, China's unique advantage is its strong village-level authority structure. The village committee, and especially, the village party leader, is respected and feared. These factors ensure that the electrician is able to do his or her job. In comparison to China's village committees, India's village Panchayats are utterly devoid of power, as well as authority, as institutions for local governance.

In India a similar experiment was tried out in Orissa, where private companies in charge of distribution first experimented with village *vidyut sanghas* (electricity cooperatives) by forming 5500 of them but are now veering around to private entrepreneurs as electricity retailers. Mishra (2004), who carried out an assessment of Orissa reforms for the IWMI-Tata programme, visited a number of these sanghas during 2003 and noted that: 'None of the village committees were operational.' These worked as long as the support organization hired to catalyse them propped them up with constant visits and organizational work; as soon as the support organization was withdrawn, the village vidyut sanghas became defunct. Mishra (2004) wrote: 'The situation today is quite similar to that [which] existed earlier before the interventions were made through the Committee.' Sanghas having failed, power distribution companies appointed three private entrepreneurs as franchisees on terms similar to those facing China's village electricians. These have resulted in sustained and significant improvements in billing and collection of electricity dues.

The Orissa experiment and the Chinese experience suggest that, in principle, it is possible to make volumetric pricing and collection of electricity charges work if private entrepreneurs are given appropriate incentives. However, in Orissa, the electricity use in agriculture is less

than 5%. If the same arrangement were to work in Punjab, Haryana or Gujarat or several other states where electricity use in the farm sector is 30% or more, farmer resistance would be greater and commensurate with the effectiveness of the volumetric pricing. And one thing that private power retailers in Indian villages would have to do without is the authority of the village party leader that helps China's village electricians to firmly pass on all costs to farmers. In the absence of such authority structures, private entrepreneurs would expect very high margins to assume the role of retailing power on a volumetric basis. This – as well as farmer propensity to frustrate metering – would raise transaction costs of metering to very high levels. If the ultimate purpose of volumetric pricing is to improve the finances of electricity utilities, I doubt this purpose would be achieved.

In a recent paper (Shah *et al.*, 2004c), we have argued that, in making an impossibly bad situation better, a more practical course available to SEBs and state governments is to stay with flat tariffs but to rationalize them through intelligent management of power supply. Farmers' needs for power are different from those of households or industries: they need plentiful power on 30–40 days of the year when crops face acute moisture stress. However, in most states, they receive a constant 8–10 h/day of poor-quality power supply throughout the year. If SEBs were to invest in understanding that their farmers are customers, it should be possible for them to supply 20 h/day of good-quality power to farmers on 30–40 days of peak irrigation need while maintaining 3–4 h/day supply on other days. In order for such an approach to work, the nature and capabilities of the power utilities have to change; so also does the thinking of donors and governments.

In sum, in improving the working of India's water economy, many policy and institutional interventions – already tried and watered down, or on the discussion table – are of little value because its predominantly informal nature makes its policing and enforcement costs prohibitive. India is not alone in devoting energies and resources to these.

In Africa several countries have, during recent years, experimented with demand management ideas such as pricing of water, instituting water withdrawal permits and restructuring regional water departments as river basin organizations. Although it may be too early to write a report on these, countries like Ghana are already having second thoughts. The concerns are of five kinds: (i) most reforms have remained largely unimplemented, especially in the informal segments of the water economy that encompass most of the users and uses; (ii) nowhere have the reforms produced evidence of improved performance of the water economy, except in countries with a large formal water economy; (iii) implementation of reforms has disrupted customary arrangements for water management that was robust enough to, at least, survive the test of time; (iv) when zealously implemented, reforms – especially water permits and water taxes – hit poor people in remote rural areas hard; and (v) 'demand management reforms' deflected national IE players from pursuing water sector priorities important to them, namely improving water infrastructure and services to their people (Shah and van Koppen, 2005).

Areas in Need of Institutional Innovation (Quadrant 2)

Rather than evolving organically from the unfolding situation on the ground – and therefore being demanded by stakeholders – many of the reforms currently being pursued in India, such as Irrigation Management Transfer (IMT), River Basin Management and metering of electricity are actually promoted aggressively by both researchers and funding agencies,[18] and are sometimes out of sync with the prevailing Indian context. By far the most frequent are situations where institutional interventions proposed would yield high productivity pay-offs if successful; but they rarely succeed because of high transaction costs.

In independent India's history, the 'communitarian ideal' – the notion that villagers will instantly come together to take over the responsibility of participatory, democratic management of virtually anything (land, water, watersheds, forests, irrigation systems, river basins) – has been behind innumerable abortive institutional interventions. What has helped fuel this enthusiasm for participatory irrigation management (PIM) by farmers are

occasional examples of such models having worked reasonably well either in the industrialized countries or in India itself, but under the tutelage of an inspired local leader or an industrious NGO. Its having worked in a few situations in exceptional conditions becomes the basis for designs of major programmes of institutional interventions, commonly bankrolled by a supportive donor.

One classic example of ideas in this genre is PIM (or its cousin IMT) which has been, for the past four decades, the ruling *mantra* for improving the productivity of irrigation systems in India. What is extraordinary about this preoccupation with PIM (or IMT) is the sway it has continued to hold on players in water IE, despite virtually no evidence of it having succeeded anywhere else except on an experimental scale, that too with facilitation of non-replicable quality and scale.[19]

The idea of farmers managing irrigation canals is not new; the British tried hard in the late 19th century to get farmers from the Indus and Ganges areas to participate in irrigation management but without much success, except in enforcing *warabandi* (rotational methods for equitable allocation of available water) in the Indus canals (Whitcombe, 1984). More recently, since 1960, WUAs (Water Users' Associations) have been tried out on small irrigation systems. Uttar Pradesh tried *sinchai samitis* (irrigation committees) way back in the early 1960s on irrigation tanks and reservoirs; following that, Madhya Pradesh too tried it on thousands of its minor irrigation tanks.

Other states have been trying to make *pani panchayats* (water councils) work. But sinchai samitis of Madhya Pradesh and Uttar Pradesh have disappeared without trace; and so have pani panchayats in Gujarat and elsewhere. Yet, Orissa recently made a law that transferred all its minor irrigation systems to instantly created pani panchayats. Gujarat introduced joint irrigation management programmes as far back as in 1983, but the 17 irrigation cooperatives lost money and became defunct. In 1991 it made another attempt, this time around with assistance from NGOs; 144 irrigation cooperatives were formed to cover 45,000 ha of irrigated area (Shukla, 2004); however, it is difficult to see precisely in what way these areas are better off than other command areas.

Indeed, a core idea of Command Area Development Agencies (CADAs) in the early 1980s was to involve farmer organizations in the management of irrigation projects. But we see no trace of CADAs or their beneficiary farmers' associations (BFAs), even in Kerala where thousands of these were formed under a 'big bang' approach in 1986. An assessment by Joseph (2001) in the late 1990s suggested that, even in this land of strong traditions of local governance, good education and high levels of public participation, BFAs were a damp squib.[20]

As in Kerala, Andhra Pradesh overnight transferred the management of all its irrigation systems to over 10,000 WUAs created by the automobile company Fiat and a World Bank loan; this 'big bang' approach to PIM has attracted all-round interest; however, now that the World Bank funds retailed to WUAs for maintenance are over, field observers are beginning to wonder precisely what the WUAs are doing better (Jairath, 2001).[21]

The central assumption underlying PIM/IMT is that, once irrigation management is transferred from remote bureaucracies to WUAs, the financial viability of the systems would improve and so would the quality and reliability of irrigation. Physical and value productivity of water and land would increase. As a result, irrigation systems would better achieve their potential for food and livelihood security for farmers in their command. PIM/IMT programmes have belied many of these expectations, even in countries like Turkey, Mexico and Philippines where they are known to have succeeded. As a result, early expectations from PIM/IMT have been increasingly moderated and IMT is now considered successful even if it just 'saves the government money, improves cost effectiveness of operation and maintenance while improving, or at least not weakening, the productivity of irrigated agriculture' (Vermillion, 1996, p. 153). The drift of the IMT discussion then, in recent times, has been more towards getting irrigation off the back of the governments than towards improving the lot of the farmers and the poor, the original goal at which much public irrigation investment has been directed over the past 50 years.

Some over-arching patterns emerge from a reading of the international experience. IMT has tended to be smooth, relatively effortless and successful where:

- The irrigation system is central to a dynamic, high-performing agriculture.
- The average farm size is large enough for a typical or a significant proportion of the command area farmers to operate like agro-businessmen.
- The farm producers are linked with global input and output markets.
- The costs of self-managed irrigation are an insignificant part of the gross value of product of farming.

These are the conditions – all of which enhance the pay-offs, reduce transaction costs or both – obtained in Mexico, the USA and New Zealand, from where emerge the resounding success stories we hear about IMT[22] (Shah et al., 2002). In South Africa the commercial farming sector, which satisfies all these conditions, took naturally to PIM through its irrigation boards; but the same logic when applied to irrigation systems serving smallholders in former homelands met with resounding failure because these met none of the conditions that irrigation boards satisfied (Shah et al., 2002).

Even where all conditions are satisfied and PIM/IMT declared 'successful', researchers have presented a mixed picture of resultant impacts. For example, an exhaustive global review carried out for IWMI of IMT impacts by Douglas Vermillion, a pioneer in IMT research, showed that impacts are significant and unambiguously beneficial in terms of cost recovery in Turkey, Mexico, the USA and New Zealand. Fee collection has improved; agency staff strength has declined. But the impact of management transfer on agricultural productivity and farm incomes is far less unequivocal even in these countries (Vermillion, 1996, p. 153). In Philippines, the Mecca of IMT and PIM, recent studies show that productivity gains from PIM have not been sustained (Panella, 1999).

None of the conditions outlined above are obtained in a typical Indian surface irrigation system. Most farmers in the command have small-holdings, subdivided further into smaller parcels. A typical major system has hundreds of thousands of smallholders, making it well nigh impossible to bring them all together to negotiate. Over 90% of the surface water irrigated area in India is under field crops yielding Rs 15,000–18,000 (US$325–400)/ha of gross value of output, compared with US$3000–7500/ha in high-value farming in industrialized countries. Irrigation systems are at the heart of the farming economy of command areas. However, the mushrooming of wells and tube wells, and booming pump irrigation markets in command areas and in the neighbourhood of irrigation tanks have reduced farmers' stakes in managing surface irrigation systems. Head-reach and tail-end farmers almost always have opposing motivations when it comes to management reform, with the former interested in preserving the status quo and the latter interested in change.

All these, together, raise the transaction costs of implementing management reform through PIM/IMT-type interventions. The prospects become worse because, almost everywhere, the agency's purpose in promoting PIM is to get WUAs to assume arduous responsibilities – maintenance, fee collection, mobilization of voluntary labour for repair and maintenance works, etc. Moreover, farmers are generally quick to figure out that PIM often means increased water fees without corresponding improvement in service quality. These reduce the perceived pay-offs from reform.

All in all, decades invested in the hope that PIM or IMT would spearhead productivity improvements in public irrigation are decades wasted. PIM has not achieved any significant success on a meaningful scale anywhere in India, and it will indeed be a great surprise if it does in the existing IE marked by hopelessly low irrigation fees, extremely poor collection and poor main system management.

There are similar institutional misadventures in other spheres. In growing regions where fluoride contamination of groundwater is endemic, governments and donors have tried setting up village-based reverse osmosis-type plants or Nalgonda-type defluoridation plants to control the growing menace of dental and skeletal fluorosis. Again, the management model chosen is communitarian, and these have invariably failed. In Gujarat, out of dozens of such plants set up during the 1980s and 1990s, not one has operated for more than a few months.

An older experiment with a communitarian model has been with inland fishery cooperatives. Numerous local water bodies controlled by irrigation departments, zilla panchayats, *taluka panchayats* (sub-district councils) and gram panchayats can potentially sustain a vibrant inland fishing enterprise and livelihood system. However, government policy has always been to give away monopoly lease rights to registered fisher-people's cooperatives. Thousands of such cooperatives are registered; but probably a very small fraction – in my surmise, less than 1 or 2% – operate as dynamic producer cooperatives as, for instance, the dairy cooperatives do in Gujarat.

In South India, which has over 300,000 irrigation tanks, a decades-old concern has been about the breakdown of traditions of maintenance of bunds and supply channels, orderly distribution of water and protection from encroachment. Several donor-supported projects first aimed at 'engineering rehabilitation' and restored tank infrastructure to their original – or even a better – condition. However, when rehabilitation of tanks again declined and needed another round of rehabilitation, planners found something amiss in their earlier approach. Therefore, in new tank rehabilitation programmes – such as the new World Bank project in Karnataka – an institutional component is added to the engineering component. But the institutional component invariably consists of registering a WUA of command area farmers. Except where such WUAs have been constantly animated and propped up by support NGOs – as in the case of the Dhan Foundation in Madurai, Tamilnadu – it is difficult to find evidence of productivity improvements in tanks because of WUAs on any significant scale (Shah *et al.*, 1998).

Besides the problem of high transaction costs of co-coordinating, negotiating, rule making and, above all, rule enforcement and improving the management of tanks – more in North India than in South India – face some special problems. One of them is of aligning conflicting interests of multiple stakeholders. Command area farmers have a direct conflict of interest with tank-bed farmers; and well owners in the neighbourhood of tanks are a potential threat to all other users because they can virtually steal tank water by pumping from their wells. Then, there are fishing contractors whose interests also clash with those of irrigators, especially during the dry season (Shah and Raju, 2001). Registering a WUA of command area farmers and hoping that this 'institutional intervention' would increase productivity of tanks is extremely naive. Improved management of public irrigation systems, tanks and fishery represents opportunities for high pay-off but has failed to be realized because the institutional models promoted have high transaction costs.

Vibrant Institutional Arrangements Ignored (Quadrant 1)

The core of New Institutional Economics is the notion that productivity of resources in an economy is determined by technology employed and institutions. And if 'institutions affect economic performance by determining transaction and transformation (production) costs', then the Indian water sector is brimming with institutional changes occurring on the margins that are doing this all the time, and yet are either glossed over (or even frowned upon) by the players in the IE. Most such institutions we explore in this section are invariably swayamb-hoo[23] (self-creating and spontaneous); they have come up on a significant enough scale to permit generic lessons. These invariably involve entrepreneurial effort to reduce transaction costs; they serve an important economic purpose, improve welfare and raise productivity; they are commonly faced with an adverse or unhelpful IE. Crucially, these constitute the instrumentality of the players of the game, and sustain as long as they serve their purpose.

The emergence of tube well technology has been the biggest contributor to growth in irrigation in post-independent India; and the spontaneous rise of groundwater (or, more appropriately, pump irrigation service) markets has done much to multiply the productivity and welfare impact of tube well irrigation. The Indian irrigation establishment is probably out of touch with the changing face of its playing field: it still believes that only 38% of the gross cropped area is irrigated, 55% of it by groundwater wells. But concerning the reality of Indian irrigation at the dawn of the millennium, the tail has begun wagging the dog.[24] IE in the Indian water sector has little or no interface with either the 75% of Indian irrigation occurring through tube wells or with the institution of water markets.

The working of groundwater markets has now been extensively studied (see Shah, 1993; Janakarajan, 1994; Saleth, 1998; Singh and Singh, 2003; Mukherji, 2004 for a good survey of the literature). These studies analyse myriad ways in which their working differs across space and time.

But common elements of groundwater markets everywhere in the Indian subcontinent are the features we listed at the start of this section: (i) they are swayambhoo; (ii) they operate on such a large scale as to account for over one-quarter of the Indian irrigated areas; (iii) water sellers everywhere constantly innovate to reduce transaction costs and create value; (iv) water markets are the instrumentality of buyers and sellers of pump irrigation service, and not of society at large or the IE; (v) as a result, water markets are unrepentant when their operation produces externalities such as groundwater depletion or drying up of wetlands; and, finally, (vi) despite their scale and significance, the IE has been blind towards the potential of water markets to achieve larger policy ends. When they take notice of their existence and role – which is seldom – water policy makers are often unable to decide whether they deserve promotion or regulation.

Much the same is the case with many other water institutions. In the previous section, I mentioned tens of thousands of fishermen's cooperatives that are lying defunct. However, pond fishery entrepreneurs have sprung up everywhere who use 'paper' cooperatives as a front for operating profitable culture fisheries. Why don't fisher cooperatives exploit the economic opportunities that these contractors are able to? The most important reason is the transaction costs of protecting their crop. Culture fishery is capital intensive but affords a high yield. In common property village or irrigation tanks with multiple stakeholders, in order to remain viable the fishermen should be able to meet many conditions. They should effectively defend their rights against poachers, and against irrigators who may want to pump tank water below the sill level during dry periods to irrigate crops, or against tank-bed cultivators who want to empty the tank so they can begin sowing.

In South Asia, fisher communities are commonly from the lowest rung of the village society. They would not only have difficulty in mobilizing capital to buy seedlings and manure but also in protecting the crop from poaching from outsiders, from the local bigwigs as well as from their own members. Fisher cooperatives, as a result, always underinvest. Reserving fishing contracts for fisher cooperatives is therefore the best formula for sustained low productivity of the inland fishery economy.

We discovered just how high the transaction cost of protecting a fish crop was when we studied who precisely the fishing contractors were in two separate studies in central Gujarat and Bundelkhand. We found that, in both the regions, the key characteristic of people who emerged as successful fishing contractors was a painstakingly cultivated image of a toughie, or a ruffian capable of enforcing his rights even by using violence. In Bundelkhand, 'Everywhere the fishing contractors involved stopped farmers from lifting water from the tank once the last five feet of water was left. They had invested in fish production and now were making sure they get their money's worth' (Shah, 2002, p. 3).

In central Gujarat, fishing contractors often have to resort to violence and even undergo a jail term to establish that they meant business when it came to defending their property right.[25] Despite this unsavoury aspect, I would not be much off the mark in suggesting that the explosive increase in inland fishery in India during the past 40 years is the result of two factors: (i) introduction of new technologies of culture fishery along with its paraphernalia; and (ii) gradual emasculation by the fishing contractors of the idealized fisher cooperatives as monopoly lease holders on water bodies. Had the cooperative ideal been enforced vigorously, India's inland fishery would not have emerged as the growth industry it is today.

How changing IE policy unleashes productive forces in an economy is best illustrated by the evolution of Gujarat's inland fishery policy over the past 30 years (Pandya, 2004). Following early attempts to intensify inland fisheries during the 1940s, Gujarat Government's Fisheries Department began supporting village panchayats to undertake intensive culture fishery in village tanks during early 1960. However, the programme failed to make headway, partly because of popular resistance to fish culture in this traditionally vegetarian state and partly because of rampant poaching from local fisher-folk that village panchayats, as managers, could not control. In a modified programme, the Fisheries Department took over the management of tanks from the panchayats to raise fishery to a produce-sharing basis; but the Department was less effective than the panchayats in checking poaching. In 1973, a special notification of the Government of Gujarat transferred inland fishing rights on all water bodies, including village tanks, to the Fisheries Department, which now set about forming fishermen's cooperatives in a campaign mode. The idea was to entrust the management to the community of poachers themselves.

In the Kheda district of Gujarat, for example, 27 such cooperatives were formed to undertake intensive culture fishing. However, these were none the better when it came to controlling poaching – including that by their own members; and the gross revenues could not even meet the bank loans. Members lost heart and cooperatives became defunct, a story that has been endlessly repeated in various fields in India's history of the cooperative movement. While all manner of government subsidies were on offer, what made culture fishery unviable were three factors: (i) a lease offered for only 3 years, a period considered too short to recoup the investment made; (ii) only registered cooperatives could be given a lease and the process of registration was transaction-costly; and (iii) rampant poaching and the high cost of policing and preventing it.

All this time, culture fishery productivity was steadily rising. Although fisher cooperatives were not doing well, culture fishery was, as entrepreneurs began using cooperatives as a front to win leases on common property water bodies. Doing this entailed significant transaction costs; office bearers of cooperatives had to be paid off, and gram panchayat leaders kept in good humour so that the lease would be renewed. Even then, whenever a gram panchayat leadership changed, the new order would terminate the contract to favour a new contractor. This dampened the contractors' interest in investing in high productivity.

In 1976, the government began setting up fish farmers' development agencies in each district to implement a new Intensive Fish Culture Programme. Terms of lease began to undergo change: private entrepreneurs were, in principle, considered for giving away leases but there was a pecking order of priority where first priority was for a Below Poverty Line (BPL) family, followed by a local poor fisherman, then a local cooperative and, if none of these were available, to any entrepreneur who bid in an open auction.

Earlier, the government had paid a puny rental to the gram panchayats for using their tanks for fish culture. Now that entrepreneurs were allowed, gram panchayats began quoting an 'upset price' derived as an estimate of the 'fishing value' of the tank, which was often 20 to 30 times the rental panchayats received earlier from the Department. Even so, as soon as leases were open to entrepreneurs, many came forward. A later change in policy gave cooperatives some discount in the 'upset price' and other benefits. In general, the IE's outlook constantly remained favourable to cooperatives and suspicious of entrepreneurs. In 2003, a series of new changes in the policy framework gave a further fillip to productivity growth: the lease period was extended from 3 to 10 years, which reduced the contractors' vulnerability to changes in panchayat leadership. It also made investment in productivity enhancement attractive. The new policy also removed the last vestiges of special treatment to cooperatives, and provided for a public auction of the lease after open advertisement.

During 1971–1998, the inland fishery output of Gujarat increased sixfold from 14,000 mt (metric tons) in 1971 to over 80,000 mt in 1998–1999 (Government of Gujarat, 2004). Considering that Gujarat had hardly any culture fishery before 1950, it must be said that the credit for this growth rightly belongs to the government's efforts. The government invested in subsidies, organizing inputs, bringing in new technology, extension and training and much else. All these played a role in expanding the fisheries economy. However, perhaps, the most important impact has been produced by two factors: (i) the changes made at the margins in the leasing policies of water bodies that have shaped the transaction costs of setting up and operating a profitable culture fishery business; and (ii) the high costs of controlling poaching, which has ensured that, besides several entrepreneurial qualities, successful fishing contractors also have to acquire and deploy muscle power.

Several less sensational examples can be offered of spontaneous institutions that operate on a large scale to serve purposes for which water establishments often promote copybook institutions such as WUAs. I briefly mentioned earlier how hundreds of defunct community reverse osmosis (RO) or defluoridation plants set up by governments and donors to supply fluoride-free drinking water to village communities have failed under community management. However, in North Gujarat, as a demand curve has emerged for fluoride-free drinking water, some 300 plants selling packed desalinated water have mushroomed in the cottage sector. Over half of these have been set up since 2001, mostly in *mofussil* (small towns) to serve permanent customers, as well as to retail water in polythene pouches.[26]

The RO cottage industry of Gujarat was quietly serving a growing demand when the 'IE' caught up with it. In 2001, the Bureau of Indian Standards (BIS) made it compulsory for cottage RO plants to achieve the ISI mark.[27] This entailed that each plant had to invest Rs 0.3–0.4 million ($6500–8670) in an in-house laboratory and pay an annual certification fee of Rs 84,000 ($1870) to the ISI. This single move immobilized the emerging RO water cottage industry; 200 operators had to close their businesses because the new announcement doubled their cost of production. Yet, setting up an in-house laboratory and paying an annual certification fee implied no guarantee of quality assurance because BIS inspectors hardly visit plants, if ever. Many customers (Indu, 2002) interviewed wondered if the ISI mark – like the AGMARK (standardized certification for agricultural food products) ghee and honey – can by itself guarantee quality unless BIS itself put its act together in the first place.

Likewise, many state governments are struggling, in vain, to cut their losses from operating mostly World Bank-funded public tube well programmes by trying to transfer these to *idealized* cooperatives registered under the Cooperative Act. If the purpose of a cooperative tube well is to enable a group of farmers to mobilize capital, to install and operate a tube well for the mutual benefit of members, such tube well groups have existed for decades in

North Gujarat. The difference is that, having been created to serve the purpose of their members, their ownership structure and operating rules are designed to minimize the transaction costs of cooperating on a sustained basis (Shah and Bhattacharya, 1993). The Government of Gujarat tried hard to transfer its public tube wells to idealized cooperatives but, thanks to the very high transaction costs relative to the pay-off facing potential entrepreneurs, the programme made no headway until 1998 when the terms of turnover were rewritten.[28]

Basically, the requirement that a cooperative be registered under the Cooperative Act was dropped; the lease period was extended from 1 to 5 years; and changes were introduced that made it possible for one or few major stakeholders to assume the role of tube well manager and residual claimant. These minor changes suddenly gave a fillip to the turnover programme and, over a 3-year period, over half of Gujarat's public tube wells, some 3500 in all, were transferred to farmer groups. An IWMI-Tata study of turned-over public tube wells (Mukherji and Kishore, 2003) showed that, within 1 year of the turnover, the performance of turned-over tube wells, in terms of area irrigated, hours of operation, quality of service, O&M and financial results improved. Two years after the turnover, it improved dramatically.

In opening this section, I talked about the significance of groundwater markets in India's irrigation. However, private provision of water services is also an important part of India's urban reality. In an IWMI-Tata study of six cities – Indore, Jaipur, Nagpur, Ahmedabad, Bangalore and Chennai – Londhe *et al.* (2004) found that municipal agencies supplied only 51% of the demand calculated at 80 l per capita per day.

In Chennai and Ahmedabad, formal organizations served only 10 and 26%, respectively, of the 'normative' demand, the balance being either self-supplied or served by informal sector players. 'Tanker markets' supply 21, 12 and 10% of the demand in Chennai, Indore and Jaipur, respectively. In Chennai, tanker operators have year-round operations and even have an association. In other cities, tanker markets emerge during the summer and quietly disappear as the monsoon arrives. Londhe *et al.* (2004) estimate that some 3000 tankers in the

six cities operate a water trade worth Rs 203 crore (US$45 million)/year. Despite being key players in urban water sectors: 'There is no record with any government department about its size, scale and modus operandi. There is an absence of any government regulation on groundwater withdrawals. Except in Chennai, municipal authorities refuse to even acknowledge the existence of such markets' (Londhe *et al.*, 2004).

Tanker markets operate much like any other market, and serve those who can pay for their services. The IWMI-Tata study estimated that 51% of consumers in the six cities are from high-income groups, 43% from middle-income groups and only 6% from low-income groups. Contrary to belief that the poorest pay the most for water, the IWMI-Tata study showed that the poorest pay the least, even when transaction costs and imputed cost of labour and time in fetching water are factored in (Londhe *et al.*, 2004).

One more case of institutions that 'planners propose and people dispose' that I want to discuss briefly concerns the world-famous Sardar Sarovar Project (SSP) on the Narmada river. SSP must be one of the world's most-planned projects. One of SSP's key planning premises was that the Project would construct lined canals with gated structures going right up to the village service area (VSA), comprising some 400 ha of command. A WUA would be organized in each VSA that would simultaneously construct the sub-minor and field channels to convey water from the *pucca* (lined minor) to the fields. When SSP water was first released to some 80,000 ha of the command just below the dam in 2001, the Project managers registered, on a war footing, WUAs as cooperatives in some 1100 VSAs. When the water was finally released, however, the village-level distribution structure was not ready in a single village.

And it will never be, as we learnt in the course of a quick assessment of farmer preparedness to receive Narmada irrigation (Talati and Shah, 2004). The perceived sum of the transaction and transformation cost[29] of constructing village distribution systems seemed by far to outweigh the benefits people expected of SSP. There was, however, a flurry of activity as SSP water began flowing into minors.

According to our quick estimates, several thousand diesel pumps and several million metres of rubber pipes were purchased by water entrepreneurs to take water to their own fields and to provide irrigation services to others.

The trend for new investments in diesel pumps and rubber pipes gathered further momentum in 2002 and 2003; and we found that village communities were none the worse for having violated the SSP planning assumption. The Government of Gujarat is, however, adamant on constructing a 'proper' village distribution system in the SSP command – never mind whether it will take 50 years to complete the canal network.[30]

The swayambhoo institutions I have discussed in this section are all driven by opportunism. However, large-scale swayambhoo institutions are often driven by more complex motives including long-term, collective self-interest. The decentralized mass movement for rainwater harvesting and groundwater recharge that the Saurashtra region of Gujarat saw from 1987 until 1998, when it became co-opted by the state government, is a good example of such an institutional development (Shah, 2000).

The movement was catalysed first by stray experiments of 'barefoot hydrologists' in modifying open wells to collect monsoonal flood waters. Early successes fired the imagination of a people disillusioned with ineffective government programmes. Soon, well recharge was joined by other water-capture structures such as check dams and percolation tanks. With all manner of experimentation going on, a kind of subaltern hydrology of groundwater recharge developed and became energetically disseminated. Religious leaders of sects like *Swadhyaya Pariwar* and *Swaminarayana Sampradaya* ennobled this work in their public discourses by imbuing it with a larger social purpose. The gathering movement generated enormous local goodwill and released philanthropic energies on an unprecedented scale, with diamond merchants – originally from Saurashtra but now settled in Surat and Belgium – offering cash, cement companies offering cement at discounted prices and communities offering millions of days of voluntary labour.

In neighbouring Rajasthan, Alwar was also undergoing similar mass action; but it was far more limited in scale, and was orchestrated by Rajendra Singh's Tarun Bharat Sangh, a grass-roots organization. Saurashtra's recharge movement was truly multicentric, unruly, spontaneous and wholly internally funded with no support from government, international donors or the scientific community – until 1998, when the Government of Gujarat became involved and proceeded to rid the movement of its quintessentially swayambhoo and voluntary character by announcing a subsidy programme (Shah, 2000; Shah and Desai, 2002).

It is difficult to assess the social value of this movement, partly because 'formal hydrology' and 'popular hydrology' have failed to find a meeting ground. Scientists want check dams sited near recharge zones; villagers want them close to their wells. Scientists recommend recharge tube wells to counter the silt layer impeding recharge; farmers just direct flood water into their wells after filtering. Scientists worry about upstream–downstream externalities; farmers say everyone lives downstream. Scientists say the hard-rock aquifers have too little storage to justify the prolific growth in recharge structures; people say a check dam is worthwhile if their wells provide even 1000 m^3 of life-saving irrigation/ha in times of delayed rain. Hydrologists keep writing the obituary of the recharge movement; but the movement has spread from eastern Rajasthan to Gujarat, thence to Madhya Pradesh and Andhra Pradesh. Protagonists think that, with better planning and larger coverage, the decentralized recharge movement can be a major response to India's groundwater depletion problem because it can ensure that water tables in pockets of intensive use rebound to pre-development levels at the end of the monsoonal season every year they have a good monsoon.

Table 5.1 offers a comparative view of a sample of six 'high pay-off–low transaction-cost' institutions that have emerged in India's water sector in recent years. If we judge institutions by their contribution to increasing productivity and welfare, all six can be considered successful. Each can be found to operate on a significant scale, thus permitting generic lessons. One notable aspect is that each institution has arisen spontaneously and flourished as an instrumentality of its players, serving a purpose important to them though not necessarily of the IE players. Each has devised

Table 5.1. Characteristics of swayambhoo water institutions.

	Decentralized groundwater recharge movement of Saurashtra	Irrigation institutions unfolding in the Narmada command and Upper Krishna basin	Urban tanker water markets	Tube well companies of North Gujarat and Gujarat's Public Tube Well Programme	Reverse osmosis (RO) plants in North Gujarat's cottage industry	Fishing contractors using cooperatives as fronts
Spread of the institution	300,000 wells modified for recharge; 50,000 check dams	Several thousand new pumps installed/year	Most Indian cities	Some 8,000–10,000 companies in North Gujarat	Around 300 plants in Gujarat	Tens of thousands of small and large tank fisheries in India
Economic contribution	Improved greatly security of kharif crops, and possibility of a rabi crop	Private investment in water distribution infrastructure; expansion of Narmada irrigation	Fill the gap between demand and supply	Create irrigation potential which individual farmers would not be able to do	Add and operate water treatment capacity to serve demand for clean water	Contributed to achieving seven- to tenfold increase in inland fishery productivity 1960–2000
Raison d'être	Improve water availability in wells for life-saving irrigation when monsoon makes early withdrawal	To profit by distributing Narmada water by lifting water from canals and transporting it by rubber pipe to user fields	To profit from supply of water in cities where public institutions cannot cope with the economic demand	To pool capital and share risks of tube well failure in creating and operating an irrigation source in an over-exploited aquifer	To profit from serving emerging demand for fluoride-free water by investing in and maintaining RO plant	Can protect fish better and therefore can invest in intensive culture fishery, which cooperatives cannot
Mode of emergence	*Swayambhoo*	*Swayambhoo*	*Swayambhoo*	*Swayambhoo*	*Swayambhoo*	*Swayambhoo*
Strategy of reducing transaction and transformation cost	Religious leaders have reduced transaction costs of cooperative action	Avoidance of making of sub-minors and field channels, reducing seepage, overcoming topography	Meet the demand as it occurs in a flexible manner	Vesting management roles in members with largest share in command area	Cultivating annual customers	Instilling fear amongst poachers
Incentive structure	Pay-off concentration	Pay-off concentration	Pay-off concentration	Pay-off concentration	Pay-off concentration	Pay-off concentration
Outlook of the 'establishment'	Sceptical, but piggybacked and lessened its swayambhoo character	Negative/neutral	Neutral/negative	Negative	Negative	Negative, but changing in states like Gujarat
Preferred alternative by institutional environment	Narmada project; scientific recharge works	Idealized WUAs	Municipal water supply improved	Idealized WUAs	Community RO plants	Registered fishermen's cooperatives

its own methods of reducing transaction costs and managing incentive structures.

Finally, each is widely viewed in the IE – by government officials, NGOs, researchers, international experts and even local opinion leaders – as a *subaltern* or inferior alternative to the mainstream notion of an institution considered ideal but that has not worked on a desired scale or in a desired manner. As a result, far from recognizing the potential of these subaltern institutions to further larger social goals, the outlook has been to ignore their existence and social value, or even to emasculate them.

Analysis and Discussion

The repertoire of institutional arrangements that operate on a large scale includes numerous 'successes' of varied types and scales produced by exceptional local leaders and industrious NGOs. By virtue of exceptional and highly scarce resources at their command – such as reputation, social status, allegiance of people, funds, goodwill, influence in the IE, skilled manpower – local leaders and NGOs are often able to drastically reduce transaction costs of fostering institutional change of a certain kind in a limited setting for a limited period. Out of hundreds of thousands of irrigation tanks in India that can produce large pay-offs from improved management, there are but a few hundred in which exceptional local leaders have established and sustained novel institutions for upkeep, maintenance, management and use of tanks to improve the welfare of the community. The IWMI-Tata Programme studied some 50 of these during 2002–2003 (Sakthivadivel *et al.*, 2004) and found that, while the architecture of institutions (as rules-in-use) varied from case to case, the common aspect of all successful tank institutions was a leader or a leadership compact which, by virtue of the sway they/it has over the community, is able to drastically reduce the transaction costs of enforcing an institutional arrangement that would neither work in their absence nor survive them.

Successful NGOs similarly create islands of excellence by reducing transaction costs *artificially* and *temporarily*. The *Sukhomajri* experiment with watershed institutions in Haryana in the mid-1980s – Vilas Rao Salunke's pani panchayats in Maharashtra, Aga Khan Rural Support Programme's irrigators' association in Raj Samadhiala, Dhan Foundation's Tank User Federations, Development Support Centre's WUAs in Dharoi command in North Gujarat, community-managed tube wells that came up in Vaishali and Deoria in Eastern UP, Anna Hazare's Ralegaon Shiddi, Rajendra Singh's profusion of *johads* in Thanagazi, Alwar district, Chaitanya's conversion of irrigation tanks into percolation tanks in Rayalaseema – all these are examples. That the transaction cost reduction in all these was *artificial* is indicated by the absence of spontaneous lateral expansion/replication of these experiments despite the high pay-offs they are seen to have produced. That it was *temporary* is evident in that many of these institutions disappeared, stagnated or declined once the 'transaction cost reducer' was removed from the scene, as in Sukhomajri, Salunke's pani panchayats and others.

A more important source of ideas – than the NGO-inspired islands of excellence – about what institutional change should occur and *can sustain* are the swayambhoo institutions that have already emerged and are thriving, as we explored earlier in the section under Vibrant Institutional Arrangements Ignored (Quadrant 1). These have found ways of reducing transaction costs in ways that are more *natural, enduring and upscalable*. This is evident in that these institutions multiply on their own, and are able to sustain and grow as long as they serve purposes important to the participants in the transactions. In my understanding, these offer six useful lessons (given under the following six headings) about how to make institutional change work in the Indian water sector.

Instrumentality

The first, and most obvious, is that institutional change which multiplies and sustains is invariably an instrument of the exchange of participants, and not of the players in the IE who often design institutional interventions. 'Opportunism with guile' is the driving force, even when high ideals and social goals are laboriously espoused as *raison d'être*. Trite as it may sound, design of incentive structures is amongst the most commonly ignored aspects in most institutional development programmes. Ideas like community-based groundwater demand management propose organizing cooperatives whose sole task would be to persuade their members to reduce their farming and incomes. Similarly, programmes to revive traditional community management of tanks commonly overlook the performance-based rewards offered to *neerkattis* (tank water distributors appointed by command area farmers) and focus primarily on generating voluntary contributions of time and effort for the greater good of the community. For institutional change to work it must serve a private purpose important to agents involved; otherwise, they will withhold participation or even work to defeat it.

Incentive diffusion or perversion

Institutions fail to emerge to take advantage of high-pay-off situations often because incentives are diffuse or even perverse, but the transaction costs of implementing change are concentrated in one or a few persons. In fishermen's cooperatives I discussed earlier, members faced perverse incentives: the cooperative stocked the pond but members stole the catch. The secretary had no incentive to make enemies by stopping poachers. When incentives became concentrated in the contractor as the residual claimant, he was willing to control poaching and invest in higher productivity. Gujarat's public tube wells had no takers until the opportunity arose for incentive concentration. That only a fraction of the surplus created by management improvement needs to be concentrated in the manager as a reward was shown 40 years ago by Amartya Sen (1966). In traditional tank institutions in South India, only a portion of the surplus output was offered to the neerkatti, who absorbed the bulk of the transaction cost of orderly distribution of tank water.

This principle is at the heart of irrigation reforms in China. Except where traditional PIM/IMT is supported by a donor loan, China's strategy of making canal irrigation productive and viable consists of changing the incentive structure facing the 'ditch manager' (Shah *et al.*,

2004a; Wang *et al.*, 2005). A pre-specified volume of water is released into a reservoir and is charged for at a certain volumetric rate. The reservoir manager's remuneration includes a fixed component and a variable component, the latter increasing with the area irrigated from the same total volume of water. Like the Chinese village electrician who is able to perform a high transaction-cost role for a fairly modest reward, the ditch manager too is able to improve water productivity for a modest bonus, if recent studies are any guide (Shah *et al.*, 2004a).

High costs of self-enforcement

Experimenting with the Indian equivalents of Chinese village electricians and ditch managers would be an interesting study. From the transaction cost viewpoint, however, there are two key differences between the Chinese and South Asian villages: first, the Chinese in general, thanks perhaps to the Confucian ethics, are more respectful to State authority compared with South Asians. Secondly, and more importantly, the village committees and the village party leader in a Chinese village enjoy far greater power and authority in the village society compared with India's gram panchayats and *sarpanch*. This has great implications for transaction costs. North (1990) suggests that: ' ... institutional setting depends on the effectiveness of enforcement. Enforcement is carried out by first party (self-imposed codes of conduct), by second party (retaliation), and/or by a third party (societal sanctions or coercive enforcement by state).' Transaction costs facing an institutional change are determined by the ease of enforcement. A Chinese village electrician or ditch manager backed by the village committee and party leader can enforce the new rules by both retaliation and recourse to coercion through the party leader.

In India, by contrast, Orissa's model of franchisees for rural billing and collection of electricity bills has attracted many entrepreneurs whose core competence is represented by their muscle power (Panda, 2002), because they have no effective local authority to either discipline them or to which they can turn to in order to defend their rights. For the same reasons, a typical culture fishery contractor has recourse only to retaliation to enforce his property right against a poacher. The high transaction cost of second-party enforcement of rules is perhaps the prime reason why entrepreneurs fail to come forward to make a business out of operating a canal or tank irrigation system.

Structures of incentives and sanction

Catalysing effective local IAs is then a matter not only of designing appropriate incentive structures that entice entrepreneurs to undertake activities with a high pay-off but also of putting into place community sanction or authority structures that: (i) enforce his/her right to do so; and (ii) establish the boundaries within which he or she operates. Here is where a community organization has a role in providing legitimacy or sanction and boundary to a service provider, thereby reducing his/her transaction cost of self-enforcement of rules. It is difficult to overemphasize this point, which is commonly overlooked in programmes of creating participatory institutions. In the much-acclaimed traditional tank management institutions, all tank management was carried out not by the community but by the neerkatti, who had the sanction and legitimacy given by the community and a reward for services that was linked to the benefits they produced for the community. A self-appointed neerkatti would find it impossible to enforce rules of water distribution amongst *ayacut* (command area) farmers.

A recent study of neerkattis by the Dhan Foundation shows that, for various reasons, many tank communities have begun withholding their sanction and questioning the legitimacy of the role neerkattis have played for centuries; as a result, the institution of neerkattis has begun to decline (Seenivasan, 2003). However, in those few tanks where we find traditional community management still working, it becomes evident that it worked through a clear specification of the 'governance' role of the community organization and the community-sanctioned, well-defined 'management' role of the neerkatti, a service provider whose rewards were linked to his performance.[31]

The value of this lesson for improving the quality of 'social engineering' is evident in the Gujarat government's public tube well transfer programme; after getting nowhere for a decade, it suddenly took off the moment entrepreneurial service providers were offered concentrated incentives coupled with some legitimacy and sanction for undertaking service provision. On these counts, I predict that such service providers have failed to come forward to provide improved water distribution in surface irrigation projects because neither concentrated incentives nor legitimacy and sanction are on offer for local entrepreneurs who would contemplate taking up such roles. Equally, the entrepreneurial service provider model too – such as the culture fishery contractor – operating without the sanction, legitimacy and boundary provided by a community organization is bound to be fragile.

Institutional environment

Finally, the IE can have a profound impact on what kind of IAs are promoted or discouraged, and what welfare and productivity impacts these produce (Mansuri and Rao, 2004); however, they *do not* have such impact because often they neither understand their working nor how to influence it. Informal pump irrigation markets, the fishing contractor and a decentralized groundwater recharge movement[32] are spontaneous and seemingly autonomous; but each of these is amenable to strong positive or negative influence from the IE.

Gujarat's cottage RO industry fell in a single swoop of the Bureau of Indian Standards; and the working of pump irrigation markets can change overnight if policies related to electricity pricing and supply to the farm sector were to change (Shah *et al.*, 2004c). Gujarat's Public Tube Well Transfer programme ploughed along without success for a decade and then suddenly took off because an actor in the IE changed the key rules of the game. And the culture fishery contractor faced drastic reduction in his transaction costs of doing business when the leasing policy for water bodies was changed at the instance of some actor in the IE. How well actors in the IE understand extant and potential institutions, their net welfare and productivity impacts and their backward and forward linkages determines how much they can influence or manage them.

Path-dependence

According to North (1990), institutional change is inherently incremental and path-dependent. It invariably grows out of its context; transposing institutional models that have worked in other, different contexts therefore seldom works in catalysing institutional change. India's state governments would probably have found it easier to manage metered electricity supply to farmers had they stayed engaged with the problems of metering rather than abandoning it the 1970s. Now that they face a huge groundwater economy based on the 'path' of flat tariff, their here-and-now options for change are tied to this path. The notion of 'path-dependence' has particular relevance to popular institutional notions, such as the Integrated River Basin Management, which have worked in highly formalized water economies in recent years. It is doubtful whether such models would work in the same way in the Indian situation, simply because by far the bulk of the Indian water economy is informal and outside the direct ambit of the IE.

Conclusion

A reader who comes to this stage of this chapter will surely remark, as did John Briscoe, World Bank's Asia Water Advisor: 'But I find very little in the chapter that would help me if I am a Secretary for Water in Gujarat, or in the Government of India, for that matter …' This response is entirely understandable; however, on the contrary, this analysis does offer useful advice for action that should always focus on the 'art of the possible'. Allan (2001) has wisely suggested that: 'The mark of effective research, advice and policy making is the capacity of those involved to know the difference between what "should" be done, and what "can" be done. This can be expressed in another way as awareness of "when" what "should" be done, "will be able" to be done'.

The upshot of this chapter is that all the things that a Secretary of Water Resources at the state or federal level is enjoined to do by the current discourse to promote improved demand management – imposing price on water resources (rather than water service),

enforcing a groundwater law, making water the property of the state and stopping unlawful diversion from nature, instituting water withdrawal permits and assigning water entitlements, managing water at river basin level – would be well nigh impossible to implement on any meaningful scale in a predominantly informal water economy such as that of India. Instead, governments of low-income countries should focus their effort on areas where they *can* produce significant impacts, which in my view are four (given under the following four headings):

Improving water infrastructure and services

This already is a high priority and will remain so for a long time, even as opinion in the rich world is turning against investments in certain kinds of water infrastructure such as irrigation projects. There are several issues to be addressed such as mobilizing capital, improving the coverage of user households – especially from poorer classes, cost recovery, and so on. The point of attack, however, is the performance of public systems, which has tended to be abysmally low, be it irrigation systems or water supply and sanitation systems.

Institutional reforms focused on incentive concentration and transaction cost reduction

Public systems' performance often responds strongly to demand for better performance not from users but from administrative or political leadership; however, such performance gains are transient, and become dissipated when demand slackens. To achieve sustainable performance improvements, institutional innovations are needed that restructure incentives and reduce transaction costs.

Honing and using indirect instruments and strategies for achieving public policy objectives

In its enthusiasm for *direct* management of water demand – through pricing, rights and entitlements, laws and regulations – the current discourse is overlooking numerous opportunities to achieve comparable aims using *indirect* instruments. True, the Secretary of Water can do little to manage water demand directly. However, in the particular situation of India, the Secretary of Energy controlling the State Electricity Board can do a great deal for groundwater demand management, through pricing and rationing of electricity to tube wells.

Undertaking vigorous demand management in formal or formalizing segments

Finally, pricing and full cost recovery, tight water law and regulations, and water rights and entitlements are definitely indicated in the predominantly formal segments of the water economy. These are to be found in cities, excluding the slums and shanty towns; and in the industrial sector where users are large and easily identifiable. It will probably take Delhi and Mumbai years before they can establish a water supply and sanitation system that can match those of Abidjan or Tunis. However, given increasing political support for management reforms, India's cities – especially, high net-worth cities like Delhi, Mumbai and Bangalore – offer by far the most fertile ground for water IE and urban governance systems for the introduction of global best practices in urban water supply and sanitation systems.

In summary, then, how formal a country's water economy is determines what kind of policy and institutional interventions are appropriate to it. In a predominantly informal water economy, where self-supply is the rule and water diversion from nature is everybody's business, regulating the actions of all water diverters is extremely costly in terms of search, information, policing and enforcement costs. As a water economy formalizes, self-supply declines and a few, visible, formal entities specialize in diverting, processing and distributing water to users; in such an economy, the range of things public policy makers can do to improve water demand management becomes much larger. The pace of formalization of a water economy is a natural response to overall economic growth and transformation of a society. This pace can be forced to a limited degree by an authoritarian state or

by investment in water infrastructure and services management. However, unless this process keeps pace with what the market can bear, it will face sustainability problems.

The current global water policy discourse focusing on *direct demand management* is misleading in two ways for developing countries like India with a highly informal water economy: (i) it is enjoining it to institute policy and institutional reforms that are good in principle but present insurmountable implementation difficulties; and (ii) in contrast, it is deflecting attention away from things that need and can be done with a better understanding of the working of the water economy, warts and all.

Endnotes

[1] Formal and informal economies are a matter of elaborate study in institutional economics. Fiege (1990) summarizes a variety of notions of informality deployed by different researchers. According to Weeks (1975), cited in Fiege (1990, footnote 6): 'The distinction between a formal and informal sector is based on the organizational characteristics of exchange relationships and the position of economic activity vis-à-vis the State. Basically, the formal sector includes government activity itself and those enterprises in the private sector which are officially recognized, fostered, nurtured and regulated by the State. Operations in the informal sector are characterized by the absence of such benefits.' According to Portes *et al.* (1987, cited in Fiege, 1990, footnote 6): 'The informal sector can be defined as the sum total of income-generating activities outside the modern contractual relationships of production.' According to Portes and Saassen-Koo (1987, cited in Fiege, 1990, footnote 6), in the formal sector activities are 'not intrinsically illegal but in which production and exchange escape legal regulation'. To most researchers, an informal economy is marked by the 'absence of official regulation' or 'official status'.

[2] In most countries, the proportion of water use in the informal sector would move in tandem with the proportion of water users. However, in countries marked by high levels of income inequality – such as South Africa or Brazil – this would not be the case. In South Africa, for instance, 95% of the water diversion and use are in the formal sector but over 99% of the users are in the informal sector.

3 The nature of the State is a crucial determinant of the level of formalization. Colonial state in British India – which lived off the land – had a huge and elaborate apparatus for land revenue administration reaching down to the village level. And since the colonial state invested in irrigation for commercial reasons, its IE evolved and maintained a firm grip over irrigated agriculture. Even today, China has a similar firm grip over its natural resources economy, thanks to the authority structure of the Communist Party and an elaborate structure of farm taxes and levies that sustain the lower rungs of its IE. However, upon independence, India all but abolished land revenue alongside the apparatus for its assessment and collection, thereby informalizing its agrarian and water economy. China is now on course to do just that. In Tanzania, during the Cold War years, Julius Nyerere had created *Mgambo*, an institution for civil defence from village youth trained in martial techniques. The Cold War over, the Tanzanian state has transformed Mgambo into a tax collection machinery. Van Koppen *et al.* (2005, unpublished report) describe how Mgambo was incentivized to undertake the recovery of a water resource fee as a kind of poll tax from rural people.

4 North (1990) defines the transaction sector as: 'that part of transactions that goes through the market and therefore can be measured' and, according to him, rapid growth in the transaction sector is at the heart of the transformation of a traditional economy into a modern one.

5 The survey estimated that approximately 36% of all rural households (which include farmers, farm labourers and households dependent on off-farm livelihoods) used some means of irrigation. Of these, 13.3% (i.e. 37% of irrigators) used their own source (well/tube well), 15.3% (i.e. 42.5% of irrigators) used shared tube wells or purchased water and 12.1% (36% of irrigators) used government-owned tube wells, canals or a river. Fewer than 2% used a locally managed irrigation source; 6.6% used more than one source, which is why the percentages fail to add up to 100. The survey also found that, of the 78,990 households interviewed, 48% reported 'no availability of community and government water resources in villages of their residence'; another 42% reported the presence of community or government sources but 'without local management'. Only 10% of households reported living in villages with access to community or government water sources 'with local management' by community or government or both (p. 44). Only 23% of all households interviewed reported depending for irrigation on a source 'other than self-owned';

30% using water for livestock rearing reported dependence on a source 'other than self-owned'.

6 Contrast this picture with a recent account by Luis-Manso (2005) of the highly formalized water economy of Switzerland: 70% of its population is urban, and the country is facing continuous reduction in industrial workers and farmers. Probably 15–20% of the Swiss population was linked to public water supply as far back as the 18th century; today, 98% of the Swiss population is linked to public water supply networks and 95% is connected with waste-water treatment facilities. Switzerland spends 0.5% of its GNP annually in maintaining and improving its water supply infrastructure, and its citizens pay an average of CHF 1.6 per 1000 l of water (CHF = US$0.786). The per capita water bill that Swiss citizens pay annually is around CHF 585, which is higher than the per capita total income of Bangladesh. All its water users are served by a network of municipal, corporate, cooperative water service providers; it has stringent laws and regulations about water abstraction from any water body, which can be carried out only through formal concessions. However, these concessions are held only by *formal* service-providing public agencies; as a result, their enforcement entails few transaction costs.

7 Scott Rozelle used this phrase recently in referring to the unexceptionable tendency of agricultural population ratios of countries to fall as their economies grow. But I think this also applies to other responses to economic development, as outlined in Fig. 5.2.

8 One commentator on an earlier draft of this chapter cited Abidjan, where a First World water supply system has operated for decades. Abidjan, however, seems to be the exception to the rule that a city's water system would rise to what its median earner is willing to pay for. If recent accounts of the travails facing global water companies like Vivendi and Thames Water – who were forced to cease trading – even in these increasingly affluent east Asian cities is any guide, we must conclude that South Asian cities have a long way to go before they can afford water supply systems of European or North American quality (see *The Economist*, 2004).

9 Societies often experience wide-ranging ideological or cultural upheavals during which customs, traditions, mores and values undergo massive change. India's Independence Movement – and the rise of the Gandhian ethos – marked one such phase in India's history. On a smaller scale, the water harvesting movement in Saurashtra under the inspiration of religious formations such as *Swadhyaya Pariwar* and *Swaminarayan*

Sampradaya too represent an L1 level change. Both these, however, have proved largely transient; besides occasional lip service paid, Gandhian ethos and ideals no longer dominate Indian psyche quite like they did during the 1940s; and Saurashtra's water harvesting movement too is now energized by the Gujarat Government's 60:40 scheme of government versus community contribution rather than the ideal of self-help the religious leaders had inspired. However, both L1 and L2 may experience rapid change in the face of rapid economic growth and transformation of a society. Since India is in the throes of such economic transformation, the pace of change at L1 and L2 levels should, in my surmise, be quicker than that suggested by Williamson.

[10] A good example is Francis Corten's work during the 1980s on reorienting the irrigation bureaucracy.

[11] A charismatic and energetic political or bureaucratic leader does often produce significant attitude and behaviour changes; however, these generally fail to last for long after the leader has been removed from the scene. In this sense, such change is not enduring.

[12] Because the law did not apply to anyone who diverted less than 1030 m^3 of water/year.

[13] Anil Shah, an illustrious former bureaucrat of the Government of Gujarat, fondly tells the story about Gujarat's groundwater bill, which was passed by the assembly in 1973. When the Chief Minister was required to sign it into the government gazette, he refused to do so because it required that every irrigation well be registered. His curt response to Mr Shah was: 'Can you imagine that as soon as this bill becomes a law, every talati (village-level revenue official) will have one more means at his disposal to extract bribes from farmers?' This is the reason there are no takers for the draft Groundwater Bill that the Ministry of Water Resources of Government of India has been tossing around to states since 1970.

[14] The Andhra Pradesh law tried harder to come to grips with rampant groundwater over-exploitation in Andhra Pradesh by emphasizing the registration of wells and drilling agencies and stipulating punitive measures for non-compliance.

[15] The 1987 Water Policy to Saleth (2004, p. 29) is '… such a simple non-binding policy statement'.

[16] Although the Network Reform Programme is a National Government programme, the government contributes only a part of the resources, the balance being contributed by the village committee. Just to give an example, Guantun village in Yanjin County of Henan got a grant of Y60,000 (US$1.00 = Y8.33) under this project for infrastructural rehabilitation. To match this, the village also contributed Y60,000; of this, 60% came from the funds from the village collective, while the remaining 40% was raised as farmer contributions by charging Y80 per person. All the power lines and other infrastructure were rehabilitated during recent years under this national programme. New meters were purchased by the township in bulk and installed in users' homes on a cost-recovery basis. A system of monitoring meters was installed too.

[17] The village electrician's reward system encourages him/her to exert pressures to achieve greater efficiency by cutting line losses. In Dong Wang Nnu village in Ci County, Hebei Province, the village committee's single large transformer that served both domestic and agricultural connections caused heavy line losses, at 22–25%. Once the Network Reform Programme began, he pressurized the village committee to sell the old transformer to the county electricity bureau and raise Y10,000 (partly by collecting a levy of Y25 per family and partly by a contribution from the village development fund) to acquire two new transformers, one for domestic connections and the other for pumps. Since then, power losses here have fallen to the permissible 12%.

[18] Saleth (2004, p. 30) asserts: ' … most of the organizational reforms, including the promotion of basin-based organizations observed in states such as Andhra Pradesh, Tamil Nadu, Orissa, and Uttar Pradesh were introduced under different World Bank-funded projects.' It is equally clear that Andhra Pradesh's irrigation reforms proceeded at a hectic pace because a World Bank loan was able to kindle interest at all levels in new resources available for maintenance work.

[19] And that too only when a mid-sized NGO invests years of effort and resources in organizing WUAs and using means to reduce transaction costs that farmers on their own would not normally possess. Some of the best-known examples of successful PIM/IMT are the Ozar on Waghad project in Nashik, Maharashtra, Dharoi in North Gujarat, Pingot and a few more medium-sized schemes in the Bharuch district. The success of farmer management in all these – and its beneficial impact – is undisputed. In each of these, however, there was a level of investment of motivation, skill, time, effort and money that is unlikely to be replicated on a large scale. In catalysing Ozar cooperatives, Bapu Upadhye and Bharat Kawale and their Samaj Pragati Kendra, and senior researchers of SOPPECOM, invested years of effort to make PIM work (Paranjapye and Joy, 2003). In Gujarat, between the Aga Khan Rural Support Programme and the Development Support Centre, Anil Shah and Apoorva Oza have

invested at least 30 years' professional staff time to organize, say, 20,000–30,000 flow irrigators into functional WUAs. My intent is not to undermine this exceptional work but to suggest that no government agency had the quality and scale of resources needed to implement an institutional intervention that could sustainably raise the productivity of the 28–30 million ha of flow-irrigated area in India over, say, 15 years.

[20] Here are some random excerpts from Joseph (2001), based on his study of the Malampuzha Project: 'It is the CADA officials who took the initiative in their formation and not the farmer groups. In most cases, membership fee of Rs5 was not paid by the farmers concerned; payment was made on their behalf by prospective office bearers, or the potential contractors of field channel lining or the large farmers in the ayacut. 86% of the Beneficiary Farmers' Associations (BFAs) were formed in these 2 years (1986 and 1987) … for making possible the utilization of funds … Only 57 Canal Committee meetings were held by the 8 Canal Committees during a span of 10 years … 43 of them were held without quorums and 35 with zero attendance of non-official members … The level of knowledge … about CCs … And their structure and functions is very low.'

[21] In a recent paper, Mansuri and Rao (2004) have reviewed a much larger body of evidence from several sectors to assess the extent to which community-based and community-driven development projects for poverty alleviation were effective, and have concluded that: (i) these have not been particularly successful in targeting the poor; (ii) there is no evidence to suggest that participatory elements and processes lead to improved project outcomes and qualities; (iii) community-based development is not necessarily empowering in practice; and (iv) 'There is virtually no reliable evidence on community participation projects actually increasing a community's capacity for collective action' (p. 31).

[22] Even in middle-income countries, huge inequalities in landholdings seem to have helped IMT. In the Andean region of Colombia where IMT has succeeded, according to Ramirez and Vargas (1999), farmers 'mostly grow crops oriented to the external markets, mainly banana and oil palm'; and while 66% of the farms have 5 ha or less, 40.3% of the land is owned by 2.8% of large farmers owning 50 ha or more. In South Africa, numerous Irrigation Boards – WUAs par excellence – have managed irrigation systems successfully for a long time; but their members are all large, white commercial farmers operating highly successful citrus and wine orchards. In Turkey, 40% of the irrigated area was in 5–20 ha holdings with a strong focus on high-value commercial crops for export to Europe. Here in Turkey, it can be argued, IMT has succeeded because, as with South African irrigation boards, in many respects there was already a 40-year old tradition of farmer participation in the maintenance of the canal system through an informal, village-level organization. Equally, irrigation fees under self-management in Turkey were 2% or less of the value of production per ha, 3.5% or less of total variable cost of cultivation and less than 6% of gross margin (Svendsen and Nott, 1997).

[23] Sanskrit for self-creating or spontaneous.

[24] A large survey, covering over 48,000 farming households throughout India during January–June 1998, suggested that over 66% of India's Gross Cropped Area under the five most important field crops (which account for over 90% of the Gross Cropped Area) is irrigated; only one-quarter of irrigated area is served by government canals. Amongst other interesting things it suggests that every fourth Indian farming household probably owns a diesel or electric pump; and the area irrigated through groundwater markets is as large as the area irrigated by all government canals (NSSO, 1999b).

[25] As North (1990) aptly notes: 'If the highest rates of return in a society are to be to piracy, the organizations will invest in knowledge and skills that will make them better pirates; if the pay offs are … to increase productivity, they will invest in skills and knowledge to achieve that objective.'

[26] An IWMI-Tata study (Indu, 2002, unpublished report) surveyed a sample of 14 such plants that served 4890 households. Reverse osmosis (RO) water in 10 and 20 l cans is delivered daily at the customer's door step; charges are levied on an annual basis (Rs 1500 (US$33) for a 10 l can daily; Rs 2500 (US$55) for a 20 l can). Plant capacities vary from 500 to 2000 l/h. In addition, most plants also retail RO water in pouches at bus stops, railway stations and crossings and market places. Consumers of pouches are typically low-income buyers; retailers are also poor youth working on commission. In sum, this institution serves a demand by transforming 800–2000 ppm TDS water into 150–300 ppm TDS water, and fluoride levels reduced to 0.25–0.50 mg/l. People had no way of ascertaining the quality, but 60 customers surveyed by Indu (2004, unpublished report) asserted that the taste of RO water was distinct. Many also claimed relief from the pain of skeletal fluorosis after adopting RO water.

[27] The seal of the Indian Standards Institution (ISI), the national agency for quality control in all manufactured products.

[28] Registering a cooperative itself meant a great hassle and cost in time and money. The policy also required that two-thirds of the command area farmers submit a written no-objection declaration for the transfer; past defaulters on water fees must first pay up their dues. In addition, several conditions specified that the violation of any of those would qualify the government to reclaim the tube well.

[29] Transformation cost would include the cost of labour and material in creating a lined sub-minor and field channels plus the cost of acquiring land. Transaction cost would basically involve persuading farmers to give up their land for making channels and to give right of way to carrying water to downstream farmers.

[30] In the North Krishna basin in western Maharashtra, a similar groundswell of numerous private irrigation service providers has created an institutional dynamic that challenges orthodox notions of how irrigation systems should be designed. The Bachawat tribunal's decision on the division of Krishna water between Maharashtra and Karnataka made Maharashtra's share contingent upon the amount of water it could develop and use by 2006. To maximize its share, the Government of Maharashtra went on a reservoir-building spree. Strapped of funds, it chose not to build canal systems; instead, it encouraged private entrepreneurs to set up numerous lift irrigation systems. In the command of one such small reservoir, Padhiari (2005) found 1200 such private irrigation service providers serving an area larger than was originally designed to be commanded. These entrepreneurs resolved most key problems that canal irrigation faces in India: while most canal projects are unable to collect even 3–5% of the gross value of crop output they help farmers produce, private service providers in the Upper Krishna basin regularly collect 25% as irrigation charge. They have a much better record of providing irrigation on demand. It is difficult to understand what this is if not Participatory Irrigation Management.

[31] This is put into bold relief in a new, unpublished case study, by Reddy *et al.*, 2004, of traditional community management institutions in a Mudiyanur tank in a system of ten tanks in the Uthanur watershed in the Kolar district. Despite sweeping socio-economic changes in its surround during recent decades, as if stuck in a time warp, the management institution of this 1200-year-old tank has still retained many of its traditional features. Its striking aspect is the fine distinction between the specialized governance role of the caste-based 'Council of Elders' (CoE), the community organization responsible for overseeing general administration of all seven villages

sharing the tank and the role of the neerkattis and *thootis* (village guards) – as management-agents of the CoEs. Most routine aspects of decision making are taken care of by inherited rules and norms that result in 'well-established patterns of behaviour' such as on crop choice, time of opening the sluice under different rainfall regimes, payments to be made to neerkattis and labour contribution in maintaining supply channels. The role of the neerkatti is to execute these routine tasks on behalf of the CoE; and his reward is a piece of cultivable, inheritable *inam* land in the command and ten bundles of hay with grains per each of the 250-odd roughly equal pieces of ayacut land cultivated. The CoE gets into the act only when conflict mediation goes beyond the authority vested in the neerkatti or when circumstances arise that require responding to a new discontinuity. As water inflow into the tank has steadily declined, the CoE decided to disallow sugarcane 20 years ago or, more recently, to make a new rule that divided the 240 acres of ayacut into three parts and irrigate one part per year in annual rotation. Helping the CoE decide whether water available can support the irrigation of a summer crop, orderly distribution of water in the ayacut without any intervention from farmers, deciding the *amount* of irrigation water to be released at different stages of crop growth, undertaking repairs and maintenance of sluices (himself), and canals and supply channels by mobilizing labour from members are amongst the tasks performed by the neerkatti. Cleaning of distributaries is carried out by farmer(s) benefiting from them; however, main canals never get cleaned of weed and silt unless the neerkatti summons all farmers to work there on a fixed day. All in all, in the smooth management of the tank, the neerkatti plays the pivotal management role; he *is* the operating system of the institution; the CoE, mostly invisible and unobtrusive, vests in him the authority and sanction to play that role on behalf of all the members. A tank management institution without a CoE or the neerkatti would be a far lesser institution.

[32] In the Vadodara district, several leases given to fishing contractors were withdrawn because the communities rejected the contractors. In one case, for instance, the contractor used dead animals as manure, a practice that offended the community. In another, the chemical fertilizers used by the contractor ended up in a drinking water well within the tank foreshore; when this was discovered, the village refused to renew the lease. Such aberrations would not occur if the contractor had to obtain the legitimacy and sanction of the community to operate.

References

Allan, J.A. (2001) *The Middle-East Water Question: Hydropolitics and the Global Economy*. I.B. Tauris, London and New York.

Bandaragoda, D.J. and Firdousi, G.R. (1992) *Institutional Factors Affecting Irrigation Performance in Pakistan: Research and Policy Priorities*. IIMI Country Paper, Pakistan, No. 4, International Irrigation Management Institute, Colombo, Sri Lanka.

Cullis, J. and van Koppen, B. (2007) *Applying the Gini Coefficient to Measure Inequality in Water Use in the Olifants River Water Management Area in South Africa*. IWMI Research Report 113, International Water Management Institute, Colombo, Sri Lanka.

Fiege, E.L. (1990) Defining and estimating underground and informal economies: the new institutional economics approach. *World Development* 18 (7), 989–1002.

Frederickson, H.D. and Vissia, R.J. (1998) *Considerations in Formulating the Transfer of Services in the Water Sector*. International Water Management Institute, Colombo, Sri Lanka.

Government of Gujarat (2004) *Gujarat Fisheries Statistics 2002–2003*. Commissioner of Fisheries, Government of Gujarat, Gandhinagar, India.

Holmes, P.R. (2000) Effective organizations for water management. *Water Resources Development* 16 (1), 57–71.

Indu, R. (2002) *Fluoride-Free Drinking Water Supply in North Gujarat: the Rise of Reverse Osmosis Plants as a Cottage Industry: an Exploratory Study*. IWMI-Tata Water Policy Programme, Anand, India.

Jairath, J. (2001) *Water User Associations in Andhra Pradesh – Initial Feedback*. Concept Publishing Co., New Delhi, India.

Janakarajan, S. (1994) Trading in groundwater: a source of power and accumulation. In: Moench, M. (ed.) *Selling Water: Conceptual and Policy Debates over Groundwater Markets in India*. VIKSAT and Pacific Institute, USA, pp. 47–58.

Joseph, C.J. (2001) *Beneficiary Participation in Irrigation Water Management: the Kerala Experience*. Discussion Paper 36, Centre for Development Studies, Thiruvananthapuram, India.

Klein, P.G. (2000) New institutional economics. In: Bouckaert, B. and De Geest, G. (eds) *Encyclopedia of Law and Economics, Vol. I. The History and Methodology of Law and Economics*. Edward Elgar, Cheltenhan, UK, pp. 456–489. http://encyclo.findlaw.com/0530book.pdf

Livingston, M.L. (1993) Normative and positive aspects of institutional economics: the implications for water policy. *Water Resource Research* 29 (4), 815–821.

Londhe, A., Talati, J., Singh, L.K., Vilayasseril, M., Dhaunta, S., Rawlley, B., Ganapathy, K.K. and Mathew, R.P. (2004) *Urban–Hinterland Water Transactions: a Scoping Study of Six Class I Indian Cities*. Working Paper, IWMI Tata Water Policy Programme, Anand, India.

Luis-Manso, P. (2005) *Water Institutions and Management in Switzerland*. College of Management of Technology, MIR Report-2005–001. http://cdm.epfl.ch/pdf/working_papers/WP05_water_ switzerland.pdf

MacDonald, D.H. and Young, M. (2001) *A Case Study of the Murray Darling Basin*: Draft Preliminary Report: CSIRO. Final for IWMI. Sydney, Australia. http://www.clw.csiro.au publications/consultancy/2001/MDB-IWMI.pdf

Mansuri, G. and Rao, V. (2004) Community-based and -driven development: a critical review. *The World Bank Research Observer* 19 (1).

Merrey, D.J. (1996) *Institutional Design Principles for Accountability in Large Irrigation Systems*. IIMI Research Report 8, International Irrigation Management Institute, Colombo, Sri Lanka.

Merrey, D.J. (2000) Creating institutional arrangements for managing water-scarce river basins: emerging research results. Paper presented at the session on *Enough Water for All, Global Dialogue on the Role of the Village in 21st Century: Crops, Jobs, Livelihoods*, August 15–17 2000, Hanover, Germany.

Mishra, S. (2004) *Alternative Institutions for Electricity Retailing: Assessment of Orissa Experiment*. Working Paper, IWMI-Tata Water Policy Programme, Anand, India.

Mukherji, A. (2004) Groundwater markets in the Ganga Meghna Brahmaputra basin: theory and evidence. *Economic and Political Weekly* XXXIX (31), 3514–3520.

Mukherji, A. and Kishore, A. (2003) *Tube Well Transfer in Gujarat: a Study of the GWRDC Approach*. IWMI Research Report 69, International Water Management Institute, Colombo, Sri Lanka.

Myrdal, G. (1968) *Asian Drama: an Inquiry into the Poverty of Nations*. Twentieth Century Fund, New York.

Narayana, P. and Scott, C. (2004) *Effectiveness of Legislative Controls on Groundwater Extraction*. IWMI-Tata Water Policy Programme, 3rd Partners' Meet, IWMI-Tata Water Policy Programme, Anand, India.

North, D.C. (1990) *Institutions, Institutional Change and Economic Performance (the Political Economy of Institutions and Decisions)*. Cambridge University Press, Cambridge, UK.

NSSO (National Sample Survey Organization) (1999b) *Common Property Resources in India. Government of India, National Sample Survey Organization*, Report No. 452 (54/31/4) 54th round, January–June 1998. Government of India, New Delhi, India.

NSSO (1999a) *Cultivation Practices in India. Government of India*, National Sample Survey Organization, Report No. 451 (54/31/3) 54th round, January–June 1998. Government of India, New Delhi, India.

NSSO (2003) *Report on Village Facilities. Government of India, National Sample Survey Organization*, Report No. 487 (58/3.1/1) 58th round, July–December, 2002. Government of India, New Delhi, India.

Padhiari, H.K. (2005) Water management in Upper Krishna basin: issues and challenges. Presented in *FPRM Research Workshop*, 19–21 October 2005. Institute of Rural Management, Anand, India.

Panda, H. (2002) *Assessing the Impact of Power Sector Reforms in Orissa. A Synthesis of ITP Studies*. Pre-publication discussion paper, IWMI-Tata Water Policy Research Programme, Annual Partners' Meet-2002, Anand, India.

Pandya, D. (2004) *History of Inland Fishery in Gujarat with Special Reference to Culture Fishery*. IWMI-Tata Water Policy Programme, Anand, India.

Panella, T. (1999) Irrigation development and management reform in the Philippines: stakeholder interests and implementation. Paper presented at the *International Researchers' Conference on Irrigation Management Reform*, 11–14 December, Hyderabad, India.

Paranjapye, S. and Joy, K.J. (2003) *The Ozar Water User Societies: Impact of Society Formation and Co-Management of Surface Water and Groundwater*. SOPPECOM paper for IWMI-Tata Water Policy Programme, IWMI-Tata Water Policy Programme, Pune, India.

Phansalkar, S. and Kher, V. (2003) A decade of Maharashtra groundwater legislation: analysis of implementation process in Vidarbha. In: Phansalkar, S. (ed.) *Issues in Water Use in Agriculture in Vidarbha*. Amol Management Consultants, Nagpur, India.

Ramirez, A. and Vargas, R. (1999) Irrigation transfer policy in Columbia: some lessons from main outcomes and experiences. Paper presented at the *International Researchers' Conference on The Long Road to Commitment: a Socio-political Perspective on the Process of Irrigation Reform*, Hyderabad, India.

Sakthivadivel, R., Gomathinayagam, P. and Shah, T. (2004) Rejuvenating irrigation tanks through local institutions. *Economic and Political Weekly* XXXIX (31), 3521–3526.

Saleth, M. (2004) *Strategic Analysis of Water Institutions in India: Application of a New Research Paradigm*. Research Report 79, International Water Management Institute, Colombo, Sri Lanka.

Saleth, R.M. (1998) Water markets in India: economic and institutional aspects. In: Easter, K.W., Rosegrant, M.W. and Dinar, A. (eds) *Markets for Water: Potential and Performance*. Kluwer Academic Publications, Boston, Massachusetts, pp. 186–205.

Saleth, R.M. and Dinar, A. (2000) Institutional changes in global water sector: trends, patterns, and implications. *Water Policy* 2 (3), 175–199.

Scott, C. and Shah, T. (2004) Groundwater overdraft reduction through agricultural energy policy: insights from India and Mexico. *Water Resources Development* 20 (2), 149–164.

Seenivasan, V. (ed.) (2003) *Neerkattis, The Rural Water Managers*. DHAN Foundation, Madurai, India.

Sen, A. (1966) Labour allocation in a cooperative enterprise. *The Review of Economic Studies* 33 (4), 361–371.

Shah, T. (1993) *Groundwater Markets and Irrigation Development. Political Economy and Practical Policy*. Oxford University Press, New Delhi, India.

Shah, T. (2000) Mobilizing social energy against environmental challenge: understanding the groundwater recharge movement in Western India. *Natural Resource Forum* 24 (3), 197–209.

Shah, T. (2002) *Who Should Manage Chandeli Tanks?* IWMI-Tata Water Policy Programme Comment, Anand, India.

Shah, T. and Bhattacharya, S. (1993) *Farmer Organizations for Lift Irrigation: Irrigation Companies and Tube Well Cooperatives of Gujarat*. Overseas Development Institute Network Paper 26, Overseas Development Institute, London.

Shah, T. and Desai, R. (2002) *Creative Destruction: is that how Gujarat is Adapting to Groundwater Depletion? A Synthesis of ITP Studies*. Pre-publication discussion paper, IWMI-Tata Water Policy Research Programme Annual Partners' Meet-2002, Anand, India.

Shah, T. and Raju, K.V. (2001) Rethinking rehabilitation: socio-ecology of tanks in Rajasthan, India. *Water Policy* 3, 521–536.

Shah, T. and van Koppen, B. (2005) Fitting water reforms to national context: a brief report on African Law Workshop. Draft paper presented at *4th IWMI-Tata Partners Meet*, 24–26 February 2005. International Water Management Institute, Anand, India.

Shah, T., Seenivasan, R., Shanmugam, C.R. and Vasimalai, M.P. (1998) Sustaining Tamilnadu's tanks: field notes on PRADAN's work in Madurai and Ramnad. Policy School Working Paper 4, Anand, India. In: Marothia, D. (ed.) *Institutionalizing Common Property Management*. Concept Publishing Co., New Delhi, India.

Shah, T., van Koppen, B., Merrey, D., Lange, M.D. and Samad, M. (2002) *Institutional Alternatives in African Smallholder Irrigation: Lessons from International Experience with Irrigation Management Transfer*. IWMI Research Report 60, International Water Management Institute, Colombo, Sri Lanka.

Shah, T., Giordano, M. and Wang, J. (2004a) Irrigation institutions in a dynamic economy: what is China doing differently from India? *Economic and Political Weekly* XXXIX (31), 3452–3461.

Shah, T., Scott, C. and Buechler, S. (2004b) Water sector reforms in Mexico: lessons for India's new water policy. *Economic and Political Weekly* XXXIX (4), 361–370.

Shah, T., Scott, C., Kishore, A. and Sharma, A. (2004c) *Energy–Irrigation Nexus in South Asia: Improving Groundwater Conservation and Power Sector Viability*. Research Report 70, International Water Management Institute, Colombo, Sri Lanka.

Shukla, P. (2004) Exposure visit to PIM project, Ahmedabad, Gujarat. *Letters* 1 (3), July.

Singh, D.R. and Singh, R.P. (2003) Groundwater markets and the issues of equity and reliability to water access: a case of western Uttar Pradesh. *Independent Journal of Agricultural Economy* 38 (1), 115–127.

Svendsen, M. and Nott, G. (1997) *Irrigation Management Transfer in Turkey: Early Experience with a National Program under Rapid Implementation*. Short report series on Locally Managed Irrigation, No. 17, International Water Management Institute, Colombo, Sri Lanka.

Talati, J. and Shah, T. (2004) Institutional vacuum in Sardar Sarovar project: framing 'rules-of-the-game'. *Economic and Political Weekly* XXXIX (31), 3504–3509.

The Economist (2004) Life is not easy for the three biggest private-sector water firms. 15–21 August, 2004.

Vermillion, D. (ed.) (1996) *The Privatization and Self-Management of Irrigation: Final Report*. International Irrigation Management Institute, Colombo, Sri Lanka.

Wang, J., Zhang, L. and Cai, S. (2004) *Assessing the Use of Pre-Paid Electricity Cards for the Irrigation Tube Wells in Liaoning Province, China*. IWMI-Tata Water Policy Programme, Anand, India and Chinese Centre for Agricultural Policy, Beijing.

Wang, J., Xu, Z., Huang, J. and Rozelle, S. (2005). Incentives in water management reform: assessing the effect on water use, production and poverty in the Yellow River Basin. *Environment and Development Economics* 10 (6), 769–800.

Whitcombe, E. (1984) Irrigation. In: Kumar, D. (ed.) *The Cambridge Economic History of India, c. 1757–c. 1970, Vol II*. New Delhi, India, reprint, 1984.

Williamson, O.E. (1985) *The Economic Institutions of Capitalism*. Free Press, New York.

Williamson, O.E. (1999) *The New Institutional Economics: Taking Stock/Looking Ahead*. Business and Public Policy Working Paper BPP-76, University of California, Berkeley, California.

Zhang, L. (2004) *The Economics of Electricity Supply to Tube Well Irrigation in North China*. Draft Report, IWMI-Tata Water Policy Programme, Anand, India.

6 Legal Pluralism and the Politics of Inclusion: Recognition and Contestation of Local Water Rights in the Andes

Rutgerd Boelens,[1] Rocio Bustamante[2] and Hugo de Vos[3]

[1]*Wageningen University and Research Centre, Wageningen, Netherlands;
e-mail: rutgerd.boelens@wur.nl;* [2]*AGUA, San Simon University, Cochabamba, Bolivia;
e-mail: vhrocio@entelnet.bo;* [3]*Freelance Researcher on Institutional Aspects of Natural
Resource Management in Latin America; e-mail: voswiz@versatel.nl*

Abstract

In the Andean countries water has become a source of intense conflicts. Powerful water-interest groups intervene in local water systems and claim a substantive share of existing water rights, neglecting local agreements. These groups are often supported by neo-liberal water reform and privatization policies. This has led to peasant and indigenous mobilization and community action, grounded in shared rules and collective rights. Attempts to formally recognize local rights systems, however, have not guaranteed concrete protection in day-to-day realities, and the 'politics of recognition' have proved problematic. Legal and policy strategies that simply aim to 'include' local and indigenous rights systems – as 'distinct sets of rules and rights' – in the national frameworks are bound to fail. This chapter outlines some important conclusions of the Latin American WALIR (Water Law and Indigenous Rights Program) and critically examines the false policy dilemma of 'incorporation *versus* autonomy'. It concludes that the rightful critique to prevailing ethnocentric and universalistic approaches must not lead to equally simplistic praise for local autonomy or to cultural relativist reification of local rights systems. Critical analysis of the power relations that underpin both customary and official rights systems is crucial in order to improve local, national – as well as international – water laws. Local water rights and identities are given shape not by isolation or policies that reduce them to folkloric practices, or by legal and hierarchical subordination, but by conscious confrontation and meaningful communication among plural legal systems.

Keywords: Andean countries, water rights, water policies, privatization, collective action, legal pluralism, cultural politics, cultural identities, recognition, autonomy, incorporation.

Introduction

In the Andean region, particularly locally, collective water management systems are key to household and community production and reproduction strategies. As such they also sustain national livelihood. They comprise a dynamic and complex set of hybrid rules, rights and organizational forms: a tremendous diver-sity of context-defined 'sociolegal repertoires' or 'normative systems' can be found that generally combine non-local rule-making patterns with local organizational arrangements, frameworks of rights and rules for water distribution, system operation and maintenance. Thus, these rules systems have not come into being in a social vacuum, nor are they limited to isolated devel-opment: alongside physical and ecological conditions, their development is interwoven with the past and present history of cultural, political, economic and technological foundations of the Andean society. Despite their crucial importance, the threats that local water control and rights systems face are huge and ever-growing in a globalizing society.

The Water Law and Indigenous Rights Program (WALIR)[1] aims to contribute to the support of local water management systems – without reifying their form and contents – and critically analyses water rights and customary management systems in comparison with current national legislation and policy. This sheds light on conflicts and negative impacts of certain legislative measures and policy decisions. As an action research, exchange and advocacy programme, the initiative especially supports activities of local communities[2] and inter-institutional platforms and networks to improve national water legislation and policies.

This chapter will elaborate some of the programme's key findings. Throughout the chapter we highlight the problems of legal and policy strategies that simply aim to 'include' or 'incorporate' local and indigenous rights systems – as 'distinct sets of rules and rights' – in the national frameworks. First, some basic features of Andean local water management dilemmas will be presented. Second, conceptual challenges of legal recognition strategies of local organizations will be analysed. Finally, it will discuss the 'politics of participation' in inclusion-oriented water law and policy strategies. The intention is not to give definite answers but to clarify important questions and dilemmas (see Fig. 6.1).

The Andean Context of Local Water Rights

Peasant and indigenous water management systems contribute fundamentally to sustaining local livelihoods and national food security in the Andes (WALIR, 2003; Bustamante *et al.*, 2004; Duran *et al.*, 2006; Gelles, 2006). In most Andean countries, smallholder and medium-scale farmers of highland communities are responsible for the major part of national food production. However, in contrast to the important role local communities have in sustaining

water management systems and food security, government policies are generally not supportive of them. Water rights in most regions of the Andes[3] are largely concentrated in the hands of a few powerful stakeholders (Bustamante, 2002, 2006; Peña, 2004; Guevara and Armando, 2006; de Vos *et al.*, 2006).

This unequal distribution has arisen not only because of historical reasons of colonial occupation and the encroachment of peasant and indigenous communities' water rights by *conquistadores* and *haciendas*, but also because of the contribution of contemporary state policies. For a long time, water policies have been focused on large-scale irrigation for *hacienda* or plantation agriculture in the lowlands and more recently on providing drinking water to the cities. Water is more and more viewed as an economic resource that has to be allocated to the most profitable economic use. The Chilean Water Policy, for example, highlights that: '... the allocation criterion for choosing between various requests will tend to be strictly economic, in practice, given that it is in the country's interest to allocate scarce water resources to those activities with the highest productivity per cubic meter ...' (ECLAC, 2005). Examples of modernist water policies benefiting predominantly the economically and politically well-to-do are abundant.

Fig. 6.1. Map of Latin America.

Illustration 1

Water rights privatization in post-colonial Andean states has ancient roots.[4] The implementation of the Choclococha project in Ica, Peru, for example, is illustrative of many of the attempts to undermine local communities' collective water rights, and formed part of privatization policies that had already started early in the 20th (and even late 19th) century, long before the current era of neo-liberalism (Oré, 1998; Mayer, 2002; Vos, 2002). Oré, for example, quotes an engineers' implementation report:

> The main reason for not having succeeded in developing irrigation in the Ica Valley up to now, through private or State efforts, is the existence of collective property systems in the pampas of Los Castillos. It is difficult to risk capital investment without having the backing in terms of security that the property rights of these valley lands will be obtained.
> (Technical Report, Ica Technical Commission, 1936, in Oré, 2005)

In the Andean region throughout the 20th century, before the arrival of neo-liberal economists and planners of the last decade, hydraulic engineers and bureaucratic policy makers in particular have fiercely promoted the destruction of collective land and water ownership. For example, the collectively owned Los Castillos valley lands in Ica legally belonged to 114 indigenous families which, according to the engineers of the Ica Technical Commission and the Peruvian Water Directorate, were considered to be the 'major obstacle for the proper execution of the irrigation project'. The ownership characteristics of communal territories counteracted the free sale and *lotización* (parcelization) of these newly irrigated fields to individual owners. Therefore, the engineers first suggested and later firmly pressed the state to enforce a law that would allow the expropriation of the Pampa de Los Castillos. The landlords of the Ica Valley, eager to appropriate these large pampas, historically owned by the indigenous communities, strongly supported the engineers' proposals. Since they were labelled by the engineers as 'the ideal owners of the irrigation area', the landlords were installed as the new land and water property owners (Oré, 2005, cited in Boelens and Zwarteveen, 2005). Since then, numerous cases of other state and private sector water management interventions have had devastating consequences for locally managed community systems in the Andean highlands.

As in other parts of the world, increasing demographic pressure and the processes of migration, transnationalization and urbanization of rural areas are leading to profound changes in local cultures, forms of natural resources management and water rights frameworks. New, powerful water interest groups intervene in local water systems and claim a substantive share of existing water rights, often neglecting local rules and agreements. Further, in the context of neo-liberal water reform, national and international elites or enterprises commonly use both state intervention and new privatization policies to undermine and appropriate indigenous and community water rights.

In the last two decades, continuing poverty and exclusion have led to massive nationwide uprisings in Andean countries. Protests have questioned privatization plans, while indigenous and peasant groups have demanded to take part in policy making. These demands aim to offset their historical exclusion and promote policies grounded in an in-depth analysis of the potential and problems of local players in issues such as water management. Increasingly, the traditional struggle for more equal land distribution has been accompanied or replaced by collective claims for recognition of territorial rights, more equal water distribution and for the legitimization of local authorities and normative frameworks for water management (Beccar *et al.*, 2002). Thereby, we see a certain shift from class-based to class-, gender- and ethnicity-based claims for water access and control rights. For example, indigenous groups are now claiming back both their water access rights and rule-making authority, especially in countries such as Ecuador and Bolivia. Thereby, prevailing racist and gender-biased water policies are profoundly questioned and under fire, and water rights (claims and definitions) become arms in a struggle for recognition and social justice (Pacari, 1998; Albó, 2002; Bustamante, 2002; Palacios, 2002; Boelens and Zwarteveen, 2005; Bustamante *et al.*, 2005; Baud, 2006; de Vos *et al.*, 2006). Struggles thus increasingly transcend sectoral demands and involve networks of larger coalitions. The Bolivian example of the 'water war' shows that collective struggles can result in participation in legislative processes.

Illustration 2

In 2000, the Central Valley of Cochabamba, Bolivia became a violent battlefield of protests against the state's plans to privatize the drinking water sector. The government signed a contract with a large foreign consortium and enacted a 'privatization support law' that gave an international company exclusive service and water rights in the district – including those of smaller systems in the peri-urban area and the rights to control aquifers. Directly after the international company had been awarded the concession, it significantly raised water fees before making any system improvement. A strong alliance of urban and rural water organizations protested: citizens protested against rising water rates, while local water committees and rural and indigenous organizations protested against infringement of their water access and control rights. After violent confrontations between these groups and the army, the government had to retract its decision and commit to amending all the law's articles to which the popular alliance objected (see Boelens and Hoogendam, 2002; Bustamante, 2002, 2006).

Following this so-called 'water war' the government, under pressure from international agencies, accepted indigenous, peasant and social organizations as participants in the elaboration of new regulations for drinking water and irrigation as part of the Consejo Interinstitucional del Agua (CONIAG) debates on the new (general) Water Law and Policy for the country. This resulted in the addressing of many previous concerns in a new 'Irrigation Law', in October 2004.

Recently, despite their generally threatened status and decline, there are growing opportunities for customary and indigenous cultures and rights systems. Most Andean countries have accepted international agreements and work towards constitutional recognition of ethnic plurality and multiculturalism. The ratification of the International Labour Organization (ILO) Convention 169 on Indigenous and Tribal Peoples in Independent Countries is an important example. The last decade's change of constitutions in the Andean countries, ratifying the multicultural and plural roots and peoples that make up the countries, is another. However, when it comes to materializing such general agreements in more concrete legislation, such as water laws and policies, many difficulties arise. Context-specific local and indigenous forms of water management (especially decision-making rights to water control) tend to be denied or forbidden (Pacari, 1998; Yrigoyen, 1998; Bustamante, 2002; Gentes, 2002, 2005; Guevara *et al.*, 2002; Palacios, 2002, 2003; Urteaga *et al.*, 2003; Boelens *et al.*, 2005). Before turning to the discussion of the 'politics of inclusion', first we will discuss some of the important conceptual and strategic problems of legal recognition.

Conceptual Problems and Strategic Challenges

Grassroots and popular action to counter rights encroachment and discrimination generally require clear messages and collectively shared goals and demands. However, programmes and platforms that aim to critically support the debate and process of recognition of local rights face several fundamental conceptual and strategic challenges. These are, for example, related to the following:

- Conceptualization of *indigenous* water rights and management rules.
- The concept of official *recognition* of local socio-legal repertoires (normative systems).
- The question of the effectiveness of *legal* (law-oriented) strategies for solving water conflicts and rights issues (Boelens, 2006).

'Indigenous' water rights: their social and political construction

Indigenous water user groups do not just struggle to reappropriate the above-mentioned water access rights, water management rules, water organizational forms and legitimate water authority: they also aim actively to construct their own counter-discourses on 'Andeanity' and

'Indianity', and the policies to regulate water accordingly. Obviously, this dynamic, strategic–political struggle for counter-identification (self-definition) is not necessarily based solely on 'local' truths, rules, rights and traditions.

In the Andes and elsewhere, the denial of contemporary forms of indigenous water management is often combined with a glorification of the past (Flores Galindo, 1988; Assies et al., 1998; Gelles, 2000, 2006; Hale, 2002; Baud, 2006). We find a folkloristic attitude towards contemporary indigenous communities. Policies are oriented towards a non-existing image of 'Indianity', a stereotype; or towards the assimilation and destruction of indigenous water rights systems.

In the Andean region, the so-called 'Indians' were invented and the concept of 'indigenousness' was constructed by various racist currents, developmentalist paradigms and romanticized narratives, and by the indigenous peoples themselves. Divergent regimes of representation constructed images or projections of 'Andean identity' or 'indigenous cultures'. These projections refer either to the backwardness of the 'Indians' populations, who therefore should be assimilated into the mainstream culture, or to neo-positive, idealized images of 'real and pure Indians', isolated from cultural interaction and defenders of original positive human values. Indigenous groups have often adopted or contributed to the creation of these stereotypes, sometimes unreflectively, sometimes with clear ideological and political purposes (Salman and Zoomers, 2003; Boelens and Zwarteveen, 2005; Baud, 2006).

Portraying Andean and indigenous cultures as 'radically different', with pre-constituted identities and static cultural properties, reminds us of the past, essentialist philosophies and ideologies: they either created the 'noble savage' or tended, on the contrary, to generate the image of the backward, ignorant and violent Indian. Essentialization and reification generally deny or dichotomize colonial and post-colonial influences. They neglect the adaptability and hybridity of local cultures and management forms, and the way they necessarily interact with – and are influenced by – others; 'otherness' can be defined only in processes of confrontation, communication and, thus, interaction. Moreover, such approaches neglect the power structures that influence the process of

cultural transformation, of both dominant and non-dominant groups. In the same way, those approaches that glorify lo andino (the Andean) tend to deny the existence of locally prevailing power structures that profoundly colour local (water) rights definitions and distributions.

This presents an important challenge to scholars, action researchers and NGOs: to refrain from naive participationism and to critically rethink intentions to support the so-called 'indigenous' knowledge, culture, rights, livelihoods and natural resources management. It also provides a challenge that water rights reform programmes need to face and shows partly how complex the objectives of rights-based empowerment initiatives are. For example, what is, or who is 'indigenous'? Is it possible to speak of specific 'indigenous' or 'Andean' cultures, communities, water management forms or socio-legal systems?

Often, what is called 'indigenous culture' combines elements from different origins – Andean, as well as modern. As mentioned by Gelles (2000, p. 12), Andean culture and identity, therefore, are:

> a plural and hybrid mix of local mores with the political forms and ideological forces of hegemonic states, both indigenous, Iberian and others. Some native institutions are with us today because they were appropriated and used as a means of extracting goods and labor by Spanish colonial authorities and republican states after Independence; others were used to resist colonial and postcolonial regimes.

'Indigenous' culture therefore has to be analysed as dynamic and adaptable to new challenges and contexts.[5] The difficulties in defining what 'indigenous' means have also led to a shift in the debate toward using the broader concept of 'collective rights'. More recently, there has been a greater acceptance of the consideration of 'local' as a concept that better suits this kind of system, in reference to the fact that these normative frameworks have hybrid, contextualized and dynamic properties (see also von Benda-Beckmann et al., 1998).

The plural and contradictory concept of 'recognition'

'Legal recognition' is another notion that poses enormous conceptual problems and challenges,

with important social and strategic consequences. 'Recognition' in contexts of legal pluralism is, by definition, many-faceted and generally
ambiguous. In another paper[6] we have
discussed the dilemma regarding 'recognition of
legal hierarchies', arguing that a distinction must
be made between analytical–academic and political–strategic recognition.

> In an analytical sense, legal pluralistic thinking
> does not establish a hierarchy (based on the
> supposedly higher moral values or degrees of
> legitimacy, effectiveness or appropriateness of a
> legal framework) among the multiple legal
> frameworks or repertoires that exist. In political
> terms, however, it is important to recognize that
> in most countries the existing, official legal
> structure is fundamentally hierarchical and
> consequently, in many fields state law may
> constitute a source of great social power – a fact
> that does not deny the political power that local
> socio-legal repertoires may have. Recognizing the
> existence of this political hierarchy and the
> emerging properties of state law in particular
> contexts offers the possibility to devise tools and
> strategies for social struggle and progressive
> change. In the discussion about 'recognition' as a
> way of giving legal pluralism a place in policy-
> related issues, both the political–strategic and
> analytical–academic aspects of recognition
> combine.
>
> (Boelens *et al.*, 2002, p. 138)

The analytical aspect of recognition concen-
trates on the academic quest of knowing how
plurality is ordered; the political aspect on
whether and how this plurality is (or is to be)
embedded in a political and legal hierarchy,
based on existing power structures that establish the power and properties of the 'recogniz-
ers' and the ones to be 'recognized'.

Thus, instead of collective and unified
claims, many questions arise in the debates and
struggles for 'recognition', for example:

- How to define and delimit the domain of the
  validity of so-called indigenous rights
  systems? i.e. who is able to make claims?
  Considering the multi-ethnic compositions
  of most Andean regions and the dynamic
  properties of local normative frameworks it
  is difficult (or impossible) to come to
  uniform definitions. Would it be better to
  define rights systems in terms of exclusive
  geographical areas, traditional territories, or
  flexible culture and livelihood domains? Do

indigenous peoples and their advocates
claim recognition of just 'indigenous rights'
(with all the conceptual and political–strategic dilemmas of the 'indigenous' concept),
or do they also struggle for recognition of the
broader repertoires of 'customary' and
'peasant' rights prevailing in the Andes? And
what precisely is the difference in concrete
empirical cases?

- Which recognition strategy is appropriate?
  Should indigenous peoples try to claim and
  defend legalization of their water *access
  rights* or try to legalize *delimited frameworks*
  of local water rights systems? Or should they
  rather claim the recognition of their autonomy to define, develop and enforce collective *water control rights*?

Since water access rights are increasingly
under threat an important strategy might be first
to claim the right to access, withdraw and
usufruct the water – and assume that water
management and control rights will follow once
the material resource basis has been secured.
Moreover, legalizing customary rights systems
can be difficult. Although there are many
dynamic, interacting and overlapping socio-
legal repertoires there are no clear-cut, indigenous socio-legal frameworks. Therefore, it does
not seem appropriate to recognize only the
explicit and/or locally formalized indigenous
property structures and water rights ('reference
rights', often, but not always, written down),
since these generally strongly deviate from the
complex, dynamic local laws and rights in day-
to-day practice.[7] Claiming *water control rights*,
instead of 'freezing' entire local rights systems in
formal law, potentially has the advantage of
granting *autonomy to develop* the rules and
normative frameworks according to a dynami-
cally changing context.

The pitfalls and challenges of 'law-oriented strategies'

One major conceptual and strategic–practical
challenge stems from the difference between
universally valid laws and context-dependent
rights systems. National (positivist) legislation
by definition claims that law must focus on
uniform enforcement, general applicability and

equal treatment of all citizens. At the same time, local and indigenous rights systems, by definition, address particular cases and diversity. How to deal with the conflict and fundamental difference between legal justice (oriented at 'right'-ness/generality) and diverse, local equity ('fair'-ness/particularity)?

Various forms of state legislation have recognized this fact when faced with losing its legitimacy in practice: official justice was perceived of as being 'unfair' for many specific cases. In many cases, a second set of principles (fairness) has been institutionalized by formulating 'special laws' (see Boelens et al., 2005). This was not to replace the set of positivist rightness rules, but to 'complement and adapt it'. In fact, it appeared that official legislation, 'justice', could often survive thanks to the 'fairness' and acceptability of common laws that had been incorporated. However, this institutionalized equity is a *contradictio in terminis*. It leads almost automatically to the ironical situation in which the set of common or customary rules, 'equity', itself becomes a *general*, formalized system and loses its pretensions of 'appropriateness', 'being acceptable' and 'doing justice' in particular cases (Schaffer and Lamb, 1981, cf. Boelens and Dávila, 1998).

These conceptual challenges raise several new questions. For example: how to avoid freezing of local dynamic systems by official recognition? Indigenous socio-legal repertoires make sense only in their own, dynamic and particular contexts, while national laws demand stability and continuity: how to avoid 'freezing' of customary and indigenous rights systems in static and universalistic national legislation in which local principles lose their identity and capacity for renewal, making them useless? How to avoid assimilation and subsequent marginalization of local rights frameworks when these are legally recognized? And how to avoid a situation in which only those 'customary' or 'indigenous' principles that fit into state legislation are recognized by the law, and the complex variety of 'disobedient rules' are silenced after legal recognition?

'Enabling' and 'flexible' legislation might solve the above problem. However, enabling legislation and flexible rights and rules often lack the power to actually defend local and indigenous rights in conflict with third parties. Thus, an important issue is how to give room and flexibility to diverse local water rights and management systems, while not weakening their position in conflict with powerful exogenous interest groups? Also, answers must be given to the question of what such legal flexibility means for 'internal' inequalities or abuses of power. If, according to the above dilemmas, autonomy of local rule development and enforcement is claimed for (instead of strategies that aim to legalize concrete, delimited sets of indigenous rights and regulations), how to face the existing gender, class and ethnic injustices that also form part of customary and indigenous socio-legal frameworks and practices? Answers to such questions necessarily point at directions where frameworks of collective rights and rule-making autonomy for local collectives are combined with the establishment of supra-local institutions and rules that need to guarantee the protection of individual and minority rights.[8] These also need to offer opportunities for second-order conflict resolution and appellation in case local conflicts cannot be solved adequately.

Another important question that arises is how to balance the strategic importance and effectiveness of legal recognition with other struggles for water rights? Considering peasant and indigenous communities' lack of access to state law and administration, this question comes prominently to the fore: is *legal* recognition indeed the most effective strategy, or would it be better and more effective for peasant and indigenous communities to defend their own water laws and rights 'in the field'? Moreover, it often is not the state law as such that sets the rules of the game in peasant and indigenous communities, but hybrid complexes of various socio-legal systems. Formal rights and rules cannot act by themselves, and it is only the forces and relationships of society that can turn legal instruments into societal practice. In particular, social and technical water engineers, lawyers and other legal advocates have often overestimated the actual functionality or instrumentality of formal law and policies in local contexts. On the contrary, their legal anthropological colleagues have sometimes tended to underestimate the power of formal law, assuming that all conflicts are settled by means of local normative arrangements, without any influence from official regulations.

Recent experience shows that legal recognition, just as legal misrecognition, tends to have an important effect on the daily lives of indigenous and peasant populations. For example, the neo-liberal water laws (in, e.g. Chile) or top-down instrumental water policies (in, e.g. Ecuador and Peru) have not only neglected customary and indigenous water-management forms but have also had concrete, often devastating, consequences for the poorest people in society. Because of the negative impact of application of official law, indigenous and grass-roots organizations have fiercely engaged in the legal battle. It is important to consider here that efforts to gain legal recognition do not *replace* but rather *complement* local struggles 'in the field'. On both levels, there is political–strategic action to defend water access rights, define water control rights, legitimize local authority and confront powerful discourses. In the next section we will elaborate on how 'recognition struggles' at the local and national level shape the complex arena in which local water rights and customary laws confront uniform policies and politics of participation.

Inclusion and Exclusion

National water policies in the Andean countries and their translation in field practice mirror the political power and cultural hegemony of a dominant stakeholder group.[9] Historically, this group has imposed rules, rights and regulations, and has controlled nation-building processes in previous centuries. As shown by Gelles (1998, 2000), state bureaucracies usually ignore indigenous models of resource management, not only because of the alleged superiority of 'modern' Western cultural forms and organization, but also because indigenous peoples are perceived as racially and culturally inferior (Gelles, 2000, pp. 9–10).[10]

Racist connotations stem from exclusionary politics from the Inca and Spanish colonial predecessors. In former days, indigenous property rights were taken away through violence, conquest, colonization and oppression. The Inca emperors and other indigenous leaders, as well as the kings, *conquistadores* and *hacendados* during the Spanish colonial period, differentiated themselves by excluding subordinated classes from resources, services and social life (Flores Galindo, 1988; Patterson, 1991; Mayer, 2002). Powerful groups were glorified through public displays, reinforcing the differentiation and social exclusion.

In the post-colonial area the opposite occurred. There was a move from 'exclusion' to 'inclusion' of indigenous peoples, from a discourse of racial (and thus 'natural' social) differentiation to one of equality (Boelens, 2006). Not the powerful authorities and landlords, but the peasant and indigenous communities and the common people, are made visible and, by means of a Foucauldian 'power of equalizing normalization'[11], they are individualized, classified and made 'cases' according to the ways that they do, or do not, fit the model. Yet, the powerful groups that benefit from this 'inclusive' power, as well as the new mechanisms and rules of subordination, now remain invisible. New irrigation legislation and state policies are often an expression of such post-colonial 'universality' and 'equality' discourses.

Throughout the Andes (as elsewhere), for example, irrigation technicians and development professionals have introduced virtually the same irrigation techniques, knowledge and norms (developed in Western research centres, universities and development enterprises). Nevertheless, they are not just 'imposed' in a top-down way. It is the indigenous peasants *themselves* who often ask for this same technology, to 'progress' and leave behind their traditional 'backward' technology, to become like the Western-oriented, *modern* farmers and to gain economic parity (cf. Escobar, 1995; van der Ploeg, 2003; Boelens, 2006).

Another clear example of normalization is found in the categorizing properties of neo-liberal market ideologies penetrating the Andean legal and policy frameworks regarding water management. Although the neo-liberal principles are imposed on Andean states by international institutions and national power groups, many of its basic concepts and dynamics have been adopted and internalized by Andean communities, penetrating and subtly transforming local management forms and often disarticulating indigenous water control. Communities are dispossessed through destruction of collective rights over resources. Recognition of private property rights has allowed rapid incorporation of land and water into the market system. Thus, the deployment of secular, rational, universally applicable irrigation models, supported nowadays by water management privatization ideologies, is a powerful means by which contemporary nation-states and private interest sectors extend their control.

Modernization and development discourses pretend to provide universal benefits but undermine balanced valuation of local beliefs and 'unruly' practices because any legitimization of these local norms calls into question both the state's and market ideology's supposed monopoly of rationality, efficiency and legitimate culture (Gelles and Boelens, 2003).

The politics of participation

The above analysis of new policies of 'integration', 'participation' and 'equality' raises some fundamental questions:

- If 'equality' is strived for, the question is: equal to what, equal to whom, equal to which model? The basic assumption in current Latin American water policies is, that 'progress' means: equality to occidental, technocentric and male-biased water management models. The concept of rational water management is interspersed with non-indigenous norms about efficiency, social security, effective organization, private ownership and economic functionality. In practice, indigenous peoples are forced to 'equalize': in other words, to adopt the norms and practices of white or *mestizo* water users, which most often run counter to local social relations and environment, and disintegrate local communities and identity.
- If 'inclusion' and 'participation' constitute the objective, the obvious question is: inclusion in what? Participation following whose objectives, visions and terms? In this respect, the Second World Water Forum (2000) concluded that: '… there is a recurrent problem for indigenous peoples, who are often constrained to deal with vital issues on terms dictated by others. Traditional knowledge is seen as inferior in current political, legal, and scientific systems and therefore their arguments are discarded time and again by courts and other institutions.'

- Regarding the important current concepts of 'integrated' water management and 'integrated' policies, there seems to be a general consensus but the underlying central question is: who does the integration? Let us have a look at some common, inclusion-oriented examples.

A first example draws on the Ecuadorian Licto Project. It illustrates problems of outside-driven integration of indigenous communities in uniform, national legislation, organizational models and engineers' designs:[12]

Illustration 3

In Chimborazo province, Licto district, the Ecuadorian State intervened in the territory of 20 Andean communities to build an irrigation system and carry out an integrated development programme. The design was formulated in the country's capital, without user involvement. It disregarded community production systems and boundaries and imposed a classic, universal blueprint. The nationwide, uniform legal recipe dictated the organization of the system, which would strengthen bureaucratic power and new, artificial leaders, and weaken existing community structures and collective action – the only way to survive in this region. The state also imposed a model in which water rules and rights were established by uniform government rationality: those individuals who had land and pay fees obtained water rights. *Indígena* and *campesino* rationality, on the contrary, says: you cannot just *buy* rights. Those who contribute with labour or organizational capacities, and participate in the meetings, *create* water access and decision-making rights. Thereby, individual rights are derived from the collective ownership of infrastructure.

When the state agency, because of financial crises and lack of capacity, did not complete system construction, the indigenous communities took over its development with the help of a local NGO. They adapted design, management and water rights to local demands and capacities. Although many had no formal education or were illiterate, the means were developed to collectively discuss the project. Through interactive design, user-to-user training and the use of scale models, the design of infrastructure and

water rights was linked. Combined literacy training and water management capacity building
strengthened the position of female water users
and female leaders, since they were to become
involved in the management of the system. And
in this region, characterized by male out-migration, they were the ones who were in charge of
creation of and maintenance of water rights in
the system. A system was developed which the
communities themselves now manage, from the
main level to the field level.

However, once the 20 indigenous communities had finished developing the system, with
clear rules and rights and strong collective
management, the state administration re-
appeared. It did not want to recognize local
management, regulations and water rights.
Simply because local rules were not sustained
by national law, they were declared 'illegal'.
The state agency intended to implement the
universal 'Decentralization and Management
Turnover' policy and 'include' the backward
user communities in modern society. In practice, however, instead of legalizing the local
system it claimed authority back, because:
'How can we hand it over if it is not in our
hands?' Many projects and policies have effects
in the field at variance with what theory
predicts, and behind official arguments a power
play is going on. Ecuadorian Water Law does
not allow for local water rights and management principles, and destroys the variety of
normative systems that *do* try to find particular
solutions for diverse contexts.

Illustration 4

Another example is the inclusion of indigenous
water communities – under certain limits and
conditions – in current global water policy
models. In Chile, all water users (including
indigenous peoples) have become 'included' in
the 1981 Water Code, dictating privatization of
water rights. While ideological studies continue
to praise the model, empirical field studies indicate the disintegration, in particular, of indigenous systems: the individualization of water
rights has increased insecurity and disorganiza-
tion – instead of decreasing insecurity, as
neoclassical theory would wish (Bauer, 1997,
1998; Hendriks, 1998; Dourojeanni and
Jouravlev, 1999; Castro, 2002).

According to Chilean legislation, decisions on
water management are weighted according to
actual possession of water rights. Right-holders
with more 'water actions' (volumetric rights per
time unit) have more decision-making power.
This contrasts with indigenous management,
where collective interests are negotiated according to the rule of 'one man, one vote'. Therefore,
the Water Code has enabled a water rights-
owning elite to effectively deny the interests of the
majority (the group of poorer users) and impose
their own playing rules (Hendriks, 1998).
Moreover, since individual water property owners
can make use of the water entirely according to
their personal interests, Chile faces the problem of
strong increase in water contamination, and individual property owners are not sanctioned for
polluting *their* property. Often, indigenous
communities and downstream cities bear the
consequences (Bauer, 1997; Dourojeanni and
Jouravlev, 1999).

Up till 2005, the 1981 Water Code did not
request water rights owners to actually make
use of their claims, or to pay concession fees.
This made hoarding and speculation of water
rights extremely attractive. When the new Water
Code was enforced in 1981, most indigenous
communities were left unaware of the need to
officially register their century-old customary
rights. One Mapuche leader said: 'The big
landowners here in the area have registered the
water rights in their names, and the Mapuches,
for not knowing about the laws of the Chilean
State, were left without possibilities to claim
their rights' (Solón, 2003). Water rights that are
not claimed, or the so-called 'unused rights',
were allocated to those who presented official
requests: powerful commercial companies,
especially mining and power-generation enterprises and landlords (van Kessel, 1992;
Hendriks, 1998; Dourojeanni and Jouravlev,
1999; Castro, 2002; Gentes, 2002).

Mapuche communities are furious about
this. As one Mapuche leader phrases his anger:

> The water sources that originate in the
> communities here have 98% of their trajectory on
> Mapuche territory, but the owner of the water is a
> landlord who lives in the city. He bought the
> water from the state, and nobody can use it. We
> cannot use it for irrigation, not even for drinking
> water, because the water has been bought. But
> the water was born in and flows through

Mapuche communities, and no one of the Mapuches was aware of the need for official recognition when this person registered the water rights on his name. No one of us was consulted and no Mapuche ever knew of the existence of this law.

(Solón, 2003)

It is not only the neo-liberal assumption that (market) information is freely available to everyone that is challenged here, but also the very basis for rights claims. Mapuche communities strongly feel that the water is theirs, because they have been using it for centuries and because it flows through their territory, whereas the Water Code demands official registration as a first basis for rights allocation (Boelens and Zwarteveen, 2005).

To counteract the negative consequences of the Water Code for indigenous communities, Chile enacted a new law in 1993: the *Ley Indígena* (Indigenous Law). Although it was meant to support indigenous populations in their defence of what was left of their territorial rights and livelihoods, in practice it was difficult to enforce. The fact that it is a 'special law' only applicable for (and within) a 'special group of the national population' (called a 'minority'), and the costly and time-consuming procedures has left most of the indigenous claims unanswered. Moreover, the Indigenous Law has proved to be extremely weak as a legal tool, whenever indigenous communities had to face the powerful Water and Mining Codes that are called upon by the country's water-owning elites.

This relates also to a recurrent problem of universal or national policy models: their validity is based on theoretical models and paradigms, but they usually fail to look at human suffering and internal contradictions *in the field.* Currently, the democratic government and civil society institutions have succeeded in changing some of the articles that most threaten the rights of indigenous communities and user collectives. Nevertheless, such attempts at changing the law towards social and environmental improvements meet with fierce resistance of powerful actors who defend their accumulated, private water rights.

The inclusion of local and indigenous rights frameworks in bureaucratic, state-oriented models or neo-liberal, market-oriented models

is not always based on brutal impositions. On the contrary, water reforms are presented as merely neutral and technical interventions aimed at better controlling and managing the water crisis. It is suggested that such interventions do not fundamentally alter or influence existing social and political relations. And to peasant and indigenous water user communities it is explained that flows of money and water follow universal, scientific laws and that human beings share the same aspirations and motives as everywhere. Such inclusion-oriented policies establish a universal rationality based on a 'natural' truth and 'objective' criteria for optimizing efficiency and water management (Boelens and Zwarteveen, 2005).

Peasant and indigenous movements point at the fact that this is a false representation of reality: the proposed water reforms are not just slight modifications that basically leave existing social relations intact, but they involve quite radical changes in social and political structures in which water management is embedded. The proposed ways in which water is to be owned, distributed and managed imply fundamental change, and so do the ways in which different water users relate to each other. If such universal modernization policies are implemented, relations are increasingly dictated by extra-communal laws, institutions and markets (Boelens and Zwarteveen, 2005).

In Bolivia, attempts to 'modernize' the water sector led to such widespread protests that the government was forced to allow real participation of protesting groups in the policy reform process. The illustration below shows not only that state politics of inclusion are contested but how the definitions of this 'inclusion' were challenged by the water user organizations. It was only through continuous pressure that their voices became officially recognized in official law.

Illustration 5

Bolivia is a country, different from other countries in the Andes, where the state has been very weak and so far almost absent in water management issues. As Larson (1992) observes, it was the colonial policy (18th century) to give private possession of land and water to individuals, in order to collect taxes. This policy was formalized by the subsequent republican government (through the so-called *Leyes de exvinculacion* in 1874). Much of the *Pueblos de Indios* (communally owned land) was assigned as private property to indigenes living in those territories and allowed a very active market by the beginning of the 20th century. This practice was resisted in the Altiplano region, especially by Aymara communities that used legal strategies to defend the communal or collective character of the land, preserved up to date in some places. As a result of the non-involvement of state bureaucracy in water resources management and regulation, water control in the rural areas, including that in peri-urban and sometimes even in urban areas, is usually autonomous and independent.

It was only very recently and as a result of external pressure that the Bolivian government started its 'politics of inclusion', aiming at a process of deregulation and privatization in the water sector, mainly relating to drinking water and irrigation. However, neo-liberal policies have been difficult to implement because they have led to conflicts and protests. After the 'water war' in the year 2000, popular resistance to uniform inclusion policies and politics led to a shift in policy-making processes, towards dialogues and consultation. Several indigenous and representative groups participated in the formulation process of the Drinking Water and Sanitation Services Act and Bylaws and the Irrigation Normative design process (2001), which recently resulted in the approval of the Irrigation Law (October 2004). Some features of this new legislation are:

- Creation of different types of rights over water resources and over water services provision.
- Recognition of indigenous and community rights to water and water services, under the legal figure of a registrar. A community receives collective and indefinite rights.
- Setting up of new institutional bodies to deal with water resources allocation, conflict resolution, water management at catchment scale, etc. regarding drinking water and sanitation (*Comisiones de Registros y Licencias*) and irrigation (SENARI and SEDERI). The new organizations will have representatives from local, indigenous and peasant water organizations.

- Respect for the local, indigenous and peasant water management norms: 'uses and customs' and local authorities.

The consultation process – even though this was not just through consensus building, but more particularly due to pressure by social action groups – meant significant progress in recognition of local, indigenous and communities' rights. However, this apparent openness of the state to 'give recognition' to customary rights places local water organizations in a dilemma. They are trying, on the one hand, to get their water rights and water control forms legally protected and, on the other, to maintain their autonomy and self-management. The question remains whether these new regulations will really empower local water organizations or only legitimize state intervention in an area where, previously, its presence was minimal.

Justice and the right to be different

The 'politics of inclusion' face fierce resistance by different social movements that demand alternative strategies of natural resources use and maintenance. While such movements are motivated by a range of concerns, such as social justice, the environment, 'right to livelihood' or ethnic identity, they all make claims for more equitable and just access to natural resources. All centre on the question of property rights, because whoever controls property rights controls the processes of resource extraction and environmental change. For example, in Ecuador and Bolivia, the countries in South America with the largest indigenous populations, well-organized social movements have been able to change national-level debates in water reform. CONAIE (1996) made its own proposal for a new water law, which included: (i) demands on resisting privatization of water resources; (ii) continued public and community control in water allocation, recognition of cultural and social rights; and (iii) representation of users, indigenous and peasant organizations, within the institutional framework for water management. In 1998, some of these proposals were recognized in constitutional reforms. However, up to now, proposed reforms of the actual water law have not been accepted in Congress.

In Bolivia, indigenous and peasant confederations also proposed an alternative water law and a new water reform agenda. This proposal emphasized social rather than just economic aspects of water and community water rights. As elaborated above, implemented in 2004, the drinking water legislation and a new irrigation law have begun to recognize some of the concerns of social organizations in relation to water rights, participation and social control. The newly installed indigenous President Evo Morales (2006) and the MAS government (*Movimiento al Socialismo*) plan to change the Water Law and install a Ministry of Water to enhance more equitable distribution of benefits and burdens, and to legally recognize and regularize the local and customary water rights of indigenous and peasant communities. In other Andean countries, indigenous, peasant and other grass-roots groups are also pressing for more equitable water rights distribution.[13]

These struggles not only concern control over water but also, and importantly, over the right to define what a water right entails. Fundamentally, the water rights struggle includes the following key issues: (i) access to water and infrastructure; (ii) rules and obligations regarding resource management; (iii) the legitimacy of authority to establish and enforce rules and rights; and (iv) the discourses and policies to regulate the resource. And it is precisely the *authority* of indigenous and peasant organizations that is increasingly being denied, their water *usage rights* that are being cut off and their control over *decision-making* processes that is being undermined.

It is, in particular, local peasant and indigenous water users' collectives that are facing both the water-scarcity crisis and the policies developed to counter that crisis. Paradoxically, it is precisely the ones with solutions – the producers of local livelihood and national food security, who developed a variety of water rights and management systems in order to adapt them to the multiple local constraints and opportunities – who are being denied and suffer most from the devastating consequences of 'modern water approaches'. But, if current cultural politics and policies of 'inclusion' constitute the problem, the solution can never

be to go back to 'exclusion'. Participation, yes, but with a different rights approach, based on the self-perception of them being right-holders and not just users, taking critically into account that peasant and indigenous communities (and other local groups) want to 'participate' on their own terms. On the one hand, there is a general demand for greater justice and equality regarding the unequal distribution of decision-making power, water and other water-related benefits and, on the other, there are the demands for internal distribution to be based on autonomous decisions, locally established rights and principles and local organizational forms for water control that reflect the diverse strategies and identities found in local communities today.

Conclusions

Triggered by population pressure, water monopolization practices, class- and ethnicity-biased intervention policies and climate change, among others, the growing scarcity of water has caused intensified conflicts over the resource in the Andean region. But it also has led to mobilization and community or inter-community action, grounded in shared rules and collective rights. In various instances, such mobilization has effectively resulted in increased recognition of local collective rights and more equitable access to the resource. Formal recognition, however, has not been proved to guarantee concrete protection of local water management systems in the day-to-day realities, against outside claims and transfer of water to economically powerful sectors. 'Inclusion' and 'integration' appear to be complex rights matters.

It is common in academic and policy circles that the question of local rights (and identity) recognition is placed as a false dilemma, 'incorporation *versus* autonomy': either accepting the universalized, liberal standards of rationality, efficiency, human rights, justice and order *or* recognizing and celebrating local diversity and rule-making whatever the outcome may be. However, the issue is not so much a matter of respecting 'otherness', if this is presented as 'isolated and radically different normative systems', entirely 'distinct rules and rights' and 'pre-constituted, static identities'. Rightful critique to ethnocentric, universalistic or rigid positivist approaches should not mislead us to reify local rules and rights autonomy, and give freeway to a cultural relativist approach or, worse, a revival of the theories that essentialize the 'noble savage', assuming that 'indigenous' is equal to 'good' and 'local' is presented as necessarily 'better' and 'more just' than national or international.[14]

This positioning not only risks the legitimization and legalization of local class, ethnic and gender injustices but also misrepresents the dynamic nature of water culture, water rights and the hybrid forms taken by water control and organization in practice. Essentializing and stereotyping of local water norms, rights and cultures, be it in law, policies, intervention strategies or theoretical reflection, deny the very existence of *interaction* among socio-legal systems and thereby equally deny the right to self-recreation (and improvement) of diverse normative systems, *always* and necessarily in contact with 'otherness'.

Opportunities for and openness to mutual critique and self-critique are essential for living law systems and their re-creation – an equally important message for formal, national law making. Critical analysis of the power relations that underpin these systems, thus, is crucial in order to improve both local, national and international water laws and rights. Local water rights and identities are given shape not by reification, isolation or folkloric policies but by confrontation and communication in an inter-legality approach. This calls for proactive, contextualized and power-critical strategies that enhance interdisciplinarity rather than multidisciplinarity, interculturality rather than multiculturalism, and inter-legality rather than 'multi-legalism'.

Endnotes

1 WALIR is a collaborative programme coordinated by the Wageningen University (WUR/IWE) and the United Nations Economic Commission for Latin America and the Caribbean (UN/ECLAC) and implemented in cooperation with counterpart institutions in Bolivia, Chile, Ecuador, Peru, Mexico, France, Netherlands and the USA (http:// www.eclac.cl/drni/proyectos/walir/). The counterparts work with a broad group of participants: institutions at international, national and local levels.

2 Local communities include not only peasants and indigenous groups but also other local organizations.

3 Bolivia is an exception to this: indigenous, peasant and local organizations have managed relatively well to keep control over their water resources.

4 This illustration is taken from Boelens and Zwarteveen, 2005 and elaborated in Oré, 2005.

5 As a reaction, in Bolivia, for instance, some social organizations claim now to be not indigenous but 'first peoples' nations', also being a clear way to challenge the state's authority (that was created only in the 19th century) over natural resources.

6 'Taking "recognition" as a point of departure implies that there is a "recognizing party" and a "party being recognized". This would put us in the kind of state-biased position in which matters are decided upon according to a state-determined hierarchy of legal systems' validity. Such a position, needless to say, would invalidate the insights derived from attention to legal pluralism. On the other hand, it is important to be aware of the possible opportunities involved in (state) recognition, taking into account and taking seriously the fact that many local groups of resource users (and right-holders), ethnic and other minorities actively aspire and strive for this form of recognition' (Boelens *et al.*, 2002; see also von Benda-Beckmann, 1996; Roth, 2003; Roth *et al.*, 2005).

7 These 'rights in action' emerge in actual social relationships and inform actual human behaviour, but are less 'tangible' (cf. Rudolf Stavenhagen and Diego Iturralde eds., 1990; Stavenhagen, 1994; von Benda-Beckmann *et al.*, 1998; Gerbrandy and Hoogendam, 1998; Bruns and Meinzen-Dick, 2000; Boelens and Doornbos, 2001; WALIR, 2002; Hendriks, 2004; van Koppen and Jha, 2005; Getches, 2006).

8 Many constitutions have set limits on customary systems, stating that they can be valid only if they do not run contrary to the official laws and regulations of the country. In other cases, the limits are set by human rights principles.

9 This section is largely based on and taken from Gelles and Boelens (2003).

10 This bureaucratic irrigation tradition has been especially powerful in countries such as Peru and Ecuador. As Lynch (1993) and Zwarteveen and Boelens (2006) have shown, its devaluation of particular water use actors extends to women, as the gender discrimination found in the field and in irrigation offices is part and parcel of the bureaucratic tradition (cf. Vera, 2004; Bennet *et*

al., 2005; Bustamante *et al.*, 2005; Zwarteveen and Bennet, 2005).

11 'This power is exercised rather than possessed; it is not a "privilege", acquired or preserved, of the dominant class, but the overall effect of its strategic positions – an effect that is manifested and sometimes extended by the position of those who are dominated' (Foucault, 1978).

12 Based on the chapter 'Recipes and resistance. Peasants' rights building and empowerment in the Licto irrigation system, Ecuador' of the book *Water Rights and Empowerment* (Boelens and Hoogendam, 2002).

13 The difference is that, in countries like Colombia and Chile, indigenous groups are considered as minorities, with special rights. Therefore, it was easier to enact and implement special laws for indigenous groups; this is not as easy in Bolivia and Ecuador, where more than 50% of the population is considered to be indigenous (see also Van Cott, 2000).

14 Paradoxically, both approaches commonly lead to subordination by incorporation and inclusion. The first aims to homogenize and 'equalize' all water rights and cultures according to the illuminating model of (neo-)liberalism and modernity, in which all actors and resources should be included. The second, although pressing for local autonomy, tends to codify, freeze and subordinate all non-official rights systems under the umbrella of national law that would 'protect local rights and cultures'.

References

Albó, X. (2002) *Iguales Aunque Diferentes*. CIPCA, La Paz.

Assies, W., van der Haar, G. and Hoekema, A. (eds) (1998) *The Challenge of Diversity. Indigenous Peoples and Reform of the State in Latin America*. Thela Thesis Publishers, Amsterdam.

Baud, M. (2006) Ethnicity, identity politics and indigenous movements in Andean history. In: *Agua y Derecho. Políticas Hídricas, Derechos Consuetudinarios e Identidades Locales*. WALIR – IEP, Lima.

Bauer, C. (1997) Bringing water markets down to earth: the political economy of water rights in Chile, 1976–1995. *World Development* 25 (5), 639–656.

Bauer, C. (1998) Slippery property rights: multiple water uses and the neoliberal model in Chile, 1981–1995. *Natural Resources Journal* 38, 110–155.

Beccar, L., Boelens, R. and Hoogendam, P. (2002) Water rights and collective action in community irrigation. In: Boelens, R. and Hoogendam, P. (eds) *Water Rights and Empowerment*. Van Gorcum, Assen, Netherlands, pp. 1–21.

Bennet, V., Dávila-Poblete, S. and Nieves Rico, M. (eds) (2005) *Opposing Currents. The Politics of Water and Gender in Latin America*. University of Pittsburgh Press, Pittsburgh, Pennsylvania.

Boelens, R. (2006) Local rights and legal recognition: the struggle for indigenous water rights and the cultural politics of participation. In: Boelens, R., Chiba, M. and Nakashima, D. (eds) *Water and Indigenous Peoples*. WALIR-UNESCO, UNESCO, Paris, pp. 38–45.

Boelens, R. and Dávila, G. (eds) (1998) *Searching for Equity. Conceptions of Justice and Equity in Peasant Irrigation*. Van Gorcum, Assen, Netherlands.

Boelens, R. and Doornbos, B. (2001) The battlefield of water rights. Rule making amidst conflicting normative frameworks in the Ecuadorian highlands. *Human Organization* 60 (4), 343–355.

Boelens, R. and Hoogendam, P. (eds) (2002) *Water Rights and Empowerment*. Van Gorcum, Assen, Netherlands.

Boelens, R. and Zwarteveen, M. (2005) Anomalous water rights and the politics of normalization. Water control and privatization policies in the Andean region. In: Roth, D., Boelens, R. and Zwarteveen, M. (eds) *Liquid Relations. Contested Water Rights and Legal Complexity*. Rutgers University Press, New Brunswick, New Jersey and London.

Boelens, R., Roth, D. and Zwarteveen, M. (2002) Legal complexity and irrigation water control: analysis, recognition and beyond. Paper for the *International Congress of Legal Pluralism*, 7–10 April 2002, Chiang Mai, Thailand.

Boelens, R., Gentes, I., Guevara, G., Armando, J. and Urteaga, P. (2005) Special law: recognition and denial of diversity in Andean water control. In: Roth, D., Boelens, R. and Zwarteveen, M. (eds) *Liquid Relations. Contested Water Rights and Legal Complexity*. Rutgers University Press, New Brunswick, New Jersey and London.

Bruns, B.R. and Meinzen-Dick, R.S. (eds) (2000) *Negotiating Water Rights*. Vistaar Publications, New Delhi, India and IT Publications, London.

Bustamante, R. (2002) *Legislación del Agua en Bolivia*. WALIR Research Vol. 2, Centro Agua, UN/CEPAL and Wageningen University, Wageningen University, Wageningen, Netherlands.

Bustamante, R. (2006) *Normas Indígenas y Consuetudinarias Sobre la Gestión del Agua en Bolivia*. WALIR Research Vol. 10, Centro Agua, UN/CEPAL and Wageningen University, Wageningen University, Wageningen, Netherlands.

Bustamante, R., Butterworth, J., del Callejo, I., Duran, A., Herbas, D., Hillion, B., Reynaga, M. and Zurita, G. (2004) Multiple sources for multiple uses: Household case studies of water use around Cochabamba, Bolivia. Available at http://www.irc.nl/content/view/full/8031

Bustamante, R., Peredo, E. and Udaeta, M.E. (2005) Women and the 'water war' in the Cochabamba valleys. In: Bennett, V., Dávila-Poblete, S. and Rico, M.N. (eds) *Opposing Currents. The Politics of Water and Gender in Latin America*. University of Pittsburgh Press, Pittsburgh, Pennsylvania, pp. 72–90.

Castro, M. (2002) Local norms and competition for water in Aymara and Atacama communities, Northern Chile. In: Boelens, R. and Hoogendam, P. (eds) *Water Rights and Empowerment*. Van Gorcum, Assen, Netherlands, pp. 187–201.

CONAIE (the Confederation of Indigenous Nationalities of Ecuador) (1996) *Propuesta Ley de Aguas*. CONAIE, Quito.

de Vos, H., Boelens, R. and Bustamante, R. (2006) Formal law and local water control in the Andean region: a fiercely contested field. *International Journal for Water Resources Development* 22 (1), 37–48.

Dourojeanni, A. and Jouravlev, A. (1999) El Código de Aguas en Chile: entre la ideología y la realidad. *Serie Recursos Naturales e Infraestructura* No. 3, CEPAL, Santiago de Chile.

Duran, A., Herbas, D., Reynaga, M. and Butterworth, J. (2006) *Planning for Multiple Uses of Water: Livelihood Activities and Household Water Consumption in Peri-urban Cochabamba. Bolivia*. Centro AGUA, Cochabamba, Bolivia.

ECLAC (United Nations Economic Commission for Latin America and the Caribbean) (2005) Circular No. 22 of the Network for Cooperation in Integrated Water Resource Management for Sustainable Development in Latin America and the Caribbean, August 2005. ECLAC, Santiago de Chile.

Escobar, A. (1995) *Encountering Development. The Making and Unmaking of the Third World*. Princeton University Press, Princeton, New Jersey.

Flores Galindo, A. (1988) *Buscando un Inca. Identidad y Utopia en los Andes*. Editorial Horizonte, Lima.

Foucault, M. (1978) *Discipline and Punish: the Birth of the Prison*. Pantheon Books/Random House, New York.

Gelles, P.H. (1998) Competing cultural logics: state and 'indigenous' models in conflict. In: Boelens, R. and Dávila, G. (eds) *Searching for Equity*. Van Gorcum, Assen, Netherlands, pp. 256–267.

Gelles, P.H. (2000) *Water and Power in Highland Peru: the Cultural Politics of Irrigation and Development*. Rutgers University Press, New Brunswick, New Jersey.

Gelles, P.H. (2006) Indigenous peoples, cultural identity, and water rights in the Andean nations. In: *Agua y Derecho. Políticas Hídricas, Derechos Consuetudinarios e Identidades Locales*. WALIR – IEP, Lima.

Gelles, P.H. and Boelens, R. (2003) Water, community and identity: the politics of cultural and agricultural production in the Andes. In: Salman, T. and Zoomers, A. (eds) *Imaging the Andes: Shifting Margins of a Marginal World*. Aksant, Amsterdam.

Gentes, I. (2002) *Estudio de la Legislación Oficial Chilena y del Derecho Indígena a los Recursos Hídricos*. WALIR Research Vol. 2, CEPAL and Wageningen University, Santiago de Chile.

Gentes, I. (2005) *Estudio Sobre Marcos Normativos Indígenas y Consuetudinarios Referente a la Gestión del Agua en Chile*. WALIR Research Vol. 5, CEPAL and Wageningen University, Santiago de Chile.

Gerbrandy, G. and Hoogendam, P. (1998) *Aguas y acequias*. PEIRAV – Plural Editors, Cochabamba, Bolivia.

Getches, D. (2006) *Indigenous Rights and Interests in Water under USA Law*. WALIR Research Vol. 6, University of Colorado at Boulder, Colorado, UN/CEPAL and Wageningen University. Wageningen University, Wageningen, Netherlands.

Guevara, G. and Armando, J. (2006) Official water law versus peasant and indigenous rights in Peru. In: Boelens, R., Chiba, M. and Nakashima, D. (eds) *Water and Indigenous Peoples*. WALIR-UNESCO, UNESCO, Paris.

Guevara, A., Vera, J., Urteaga, P., Vera, I. and Zambrano, G. (2002) *Estudio de la Legislación Oficial Peruana sobre la Gestión Indígena de los Recursos Hídricos*. WALIR Research Vol. 2, Wageningen University, Wageningen, Netherlands and UN CEPAL/ECLAC, Santiago de Chile, pp. 101–131.

Hale, C.R. (2002) Does multiculturalism menace? Governance, cultural rights and the politics of identity in Guatemala. *Journal of Latin America Studies* 34 (3), 485–524.

Hendriks, J. (1998) Water as private property. Notes on the case of Chile. In: Boelens, R. and Dávila, G. (eds) *Searching for Equity*. Van Gorcum, Assen, Netherlands, pp. 297–310.

Hendriks, J. (2004) Legislación de aguas y gestión de sistemas hídricos en la región Andina. WALIR Position Paper, *International Workshop on Collective Rights, Local Water Management and National Legislation*, October 2004, WALIR, Quito.

Larson, B. (1992) *Colonialismo y Transformación Agraria en Bolivia – Cochabamba 1500–1900*. CERES/HISBOL, La Paz.

Lynch, B.D. (1993) The bureaucratic tradition and women's invisibility in irrigation. *Proceedings of the 24th Chacmool Conference*, University of Calgary Archeological Association, Alberta, Canada, pp. 333–342.

Mayer, E. (2002) *The Articulated Peasant: Household Economies in the Andes*. Westview Press, Boulder, Colorado.

Oré, M.T. (1998) From agrarian reform to privatization of land and water: the case of the Peruvian coast. In: Boelens, R. and Dávila, G. (eds) *Searching for Equity*. Van Gorcum, Assen, Netherlands, pp. 268–278.

Oré, M.T. (2005) *Agua, Bien Común y Usos Privados. Riego, Estado y Conflictos en La Achirana del Inca*. Wageningen University, Wageningen, Netherlands and WALIR and PUCP, Lima.

Pacari, N. (1998) Ecuadorian water legislation and policy analysed from the indigenous-peasant point of view. In: Boelens, R. and Dávila, G. (eds) *Searching for Equity*. Van Gorcum, Assen, Netherlands, pp. 279–287.

Palacios, P. (2002) *Estudio Nacional de la Legislación Oficial y los Marcos Normativos Consuetudinarios Referente a la Gestión Indígena de los Recursos Hídricos*. WALIR, CEPAL, Lima and Wageningen University, Wageningen, Netherlands.

Palacios, P. (2003) *Estudio Sobre Marco Normativos Indígenas y Consuetudinarios*. WALIR Research Vol. 2, CEPAL, Quito and Wageningen University, Wageningen, Netherlands.

Patterson, T. (1991) *The Inka Empire: the Formation and Disintegration of a Precapitalist State*. Berg Press, New York.

Peña, F. (ed.) (2004) *Los Pueblos Indígenas y el Agua: Desafíos del Siglo XXI*. El Colegio de San Luis, WALIR, IMTA. Obranegra Editores, Mexico D.F. and Bogotá.

Roth, D. (2003) Ambition, regulation and reality. Complex use of land and water resources in Luwu, South Sulawesi, Indonesia. PhD dissertation, Wageningen University, Wageningen, Netherlands.

Roth, D., Boelens, R. and Zwarteveen, M. (eds) (2005) *Liquid Relations. Contested Water Rights and Legal Complexity*. Rutgers University Press, New Brunswick, New Jersey and London.

Salman, T. and Zoomers, A. (eds) (2003) *Imaging the Andes: Shifting Margins of a Marginal World*. CEDLA, Aksant, Amsterdam.

Schaffer, B. and Lamb, G. (1981) *Can Equity Be Organized? Equity, Development Analysis and Planning*. Institute of Development Studies, Sussex University, Brighton, UK.

Solón, P. (2003) *La Sangre de la Pachamama*. Documentary. Fundación Solón, La Paz.

Stavenhagen, R. (1994) Indigenous rights: some conceptual problems. In: Assies, W. and Hoekema, A. (eds) *Indigenous Peoples. Experiences with Self-Government*. University of Amsterdam and IWGIA, Copenhagen.

Stavenhagen, R. and Iturralde, D. (eds) (1990) *Entre la ley y la Costumbre. El Derecho Consuetudinario Indígena en América Latina*. Instituto Indigenista Interamericano e Instituto Interamericano de Derechos Humanos, Mexico City.

Urteaga, P., Vera, I. and Guevara, A. (2003) *Estudio Sobre las Reglas y Regulaciones Indígenas y Consuetudinarias para la Gestión de los Recursos Hídricos en el Perú* (draft version). WALIR, Wageningen University, Wageningen, Netherlands and UN CEPAL, Lima/Santiago.

Van Cott, D.L. (2000) *The Friendly Liquidation of the Past: the Politics of Diversity in Latin America*. Pitt Latin American Series, Pittsburg University Press, Pennsylvania.

van Kessel, J. (1992) *Holocausto al Progreso: Los Aymarás de Tarapacá*. Hisbol, La Paz.

van Koppen, B. and Jha, N. (2005) Redressing racial inequities through water law in South Africa: interaction and contest among legal frameworks In: Roth, D., Boelens, R. and Zwarteveen, M. (eds) *Liquid Relations. Contested Water Rights and Legal Complexity*. Rutgers University Press, New Brunswick, New Jersey.

van der Ploeg, J.D. (2003) *The Virtual Farmer*. Van Gorcum, Assen, Netherlands.

Vera, J. (2004) 'Cuanto más doy, más soy …' Discursos, normas y género: la institucionalidad de las organizaciones de riego tradicionales en los Andes del sur peruano. In: Peña, F. (ed.) *Los Pueblos Indígenas y el Agua: Desafíos del Siglo XXI*. El Colegio de San Luis, WALIR, IMTA. Obranegra Editores: Mexico D.F. and Bogotá.

von Benda-Beckmann, F. (1996) Citizens, strangers and indigenous peoples: conceptual politics and legal pluralism. *Law and Anthropology* 9, 1–43.

von Benda-Beckmann, F., von Benda-Beckmann, K. and Spiertz, J. (1998) Equity and legal pluralism: taking customary law into account in natural resource policies. In: Boelens, R. and Dávila, G. (eds) *Searching for Equity*. Van Gorcum, Assen, Netherlands, pp. 57–69.

Vos, J. (2002) Metric matters. Water control in large-scale irrigation in Peru. PhD thesis, Wageningen University, Wageningen, Netherlands.

WALIR (Water Law and Indigenous Rights Program) (2002) *Indigenous Water Rights, Local Water Management, and National Legislation*. WALIR Research Vol. 2, CEPAL-United Nations, Santiago de Chile and Wageningen University, Wageningen, Netherlands.

WALIR (Water Law and Indigenous Rights Program) (2003) *Análisis de la Situación del Riego en la República del Ecuador*. Misión de Consultoría (Hendriks, Mejía, Olazával, Cremers, Ooijevaar, Palacios), CONAM-BID, Quito.

Yrigoyen, R. (1998) The constitutional recognition of indigenous law in Andean countries. In: Assies, W., Van der Haar, G. and Hoekema, A. (eds) *The Challenge of Diversity*. Thela Thesis Publishers, Amsterdam.

Zwarteveen, M. and Bennet, V. (2005) The connection between gender and water management. In: Bennett, V., Dávila-Poblete, S. and Rico, M.N. (eds) *Opposing Currents. The Politics of Water and Gender in Latin America*. University of Pittsburgh Press, Pittsburgh, Pennsylvania, pp. 13–29.

Zwarteveen, M. and Boelens, R. (2006) Rights, meanings and discourses. Gender dimensions of water rights in diverging regimes of representation in the Andes. In: Lahiri-Dutt, K. (ed.) *Fluid Bonds: Views on Gender and Water*. Stree, Kolkata, India.

7 Water Rights and Rules, and Management in Spate Irrigation Systems in Eritrea, Yemen and Pakistan

Abraham Mehari,[1*] Frank van Steenbergen[2**] and Bart Schultz[1,3***]

[1]UNESCO-IHE, Netherlands; *e-mails: abrahamhaile2@yahoo.com; a.meharihaile@unesco-ihe.org; [2]MetaMeta Research, 's-Hertogenbosch, Netherlands; **e-mail: fvansteenbergen@metameta.nl; [3]Civil Engineering Division, Rijkswaterstaat, Utrecht, Netherlands; ***e-mail: b.schultz@unesco-ihe.org

---

Abstract

Spate irrigation is a system of harvesting and managing flood water. In spate irrigation, flood water is emitted from *wadis* (ephemeral streams) and diverted to fields using earthen or concrete structures. By nature, flood water is unpredictable in occurrence, timing and volume, which puts special challenges to the farmers who use, co-share and co-manage the resource. Primarily based on the research conducted in spate irrigation systems in Eritrea, Yemen and Pakistan, this chapter discusses the interlinkage between local flood water management and water rights and rules, and the enforcement mechanisms in place. It assesses how formal national/provincial land and water laws affect local flood water management and argues that what matters most are the local rules for cooperation and sharing the resource and, hence, that formal water and land rights for spate irrigation should recognize local water rights and management.

Keywords: customary practices, enforcement, flood water management, irrigation management transfer, local organizations, spate irrigation, water rights and rules, Eritrea, Yemen, Pakistan.

Introduction

This chapter describes the water rights and rules in spate irrigation and discusses their role in water management. There are three ways in which this chapter contributes to the central theme of this volume. The first is by analyzing the complexity and robustness of local water rights. Spate irrigation water rights, which are different from perennial irrigation water rights, are not fixed quantities or entitlements. Instead, they are operating rules that respond to a variety of circumstances, which are at the core of spate irrigation. We emphasize this point to move away from naive and simplistic understanding of formal water rights, where water rights are seen as mechanisms to create distinctive ownership. In this naive understanding – that can be traced back to the work of Douglas North on early land rights (North and Thomas, 1977) and the subsequent work in the field of New Institutional Economics – property rights are seen as the main institution to claim entitlements.

At policy level, water rights reform is often simplified as the intervention that will either help protect weaker interests on the strength of the property claim or, alternatively, help achieve better economic efficiency by facilitating trade and exchange of rights. The point made in this chapter is that water rights in spate irrigation (as in other fields of water management) are inseparable from the way water management is organized and that the rights are part of a bundle of responsibilities to the common group. Water rights are not something that precedes water management or can be used in isolation to change water management and water distribution.

The second way this chapter contributes to the central theme of this volume is by recognizing that water rights and water allocation in spate irrigation rules differ between societies, although there are also cross-cutting similarities. In this chapter we hope to provide some examples from Eritrea, Yemen and Pakistan (see Fig. 7.1). It is important to understand not only that water rights are the product of the resource system (the spate irrigation system) but that there are higher forces at work (e.g. the presence of politically

and financially influential farmers) that determine what rules and rights have to be implemented.

The last is by discussing how water rights change in the course of developing infrastructure, particularly in spate irrigation. Rights relate very much to operational rules, and these rules change with changing infrastructure – with different possibilities for upstream control and different common maintenance requirements.

This chapter is divided as follows. First, it discusses the different operational rules and practices – giving examples from different societies. Then, it discusses the way local organizations and institutions have enforced (with various degrees of effectiveness) these water rights and rules, and have even tried to codify them. Next, it discusses how some of the water rights and rules have changed over the past decades under the influence of particular external investment programmes. To start with, however, we want to describe briefly what spate irrigation is.

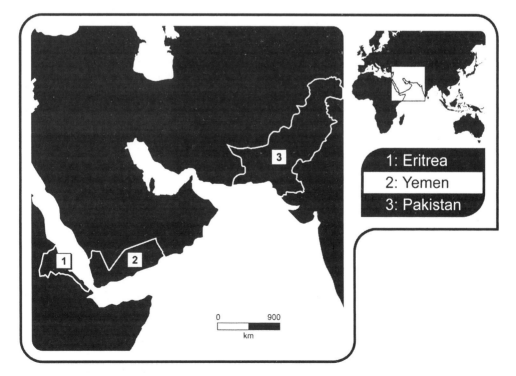

Fig. 7.1. Map study areas.

Spate Irrigation

Spate irrigation is a resource system, whereby flood water is emitted through normally dry wadis and conveyed to irrigable fields. It is a pre-planting system, where the flood season precedes the crop production period. In most spate irrigation systems in Eritrea, Yemen and Pakistan the major floods occur between June and September, which is the time of heavy rainfall in upper catchments; and crop growth takes place between October and February exclusively, depending on the water stored in the soil. To establish a spate irrigation system, there should be a mountainous or hilly topography that generates run-off; and adjacent, low-lying fields with deep soils able to store ample moisture for the crops during periods of no precipitation (Mehari et al., 2005).

Spate irrigation systems support livelihoods of often the poorest segments of the rural population in the Middle East, west Asia and North and East Africa (van Steenbergen, 1997). An estimate of the land coverage of spate irrigation systems in some countries, which the authors compiled from various sources, is presented in Table 7.1. Apart from the names of countries listed in Table 7.1, the existence of spate irrigation is reported in Chile, Bolivia, Iran, Afghanistan, Mauritania, Senegal, Ethiopia and Kenya; but there is no reliable estimate of its land coverage.

In spate irrigation systems uncertainty is a given. The unpredictability in timing, volume and sequence of flood water is the main cause of uncertainties and risks in crop production under spate irritation systems. It can also, in theory, confuse cooperation and create a free-for-all competition. Water rights and water distribution rules in spate irrigation, however, regulate access to water and – when enforced – minimize conflict. Water rights and water distribution rules also define the likelihood of irrigation for different areas and, hence, serve as the key to the collective maintenance and rebuilding of diversion infrastructure. In particular, where flood water users depend on one another for maintaining flood channels and (re)constructing diversion structures, and this work is substantial, agreement on how water is distributed is a precondition for cooperation. Water distribution rules will also make it easier to predict which land will be irrigated. As such, they encourage pre-flooding land preparation, which is important for adequate water storage and moisture conservation and key to high yields.

Water Rights and Rules in Managing Unpredictable Flood Water

To manage the unpredictable nature of flood water and reduce the risk of conflicts, several categories of water rights and rules are in place in different spate irrigation systems. The most common and widely applied rights and rules (Mehari et al., 2003; van Steenbergen, 2004) relate to the following:

- Demarcation of land that is entitled to irrigation.

Table 7.1. Spate-irrigated versus total irrigated area in selected countries.

Country	Year of data collection	Total irrigated area (ha)	Spate-irrigated area (ha)	Total irrigated area covered by spate irrigation (%)
Algeria	1997	560,000	70,000	13
Eritrea	2005	28,000	15,630	56
Libya	1997	470,000	53,000	11
Mongolia	1993	84,300	27,000	32
Morocco	1997	1,258,200	165,000	13
Pakistan	2005	17,580,000	1,450,000	8
Somalia	1984	200,000	150,000	75
Sudan	1997	1,946,000	280,000	14
Tunisia	1997	481,520	98,320	20
Yemen	2003	485,000	193,000	40

- Breaching of bunds.
- Proportion of the flood water going to different canals and fields.
- Sequence in which the different canals and fields are irrigated.
- Depth of irrigation that each field is entitled to receive.
- Access to second (and third) water turns.
- Distribution of large and small floods.

These categories of water rights and rules are discussed below, with some relevant illustrative examples from Eritrea, Yemen and Pakistan.

Rights and rules on land demarcation

Demarcation rights and rules are common in the lowland spate-irrigated areas in Eritrea, Yemen and Pakistan, where water is scarce and land is abundant; yet, they are almost non-existent in the central highlands of the countries where water is relatively more plentiful than land. Demarcation rights and rules define the boundary of the area entitled to irrigation and set priorities to access to water depending on the year of establishment of the different fields. Instead of merely regulating seasonal water supplies, the demarcation rules also predict what will happen when changes in the entire system occur. Spate systems are dynamic. Among others, changes in the course of rivers, breaching, silting up or scouring of canals and rising of fields above irrigable command levels are frequent and can occur on a yearly basis. Demarcation rules are conservative because, in the wake of these changes, they try to re-establish the prior situation. They often protect the prior rights of downstream landowners by restricting or even prohibiting new land development upstream, which could have resulted in the diversion of flood water to new territories and a redefinition of the group of shareholders.

To cite an example: in the Wadi Laba, Eritrea, about 1400 ha (besides the annually irrigated 2600 ha) were distributed in 1993 in the upstream Sheeb-Kethin area. The concerned farmers were, however, clearly informed that they would have to abide by the demarcation rule: new fields could be allocated water only after all the previously established fields had received the quantity of water granted to them by the other various rules. Due to the strict adherence to this rule, only 50 ha of the 1400 have been established so far and the water right of downstream farmers has been preserved. In Eritrea, fields are considered to be fully established when they accumulate a minimum depth of about 10 cm of alluvial sediments. With a mean annual sediment deposition of about 3 cm, this would require at least three flood seasons.

Rights and rules on breaching of bunds

Rights and rules concerning the breaching of the bunds of diversion and distribution structures and fields are widely applied in areas where the entire river bed is blocked by earthen bunds, and access of water to downstream canals and fields depends on the breaking of these immediate upstream structures. In many cases, the earthen and brushwood bunds are constructed in such a way that they breach during large flood (> 100 m^3/s) events. This prevents damage to many upstream structures and fields while increasing the probability of irrigation of the downstream fields.

In several spate irrigation systems in Eritrea, Yemen and Pakistan there are rules on when farmers can breach bunds: for instance, once the area served by an upstream bund is fully irrigated or when a certain period of the flood season has lapsed. Boxes 7.1 and 7.2 present examples of some of such rules from Eritrea and Pakistan, respectively.

Rights and rules on flood water division

The rights and rules on flood water division guide the distribution of water among different canals. In the indigenous systems in Eritrea, both proportional and rotational distributions of flood water are practised among the main and branch canals. During medium (25–50 m^3/s) and medium–large (25–100 m^3/s) floods, proportional distribution is used. This has a dual purpose. First, it irrigates two or more different areas at the same time. Second, by dividing the flow, it minimizes collateral damages such as destruction of structures and

---

Box 7.1. Rights and rules on breaching bunds in the Wadi Laba and Mai-ule, Eritrea, established in 1900 (from authors' survey, 2003).

- In July and August, the peak flood months, if the large floods do not breach the upstream *agims* and *musghas* (diversion and distribution structures, respectively), the upstream farmers have the obligation to allow the downstream farmers to breach them purposely to allow the flow of water to their fields. July and August floods are considered to be rich in nutrients, and all farmers are entitled to have a share. It is the responsibility of both the downstream and upstream farmers to maintain the structures to increase the probability of diverting the next flood(s).
- In September, where floods are assumed to be low in nutrients and marginally important for crop production, the upstream farmers are not obliged to allow the breakage of their bunds.
- If an upstream field receives an irrigation depth up to knee height (about 50 cm – see rule on depth of irrigation), the landowner of the immediate downstream field has the right to breach the relevant bund and irrigate his field. If the downstream field holder is not on site during the irrigation period, the upstream farmer is not obliged to breach his bund.

Box 7.2. Rights and rules on the Nari system, Kacchi, Pakistan, prepared in 1917 on revision of the old rules (from authors' compilation, 2004).

- From 10 May to 15 August, the landowners of the Upper Nari are allowed to make *gandas* (earthen bunds) in the Nari river.
- When the land served by one ganda in the Upper Nari is fully irrigated, the landowners in that ganda must allow landowners of the next ganda to breach it.
- After 15 August, the landowners of the Lower Nari are allowed to make a ganda in the Nari river. Landowners in the Upper Nari are not allowed to irrigate their land during this period or let the water go waste. Water is not allowed to go waste to the low-lying areas east and west of the Nari river. Guide bunds will prevent water flowing to these areas. All landowners will contribute towards these bunds, with farmers in the Lower Nari paying twice the amount per hectare in case bunds on the upper Nari are broken.

---

erosion of field bunds. During small and small–medium floods (< 25 m³/s), rotational distribution is the choice. The flow of these floods, if divided, may not have the strength to reach the most upstream fields. The rights and rules in the Nari system in Pakistan are given in Box 7.2.

In many indigenous spate irrigation systems, flow division is made flexible in order to adjust to changing bed levels of the wadi and the canals, and to variations of the flow. One example of a flexible flow division is the Wadi Laba indigenous distribution structure. The structure is constructed from earthen material. Its downstream section is reinforced with brushwood that can be easily moved in and outwards to change its orientation as needed. The structure divides the flow from the wadi to two main canals – Sheeb-Kethin and Sheeb-Abay. The management of the structure is the sole responsibility of the farmer leaders of the five main canals in Laba. Prior to each anticipated flood event, all five leaders gather on the site. Taking

into account the size of the different areas irrigated in the previous floods, they make a collective decision on how to adjust the structure so that the flows to each area are fair.

Rights and rules on sequence

The rights and rules on sequence supplement the rights and rules on the division of flood water. They describe the route that water follows within the area entitled to irrigation by clearly spelling out which main and branch canals have priority right to water, and which fields are entitled to receive water first. The sequence usually adjusts to the level of the floods. In the indigenous Wadi Laba and Mai-ule spate irrigation systems of Eritrea, the underlining rule is: upstream canals and fields have absolute priority right over small, small–medium and medium floods; and the downstream canals and fields have an equal priority rights over medium–large and large floods. This rule has created a perception of fairness of water distribution among the farmers and strengthened the degree of cooperation between them. Most of the indigenous structures are constructed from earthen and brushwood materials. They are susceptible to frequent destruction by flood water. The downstream and upstream farmers depend on one another for timely maintenance of the structures.

In the indigenous spate irrigation systems in the Tihama Plain, Yemen, the fundamental sequence rule, locally called *al aela fil aela* (this Arabic phrase, when literally translated, means 'the top is always at the top'; in this case, at the top of the list to obtain water) grants an absolute priority right to the upstream farmers regardless of the size of the flow. The downstream farmers are not, however, denied the right to surplus water after the upstream farmers have withdrawn a sufficient quantity of water in accordance with their right. This rule might seem very unfair to the downstream farmers and might give the impression that the upstream farmers have been utilizing almost all the flood water. That has not usually been the case. The indigenous structures have frequently been breached by large floods providing ample water to the downstream farmers, which in some years was more than the quantity of water received by the upstream.

Rules on depth of irrigation

The rules on depth of irrigation are not common in spate-irrigated areas in Pakistan, but are standard practices in Eritrea and Yemen where the field-to-field water distribution system is practised. In this distribution system, a farmer takes his turn as soon as his neighbour completes the inundation of his land. He does so by breaking a relevant section of the bund surrounding the field of the upstream landowner. In this practice, fierce competition usually arises among neighbours, which in many cases leads to conflicts. Probably, the rules on water depth were introduced mainly to mitigate such conflicts. In contrast, when each field (usually of very large size) is fed by its own separate intake, as is the case in many spate irrigation systems in Pakistan, such conflicts are

rare, which might be the reason why the rules on the depth of inundation are unusual.

The rules on depth of irrigation could be viewed as complementary to the rights and rules on sequence because they quantify the amount of water a certain field could receive during its turn. In Eritrea (Wadi Laba and Mai-ule) and Yemen (the Tihama Plain), the rule on irrigation depth states that each field is entitled to a depth of a knee height (about 50 cm) at each turn. When the rule was first introduced 100 years ago, the farmers attempted to ensure its implementation by limiting the height of the field bunds to around 50 cm. With time, however, this became impractical. The sediments deposited in the fields are the only sources for maintaining the field bunds. Nevertheless, the degree of damage done to the bunds is not the only factor that determines the amount of sediments to be removed from the fields. Even when there is no maintenance work to be done, certain quantities of sediments need to be removed from some fields in order to keep the field level within that of the irrigable command area of the concerned structures and canals. The excavated sediments are re-deposited in the only convenient disposal places – the field bunds. This has resulted in irregularities in the height of many field bunds. In Wadi Laba and Mai-ule, and in the Tihama, the height of field bunds ranges from 0.30 to 1.0 m.

The farmers explained that the rule on breaking bunds, when first introduced a little over a 100 years ago, referred only to the breaking of the bunds of the diversion and division structures. It was only 10 years later that it was modified to include the breaking of field bunds, when the farmers realized that it was impractical to standardize and limit the maximum height of field bunds to 0.50 m.

Rules on second turns

Although several crops, such as sorghum, wheat and cotton, can survive on one turn of water application, they yield significantly higher returns when irrigated more than once. In the case of sorghum, which is the main crop in Wadi Laba and Mai-ule systems in Eritrea, the farmers informed that with one, two and three irrigation turns they could harvest a maximum of 1, 2.5 and 3.5 t/ha, respectively. Hence, to ensure that the majority of the fields receive at least one turn, thus guaranteeing that most of the households earn the minimum possible yield of food crops, a rule was introduced in the 1920s that defined the access to second turns. This rule states that, regardless of its location, the type of crop grown in it and the social and economic status of its owner, a field is allowed a second turn only after all the other fields that are entitled to irrigation (in line with the rule on demarcation) have received one turn. This rule has, however, some practical shortcomings. The degree to which it is possible to honour it depends on the size of the flood. If the floods are small with no strength to reach the dry fields (especially under the prevailing field-to-field system), the only option would be to apply them to the area, which is already irrigated.

In Wadi Tuban, Yemen and Rod Kanwah, Pakistan, the rules on second turns are different from those in Wadi Laba and Mai-ule; they limit the access to second turns only for the most important subsistence crops – wheat in Pakistan and red sorghum in Yemen.

Rules on large and small floods

Finally, the water distribution may differ according to the size of the floods. One example given is the automatic flow division when floods are large and able to breach the bunds in the various flood channels. In other systems there are explicit rules on how to accommodate small and larger floods. Small floods tend to be diverted to the upper sections of the command area, if only because small floods are not likely to travel that far. A rare example of explicit rules dealing with floods of different sizes concerns the Irrigation Plan for Wadi Tuban in Yemen (see Box 7.3).

Enforcement of Water Rights and Rules

The type of enforcement strategies and the degree to which the water rights and rules can be enforced vary, depending mainly on the social structure of the communities and the level of the overall governance in the area. In the spate systems in Eritrea, Yemen and Pakistan, the enforcement of water rights and rules can be related to the following three factors:

- Local organizations and institutions.
- Relationship between water rights and rules, and maintenance.
- Codification.

Local organizations and institutions

For 600 years until the 1970s, the enforcement of the water rights and rules in many spate systems in Yemen had been the responsibility of the local *Sheikhs al-wadis* who were appointed by, and who worked under, the direct and strict instructions of the local Sultans. Sheikhs in Arabic usually refers to religious leaders. In this case, however, Sheikhs means chiefs, who may or may not have any religious ranks. Hence, Sheikhs al-wadis refers to 'chiefs of the wadis'. 'Sultans' is also an Arabic word and, as used here, means roughly 'supreme leaders'.

Many communities comprising several tribes in the Tihama Plain, Yemen, had depended on spate irrigation for their livelihood. The Sheikhs and Sultans who had the leading role in the enforcement of the water rights and rules always belonged to the tribe that had the largest number of members, the most powerful in terms of material and capital wealth and believed to be the most native in the area. Sheikhs and Sultans were very respected and feared leaders. Their leadership was passed to the eldest son on a hierarchical basis. In the Muslim spate irrigation communities in Yemen, a female had no right to be a Sultan or a Sheikh.

In Yemen, there were no other people or institutions that could challenge the ruling of the Sultans and Sheikhs regarding the implementation of the local water rights and rules. They had the final word, which all members of all the tribes within the concerned communities had to abide by, either willingly or unwillingly. Many of the interviewed elderly farmers in Wadi Tuban, Zabid, Mawr and Siham explained that the Sheikhs and Sultans were authoritarian, but gave them credit for their effectiveness in safeguarding the water rights of the downstream

---

Box 7.3. Water allocation rules for Wadi Tuban, Yemen (from authors' survey, 2004).

To ensure efficient use of spate water, irrigation is planned as follows:
- When the spate flow is small (5–15 m^3/s), priority is given to the canals in the upper reach of the wadi.
- When the spate flow is moderate (15–25 m^3/s), priority is given to canals in the middle reach of the wadi.
- When the spate flow is large (25–40 m^3/s), the flow is directed to either Wadi Kabir or Wadi Saghir in the lower reach of the delta, depending on which one has the right to receive the spate water.
- When the spate flow exceeds 40 m^3/s, the flow is divided equally between Wadi Kabir and Wadi Saghir.

---

farmers. To exemplify, in Wadi Tuban, Yemen, the Sheikh-al-wadi had the full power to impose sanctions on upstream farmers who took water in violation of the rules and/or without his permission. The sanctions, which were frequently applied upon approval by the Sultan, included the following:

- The farmers concerned were not allowed to grow any crop on their fields, and the immediate downstream farmers had the right to grow crops on the irrigated fields of their upstream neighbours.
- If crops were already being cultivated, the yields had to be given to the immediate downstream farmers.

The interviewed farmers informed us that, due mainly to the high degree of heterogeneity in the level of power of the tribes, conflicts in the Tihama Plain were very intense and serious. The Sultans and Sheikhs were not able to prevent the occurrence of such conflicts, but they were often successful in settling them.

Following huge investments in the 1970s in structurally modernizing the indigenous spate irrigation systems in Yemen in general and in the Tihama Plain in particular, and the introduction of formal government rules and the collectivization of agriculture in south Yemen, the task of managing the spate irrigation systems was transferred from the Sultans and Sheikhs to government employees and staff in agricultural cooperatives who, over the years, had to face reduced funding inflows and erosion of authority. The majority of the interviewed farmers also spelled out that, after the reunification of southern and northern Yemen, the central government further diminished the role of the cooperatives without putting in place an alternative institution that could better handle the spate irrigation management, effectively creating a governance vacuum. Al-Eryani

and Al-Amrani (1998), in support of this assertion, stated that due to the decline in the role of the cooperatives in the management of spate irrigation systems, a worrying vacuum was left that resulted in more conflicts between the upstream and downstream users.

The social structure of the Wadi Laba and Mai-ule communities in Eritrea differed significantly from that of the Tihama communities in Yemen. The Wadi Laba and Mai-ule communities did not comprise a dominant tribe and had no Sultans or Sheikhs with absolute authority to enforce water rights and rules. Almost all members of the communities in the Wadi Laba and Mai-ule were largely homogenous in terms of land ownership, and material and capital wealth. Each of their landholdings ranged from 0.5 to 2.0 ha, with the majority of the households owning 1 ha. Nearly all were poor, living from hand to mouth.

For 100 years, till 2001, the authority of enforcing the water rights and rules in the Wadi Laba and Mai-ule was shared among the farmers' organization and the government institutions – the local administration and the local Ministry of Agriculture. The farmers' organization came into being around the 1900s and its key players were the *Teshkil* (plural: Teshakil), *Ternafi* (plural: Ternefti) and *Abay-Ad* (village elders). Teshkil is a local term that means a 'subgroup leader'. The Teshkil commanded a group of 20 to 40 farmers who usually irrigated through one branch canal. The Teshkil was responsible for implementing all the water rights and rules that applied to the farmers within his command. It was only on his request or on the request of a group of farmers unsatisfied with his judgement in, for example, resolving some conflicts, that the respective Ternafi could interfere. Ternafi is also a local term that refers to a 'group leader'. The Ternafi had the authority to enforce rules and rights that governed the sharing of water among two or more groups of farmers led by a Teshkil.

When conflicts arose between upstream and downstream farmers due to, for instance, the improper location and/or adjustment of a certain structure, and the Ternafi failed to satisfactorily solve them, he could request the Abay-Ad as a first step and the local administration as the last chance for mediation. The Abay-Ad were a group of old men widely respected for their skill and impartiality in solving conflicts. Two or more Teshakil could also make the same request if the Ternafi did not do so. In solving conflicts, the local administration visited the site with experts from the local Ministry of Agriculture and gave a verdict, which was final and binding.

The concerned farmers elected the Teshakil and Ternefti. There was no time limit on the number of terms and years they could serve. If most farmers concluded that they were not performing well, however, they could remove them from their power by a simple majority vote. As was the case in Yemen, in the Muslim communities in the Wadi Laba and Mai-ule females were not allowed to have any leadership position or to participate in any decision making in issues that affected the water management in spate irrigation systems. The cultural and social beliefs that led to such a restriction in women's participation are still in place.

Unlike the Sultans and Sheikhs, the Ternefti and Teshakil had no power to impose harsh sanctions against those who violated the rules. Nevertheless, the farmers' organizations in the Wadi Laba and Mai-ule were able to successfully enforce the water rights and rules, protect the rights of the downstream farmers and minimize conflicts. Among the factors that led to this achievement are: (i) the existence of the homogenous society that strongly believed in equity of water distribution; (ii) the fact that the Ternefti and Teshakil were democratically elected and were largely viewed as 'accountable' by their customers – the farmers; and (iii) the unambiguous sharing of responsibilities between the leaders of the farmers' organization and those in the government institutions.

Here, 'accountable' means that the farmer leaders effectively understand and represent the specific interests of the farmers. The degree of 'accountability' of any farmers' organization leaders greatly depends on the following:

- The nature of the relationship of the farmers' organizations with the respective government institutions involved in the management of the system.
- The nature of the farmers' organizations themselves.

The nature of the relationships between farmers' organizations and the government institutions ranges from 'autonomy' to 'dependence' in both the 'financial' and 'organizational' dimension (Hunt, 1990). The more autonomous the farmers' organizations the less their leaders are influenced by higher officials in the government offices and the more accountable they are to their customers – the local farmers. The farmers' organizations in the indigenous Wadi Laba and Mai-ule systems could be considered fully autonomous in the 'organizational dimension' – the 'organizational control of water' – as they were entirely responsible for making all decisions on how water should be shared, and it was only on their request that government institutions interfered. They could also be assumed as largely autonomous in the 'financial dimension', because most of the maintenance work of the indigenous structures had been largely accomplished by mobilizing the human labour and draught animals of the local communities. The government institutions provided only some materials such as shovels and spades – even that on request from the organizations.

The 'nature of farmers' organizations' refers to how inclusive the organizations are of the various wealth groups and the male and the female gender members of the community, and how representative their leaders are. There was no big gap between the rich and the poor in the Wadi Laba and Mai-ule communities and hence the wealth category did not apply. As stated earlier, the female members of the society, although allowed to be members of the organizations, did not have decision-making voices and they were not allowed to elect or be elected. This exclusion of the females did not, however, affect the accountability of the organizations and their leaders as far as their activities in enforcement of water rights and rules were concerned. The household heads, usually the

men, were fully represented in the organizations, and it was they who actually owned the
land and who made all the decisions on the
behaviour of all the household members. Even
in the case of the fewer than 5% female-headed
households in Wadi Laba and Mai-ule
(widowed or divorced women), it was the close
male relatives of the women who served as
representatives of the households in making all
the necessary decisions.

Relationship between water rights and rules and maintenance

The links between the water rights and rules,
and the organization and execution of maintenance tasks can be categorized into three
aspects. To start with the first aspect, in many
spate irrigation systems, the right to flood water
is tantamount to one's contribution to maintenance of main and branch canals and structures. If one fails to contribute, one can simply
not be allowed to irrigate one's field. This was a
common practice in the indigenous systems in
the Tihama, Yemen, but non-existent in many
of the indigenous systems in Eritrea. As
mentioned earlier, in Eritrea, most of the
communities engaged in spate irrigation were
homogenously poor and their livelihood
depended entirely on their spate-irrigated
fields. There was a strong belief in the society
that prohibiting a certain field access to water,
because its owner – the household head – had
failed to report for maintenance duty, was not
the right decision. Such an action was viewed
as depriving the whole family of their very basic
food for a mistake perpetrated by one of its
members – the household head. Hence, in the
indigenous Wadi Laba and Mai-ule systems,
contributing labour was not a prerequisite for
preserving one's water right.

The second aspect of the link relates to the
water rights and rules, and 'the critical mass' –
the minimum amount of labour and materials
needed for maintenance. In the indigenous
Wadi Laba and Mai-ule and the Tihama spate
irrigation systems, the maintenance task was
largely dependent on human labour and
draught animals. In such a situation, a large
task force was required, which could only be
made available through strong cooperation

between upstream and downstream farmers.
That tail-end farmers were only interested in
sharing the burden of maintenance, if not for
the fact that they were systematically deprived
of their water right, made 'the critical mass
factor' vital for serving as a check on too large
an inequity in water sharing.

To come to the third aspect of the link,
water-sharing rights and rules – in particular the
rules on demarcation – help to identify the
group of farmers entitled to flood water and
who have an interest in jointly undertaking the
necessary maintenance job. Without the
demarcation rules, it is very difficult to form a
group of partners, making the organization and
cost sharing of the recurrent maintenance work
problematic.

The significance of the 'critical mass' has
considerably diminished in many systems in the
Tihama and may be affected in the Wadi Laba
and Mai-ule systems in Eritrea, mainly due to
the structural modernization of the indigenous
structures and mechanization of the maintenance, usually undertaken by government institutions. This is elaborated in the section on
'modifying/changing water rights and rules'.

Codification of rules

In all the spate irrigation systems in Eritrea,
whether in the relevant government institutions
or the farmers' organizations, there are no
complete records of water rights and rules. In
most cases, however, the rules and rights are
presented in plain, unambiguous language,
which has helped to disseminate them easily
and correctly among large (greater than 3000
households) communities by word of mouth. In
Wadi Zabid, the Tihama Plain in Yemen, the
renowned Islamic scholar, Sheikh Bin Ibrahim
Al-Gabarty, is believed to have first recorded
the rules and rights for distributing flood water
about 600 years ago. Rights and rules on flood
water distribution in the Suleman range in
Pakistan were codified by the revenue administration during the period of the British rule in
1872. The documents, which are still available
in a register, the *Kulyat Rodwar*, contain a list of
all villages responsible for contributing labour
for maintenance of the various bunds. The
document also identifies a special functionary

who was responsible for enforcing the rules. The Kulyat Rodwar and the rights and responsibilities contained therein have not been updated, but the creation of these functionaries serves to keep the system flexible, as it allows the build-up of an institutional memory of 'jurisprudence'.

There is a large added value in codifying water rights and rules into written documents such as laws and regulations. It could serve at least as a basis for clarifying disagreements in interpretations and introducing a neutral factor in any dispute. The continued use made of the Kulyat Rodwar registry in Pakistan is a proof of the importance and relevance of codifying. Yet, codifying water rights and rules may not as such be sufficient to ensure that they are observed or to mitigate conflicts. The ubiquitous disputes in Wadi Zabid, where powerful parties stand accused of violating the water rights and rules in spite of the presence of the more than six-centuries-old records, and the barely existent vehement conflicts in Wadi Laba and Mai-Ule, although none of the rules and rights are codified, all illustrate the point.

Modifying and Changing Water Rights and Rules, and Implications

If water rights and rules in spate irrigation systems are to continue to deliver, they must necessarily adjust to new situations created by various factors – new land development, changes in crop pattern, structural modernization (infrastructural investment), shift in power relations and change in levels of enforcement.

In this section, with the help of examples from Eritrea, Yemen and Pakistan, we discuss the consequences of tailoring some of the water rights and rules and the managing organizations in response to some of the mentioned factors, and a failure to do so.

To start with the case from Eritrea, in the Wadi Laba, due to an increase in the number of inhabitants the land under spate irrigation increased from about 1400 ha to nearly 2600 ha between 1900 and 1990. As a result, the farmers explained that for 20 years (1960–1980) they consistently witnessed that, even during the best flood seasons, their existing rules failed to guarantee that all the fields

received at least a single turn. To deal with this new reality, by around the mid-1980s the farmers had added a phrase to the 'water right on sequence' – as 'in a new flood season, dry fields first'. Its full interpretation is that, regardless of the location of the fields, in a new flood season the fields that did not get a single irrigation turn in the previous flood season are irrigated once before any of the other fields get a single turn. An overwhelming majority of the interviewed farmers seemed content with the degree of the impact this modification had in preserving the perception of the fairness of water distribution that had existed prior to the land expansion.

To provide another example from Wadi Laba, the structural modernization that was completed in 2001 replaced the flexible, main indigenous structure with a rigid, permanent weir, and many other secondary earthen distribution structures with gabion (cylindrical baskets filled with earth, rubble, etc.). The modern structures necessitate a different type of maintenance. They do not depend on labour and the collection of brushwood, but instead require earthmoving machinery such as loaders, bulldozers and trucks which, in turn, call for different organizations, managerially, financially and technically. The main factor in the past that was key to the enforcement of the water rights and rules during the indigenous systems was 'the critical mass' – the need for a large number of farmers who would work on collective maintenance.

There is a risk that the different maintenance requirements will change the way that water distribution is organized. Though it is too early to say, in the 2003 flood season the authors witnessed 15 occasions when the upstream farmers utilized large floods and irrigated their fields two to three times before downstream fields got a single turn. This caused a lot of conflicts. The 300 ha furthest downstream did not receive a single turn in 2002 and 2003. The earlier rule on sequence and large and small floods was not applied, partly because the new infrastructure attenuated the floods and effectively reduced the number of big floods, which were the ones that had previously served the tail areas.

Over 30 years of management of spate systems by large government irrigation institutions in Yemen have proved that such institutions have difficulty in handling the task all by themselves. Some of the factors include: (i) poorly defined sharing of responsibilities and the long communication lines, which lead to a slow decision-making process; (ii) lack of adequate funding; and (iii) little 'accountability' towards the bulk of users. More than anything, the chronic underfunding of maintenance and the loss of vigour in the operation and maintenance departments were the undoing. It left a vacuum where it was not clear who was responsible for water distribution, with no one doing the hard work of timely maintenance.

If the relatively fair distribution of the flood water that existed prior to modernization is to be preserved and the economic homogeneity of the Wadi Laba communities largely conserved, the farmers' organizations in Wadi Laba and Mai-ule, which have run the system for 100 years and have a good knowledge of flood water management practices, must continue to take the lead role. To perform this task, the farmers' organizations need to have financial and organizational autonomy, and hence their accountability. Great strides have been made with the establishment of the Wadi Laba and Mai-ule farmers' organization (also commonly called the Sheeb Farmers' Association), with almost full membership of all farmers in the area and the universal endorsement of its by-laws. The leadership of this new organization is very much based on the time-tested system of Ternefti and Tesahkil. The main challenges in the coming period are the internal organization, the water distribution, the acquisition of adequate funding (also in the occasional disaster year), the running of earthmoving equipment and the operational fine-tuning of the modernized system. In addition, there are issues concerning some national and provincial laws that need to be considered. These are discussed below.

For the past 100 years, till 2001, the Wadi Laba communities did not rely on national or provincial laws and policies to manage their indigenous spate irrigation systems; nor did they bother to clarify what impact those policies and laws could have had on flood water management. Since the structural modernization in 2001, however, some farmers and their leaders are frequently asking this question: after the huge financial investments, will the govern-

ment still allow us to continue to own and utilize 'our' land and flood water? The urgency of receiving a reply to this question emanates from the perceived fear of the farmers that the government may implement the '1994 Land Proclamation' to dispossess them of the land they had considered theirs for decades. In Eritrea in general, and in the Wadi Laba and Mai-ule spate-irrigated areas in particular, owning or having land usufructuary right is a prerequisite to securing a water right for agricultural production.

For decades, the farmers in Wadi Laba and Mai-ule have practised the traditional land tenure system, the *Risti* (literally translated, inherited land from the founding fathers). Under this tenure system, ownership of land in a certain village or villages is vested on the *Enda* (plural: Endas) – the extended family that has direct lineage to the founding fathers of the village(s). The system is highly discriminatory against women. Besides, as it allows partition of the land through inheritance, it may also cause land fragmentation and render the farm plots economically non-feasible. However, the major tenets of the Risti (see Box 7.4) collectively provide a strong sense of land, and hence water, security to the eligible landholders.

The 1994 Land Proclamation refers to the Risti and the other indigenous tenure systems as obsolete, progress-impeding and incompatible with the contemporary demands of the country. Thus, one of its stated objectives is to replace/reform the traditional tenure system with a new, dynamic system. Most of the provisions of the Proclamation (see Box 7.5) are important milestones, particularly in the provision of gender equity and preservation of the economic viability of the arable land. When some of its provisions are read against the background of the Risti, however, they seem to have given too much power to the government at the expense of the farmers' organizations. This power shift may create (as seems is the case in Wadi Laba and Mai-ule) tenure insecurity.

The provision of the Land Proclamation that grants the government absolute power and right of land appropriation is the one frequently singled out by almost all the interviewed Wadi Laba and Mai-ule farmers who expressed fear and nervousness with respect to their land and water security. The majority of the farmers

---

Box 7.4. The main tenets of the Risti land tenure system in Wadi Laba and Mai-ule (from authors' survey, 2004).

- The Enda holds a lifetime ownership of land within the territories of its native village(s). The land is distributed equally among the male Enda members. Only widowed women are allowed to own half of the parcel of land granted to men.
- An individual member of the Enda has the right to utilize his plot for the production of whatever crops he wants. He has also an absolute right to bequeath his land to his sons, lease or mortgage it. He can sell the land, however, only with the consent of the extended family – mainly the father, grandfather and the first cousins.
- The village assembly, the *Baito*, together with the Wadi Laba and Mai-ule farmers' organizations, are responsible for screening those eligible for the Risti land, distributing the available land equally among the eligible and carrying out other related land administration tasks. They, however, have neither the right nor the power to confiscate land allocated to a verified Enda member.

Box 7.5. Some of the provisions of the 1994 Land Proclamation (from authors' compilation, 2004).

- The Government of the State of Eritrea is the sole owner of all land of that country.
- All citizens of Eritrea above the age of 18 are eligible to usufructuary right regardless of sex, race, clan, Enda or beliefs. Any individual may lease his/her usufructuary right over the land in whole or in part, but under no circumstance can he/she sell the land.
- To preserve the economic viability of farmlands, partition of land through inheritance is prohibited.
- A land administration body (LAB) – consisting of a representative of the Government's Land Commission (GLC), members of the village assembly and farmers' organization leaders and different local government bodies – is responsible for classifying land and distributing it equally to the eligible by virtue of the proclamation and to those who make a living by farming. The LAB is a subordinate executive body with respect to land distribution and it carries its functions under strict orders and directives from the GLC.
- The government or its appropriate government body has the absolute right and the power to expropriate land that people (regardless of their clan, Enda, race, sex, beliefs) have been settling on or have been using for agricultural or other activities, for purposes of various development and capital investment projects aimed at boosting national reconstruction or other similar objectives. This provision further states that compensation will be given whenever land is confiscated, but it does not elaborate what such compensation will be, who decides on the nature of such compensation or whether or not the individual landholder or the farmers' organizations that represent him can challenge any compensation arrangements made by the GLC.

---

believe that the government would alter the cropping pattern, from the current entire focus on food crops to high-value cash crops, to boost national production and recover the huge (about US$4 million) investments made for the modernization of the Wadi Laba and Mai-ule systems. In an attempt to justify this assertion, the farmers point to the continuous push that they claim is being made by the local government and the local Ministry of Agriculture to introduce a cotton crop, despite their reservations. The farmers foresee that in the near future their status will be changed from landowners (users) to daily labourers under government payroll. They contend that, although they trust the government will do all it can to provide reasonable compensation should it confiscate

their land, no compensation will have a comparable value, as they attach a lot of pride to the land they currently own. The farmers argue that they should be the ones to decide whether or not to hand over their land once the government reveals its compensation plans.

The farmers' analyses of the postmodernization situation of their irrigation systems, although it seems to have evolved from a genuine perception of land and hence water insecurity, may as well end up being just a logical speculation. The government has clearly stated that the objective of modernizing the Wadi Laba and Mai-ule systems is to improve the living standards of the concerned communities; and that it will ultimately entrust the operation and management responsibility of the systems to the farmers' organizations. If this noble objective is to be translated into reality, however, real and active farmers' participation throughout the ground-laying process and activities (this has yet to properly start) for the management transfer are vital.

Nevertheless, such farmers' participation may not be achieved unless the land and water insecurities perceived by the farmers – justified or not – are addressed. We believe that the introduction of some complementary (to the Land Proclamation), easily understandable provincial/sub-provisional laws may be useful toward this end. Among others, these may spell out: (i) in the postmodernization era, what kind of land and water user rights do the spate irrigation communities have? (ii) What decision-making power do these user rights bestow on the farmers' organizations as far as the cropping system, modifying/changing water rights and rules, and other important land and water utilization activities are concerned? (iii) Do the farmers' organizations and the communities as a whole have any new obligations they need to fulfil if they are to retain these rights? And (iv) if yes, what are they?

Another related issue that needs to be given due consideration is the legality of the Wadi Laba and Mai-ule farmers' organizations. Although these organizations are officially recognized at the sub-provincial level – official in a sense that the sub-provincial local government and the Ministry of Agriculture acknowl-edge the organizations as important partners in the management of the irrigation system – these organizations cannot yet be considered as having full legal status. Their establishment and existence are not supported by any official decree or law, nor do they have the legal authority to, for instance, make direct contacts with donor agencies, own property such as machinery or operate independent bank accounts. We presume that it is useful to intro-duce national/provincial laws that strengthen the legality of the organizations and provide them the authority they need to cope with the new management challenges of the modern-ized systems.

Regarding the example from Yemen, in the spate irrigation systems of Wadi Zabid, Siham and Mawr, the structural modernizations carried out in the 1970s replaced the indigenous earthen and brushwood structures with concrete weirs. This resulted in almost complete control of the flood water by the upstream users. Although the al aela fil aela rule granted an absolute priority right to the upstream farm-ers, as stated earlier, it did not usually cause unfairness of water distribution during the indigenous systems. This was because the indigenous structures were frequently washed away delivering water downstream. In contrast, the weirs seldom breach. Hence, applying the al aela fil aela rule effectively led to the 'capture' of the flood water by the upstream lands.

Due mainly to the vacuum of governance created after the fall of the Sultans and Sheikhs, who were replaced by 'weak' local govern-ments, the al aela fil aela rule was not modified to meet the demands of the new reality. Instead, the upstream farmers strictly applied it. Moreover, encouraged by the abundance of water furnished to them and the absence of any effective countervailing power, the upstream farmers shifted from the cultivation of food crops to the more water-demanding but highly profitable banana crop on the basis of conjunc-tive use of groundwater and spate flow. This further reduced the amount of water that could have reached downstream. The local govern-ment did not interfere to stop this change in the cropping pattern. The ultimate consequence is that many of the downstream fields are now abandoned and their owners are earning their living on a crop-sharing arrangement by serv-ing as daily labourers in the fields of the now rich upstream landlords. In Wadi Zabid, where the crop-sharing arrangement is more common, the tenants perform all the labour (from planting till harvest) for a return of one-quarter of the harvest in kind.

The term 'weak' here refers to a local government lacking in-depth knowledge of: (i) local water rights and laws and approaches and strategies to enforce them; (ii) accountability to the poor segments of the farmers; and (iii) the power to correct some unfair land and water utilization decisions taken by some individuals or communities.

Regarding the example from Pakistan, in Anambar Plain in Balochistan, one of the intro-duced modern weirs significantly changed the indigenous water distribution system. The weir was constructed to divert spate flows to upstream fields. It performed this function, but it also considerably reduced the base flow to the downstream fields. This deprived the downstream farmers of their basic access to water granted to them by the water rules that had been implemented for years. Essentially, the design was made with a major oversight as to the prevailing water distribution rules. Hence, the weir became the main cause for many tensions and conflicts. Unlike in the Yemen case, the upstream community, faced with an equally socio-economically powerful downstream community, did not manage to maintain the water control power offered to it by the weir and did not shift from food crops to highly profitable commercial crops. As conflicts became unbearable, the two communities – in harmony – reached a mutual agreement: they purposely blew up the weir and returned to their indigenous structures and water-sharing arrangement.

Conclusions

Water rights and rules mitigate unpredictable flood water supplies to a large extent by introducing a series of interdependent, flexible regulation mechanisms that define acceptable practices on how water should be shared during each flood occurrence. They play the following roles: (i) protecting the rights of the farmers entitled to flood water; (ii) defining the type of water-sharing system and the sequence that should be followed in the event of different flood sizes; (iii) limiting the amount of water a certain field receives at each turn; and (iv) outlining which field, and when, is entitled to a second turn.

Collectively, the water rights and rules create a perception of fairness of water distribution between the upstream and downstream farmers, thus generating an atmosphere of cooperation between them. This, in turn, enables the attainment of the 'critical mass' needed for accomplishing the important component of the flood water management – timely maintenance of the indigenous structures. To perform these tasks, however, the water rights and rules must be observed by the majority of the farmers. This can be achieved only when there are local organizations accountable to most farmers and which apply enforcement approaches that take

into account the social structure of the concerned communities.

The water rights and rules are drafted and implemented in a way that meets the flood water management needs in a given situation. They need to be constantly tailored, and the enforcement organizations and the strategies they use are adjusted to cope with changes in events over time, if the above-stated achievements are to be sustainable. Should this not be done, as was the case in some systems in Eritrea, Yemen and Pakistan, the water rights and rules can end up being frequently violated and become sources of unfairness of water distributions and conflicts that, in turn, could result in the following:

• Pave the way for disintegration of the long-established local farmers' organizations; and cause the creation of a gap between the poor and the rich in what were rather wealth-wise homogenous societies.
• Accelerate the downfall of downstream farmers, leaving them unprotected against the illegal capture of the flood water by upstream farmers.
• Result in deliberate destruction of investment.

In general, national and provisional policies and laws have hardly any direct impact on the flood water management in the spate irrigation systems. The water distribution and maintenance are carried out according to local water rights and rules and they are sufficient. Where national legislation could become helpful, however, is in providing farmers' organizations with legal recognition and legal authorities with the means to perform activities that would enable them to be financially and organizationally autonomous. This requires more than legislation, however – it also necessitates sincere efforts to support the local organizations and graft them on to earlier local organizations and avoid the creation of dual structures (traditional and formal).

Acknowledgements

The authors of this chapter would like to thank the Netherlands Organization for International Cooperation in Higher Education (NUFFIC), the Netherlands; the Centre for Development and Environment (CDE), Berne University, Switzerland; and the UK Department for International Development (DFID) for funding this study. Special gratitude also goes to the staff of the Eastern Lowland Wadi Development Project (ELWDP) in Eritrea and the Irrigation Improvement Project (IIP) in Yemen for their logistical and advisory support during the field surveys; and the spate irrigation communities in Eritrea, Yemen and Pakistan for their unreserved participation in the interviews and group discussions that generated valuable information. Reference is also made to the Spate Irrigation Network (http://www.spate-irrigation.org), a network of practitioners and researchers in the field of spate irrigation.

References

Al-Eryani, M. and Al-Amrani, M. (1998) Social and organizational aspects of the operation and maintenance of spate irrigation systems in Yemen. In: Diemer, G. and Markowitz, L. (eds) *Proceedings of the World Bank Sponsored Participatory Irrigation Management (PIM) Seminar*, November 1998, Hudaidah, Yemen.

Hunt, R.C. (1990) Organizational control over water: the positive identification of a social constraint on farmer participation. In: Young, R.A. and Sampath, R.K. (eds) *Social, Economic, and Institutional Issues in Third World Irrigation Management*. Westview Press, Boulder, Colorado, pp. 141–154.

Mehari, A., Schulz, B. and Depeweg, H. (2003) Water sharing and conflicts in the Wadi Laba Spate Irrigation System, Eritrea. Available at http://www.spate-irrigation.org (accessed 25 September 2005).

Mehari, A., Schulz, B. and Depeweg, H. (2005) Where indigenous water management practices overcome failures of structures. *ICID Journal of Irrigation and Drainage* 54, 1–14.

North, D.C. and Thomas R.P. (1977) The first economic revolution. *The Economic History Review* 30, 229–241.

van Steenbergen, F. (1997) Understanding the sociology of spate irrigation: cases from Balochistan. *Journal of Arid Environments* 35, 349–365.

van Steenbergen, F. (2004) Water rights and water distribution rules. Available at http:www.spate-irrigation.org (accessed 25 September 2005).

8 Local Institutions for Wetland Management in Ethiopia: Sustainability and State Intervention

Alan B. Dixon[1] and Adrian P. Wood[2]

[1]*Department of Geography, University of Otago, Dunedin, New Zealand;*
e-mail: alan.dixon@geography.otago.ac.nz; [2]*Centre for Wetlands, Environment and*
Livelihoods, University of Huddersfield, Huddersfield, UK;
e-mail: a.p.wood@hud.ac.uk

Abstract

Locally developed institutions that include rules and regulations, common values and mechanisms of conflict resolution are increasingly regarded as adaptive solutions to resource management problems at the grass-roots level. Since they are rooted in community social capital rather than in external, top-down decision making, they are seen as being dynamic, flexible and responsive to societal and environmental change and, as such, they promote sustainability. Within this context, this chapter examines the case of local institutions for wetland management in western Ethiopia. It discusses how the structure and functioning of these institutions have evolved in response to a changing external environment, and the extent to which this has facilitated the sustainable use of wetlands. It is suggested that these local institutions do play a key role in regulating wetland use, yet they have, uncharacteristically, always relied on external intervention to maintain their local legitimacy. Now there are concerns that the institutional arrangements are breaking down due to a lack of support from local administrative structures and current political ideology. This has major implications for the sustainable use of wetland resources and food security throughout the region.

Keywords: community, local, institutions, natural resources management, social capital, sustainability, wetlands, state, Ethiopia.

Wetlands, Local Institutions and Sustainability

Wetlands are becoming increasingly recognized as important natural resources in developing countries because of their ability to fulfil a range of environmental functions and produce a number of products that are socially and economically beneficial to local communities (Dugan, 1990; Silvius *et al.*, 2000). Wetlands act as sponges during dry periods of the year; they regulate run-off and recharge groundwater

resources, and they purify water supplies. Their capacity to store water means they are able to support livelihood strategies, such as fishing, pastoralism and agriculture, as well as providing craft materials, clean drinking water and medicinal plants. People's long association with wetlands means that indigenous systems of wetland management and utilization are to be found throughout the developing world.

In recent years, however, much attention has been focused on the need for the 'wise use of wetlands' in the context of an increase in wetland exploitation and development, fuelled by socio-economic, political and environmental change. In many parts of Africa in particular, agricultural use of wetlands has increased as more and more people have been forced to seek new livelihood strategies, as a result of environmental degradation of other farmlands and population pressure. Government policies that have failed to recognize the significance of local wetland management practices, and indeed the wider value of wetlands, have also stimulated the intensification of wetland agriculture, in an attempt to create more economically productive land. Consequently, a key concern in the long term is that the carrying capacity of wetlands, in terms of the exploitation of products and functions, will be exceeded, resulting in degradation and loss of livelihood benefits for all.

Whilst a Malthusian perspective would argue that such degradation is unavoidable, alternative perspectives in recent years have drawn attention to the ability of local people themselves to adapt their natural resources management (NRM) systems to changes taking place, enabling resources use to remain sustainable (Boserup, 1965; Tiffen et al., 1994). At the core of this adaptive capacity is social capital; commonly interpreted as the shared norms and values, knowledge, institutions and networks intrinsic to a specific community (Pretty and Ward, 2001). Social capital includes the processes of communication and innovation, mechanisms through which new knowledge and practices evolve and which facilitate adaptation. Social capital also constitutes the space in which community-based 'traditional' or 'local' institutions exist (Shivakumar, 2003).

Such institutions, particularly those concerned with NRM, provide the rules and regulations for resources exploitation; they are effective in mobilizing human resources; they are involved in conflict resolution; and, perhaps fundamentally, they have been linked to equitable and sustainable NRM (Uphoff, 1992; Blunt and Warren, 1996; Manig, 1999; Hulme and Woodhouse, 2000). Uphoff (1992) in particular, suggests that local institutions may be particularly successful in NRM where the resources in question are known and predictable, rather than shifting and variable, and where the users are an identifiable group.

Many wetlands in developing countries clearly fit these criteria, in that they usually have a discrete community depending upon their various products and services. Hence, it could be argued that local institutions potentially have a key role to play in facilitating the adaptive and sustainable management of wetlands throughout the developing world.

Mazzucato and Niemeijer (2002) propose local institutions as the 'missing link' in development and adaptation at the people–environment interface. They are regarded as Boserupian adaptations to resource depletion and, through the networks, knowledge, rules and social cohesion associated with them, they mediate people's relationship with the environment (Leach et al., 1999; Manig, 1999; Mazzucato and Niemeijer, 2002). Their overall effect can lead to adjustment and adaptation rather than to environmental degradation and, hence, they can be facilitators of sustainable NRM. Their strength, according to Shivakumar (2003), lies in their indigenous nature, in that they represent 'home-grown' solutions to problems, based on collective understanding. Since they are based on indigenous knowledge, and rooted in social capital that has evolved over generations in a specific culture or environment, they are often regarded by local communities as having greater credibility and legitimacy than external institutions.

As empirical evidence of a relationship between local institutions and sustainable community- based natural resources management (CBNRM) has emerged (Ostrom, 1990; Blunt and Warren, 1996; Hinchcliffe et al., 1999; Pretty and Ward, 2001; Mazzucato and Niemeijer, 2002), so has interest among development practitioners. In an era when rural development and CBNRM have been dominated by the ideals of participation, local institutions have often been regarded as potential short cuts to development; they represent ready-made power structures through which policies can be formed and development initiatives implemented, and they have been taken as models for grass-roots development which can be replicated elsewhere (Warren et al., 1995; Blunt and Warren, 1996; Howes, 1997; Koku and Gustafson, 2001; Guri, 2003; Watson, 2003). The empowerment of local institutions has become a key policy objective, not only in CBNRM projects but also in the context of a shifting focus on development issues such as governance, decentralization and civil society.

One critical area of debate, however, centres on the extent to which local institutions themselves are sustainable: whether they can continue to function and support sustainable NRM strategies in the context of rapid change or, in the case of the above, increased intervention from external institutions. At best, such intervention may involve a participatory NGO seeking to empower community relations and the functioning of the institution, whilst in the worst case scenario, intervention may be government driven, top-down and prescriptive in nature, seeking to replace such institutions with state structures.

Watson (2003) describes the failure of NGOs to facilitate sustainable livelihoods, through establishing local, indigenous-based NRM institutions in Borana, Ethiopia. By underestimating both the complexity of resources use and the power relations within the indigenous *Gadaa* system, NGO intervention has resulted in a perceived devaluing of the existing institutions among their members. Government intervention in the operations of local institutions, meanwhile, has also been recognized as a threat to their legitimacy, credibility and effectiveness (Richards, 1997; Serra, 2001, unpublished), although much depends upon the nature and extent of the intervention, and the resources in question.

One of the enabling conditions for the sustainability of common-pool resource institutions, cited in a review by Agrawal (2001, p. 1659), is that 'the government should not undermine local authority in the functioning of local institutions' (although 'supportive external sanctioning institutions' are considered a prerequisite). Moreover, Rasmussen and Meinzen-Dick (1995) cite the work of Wade (1988), Ostrom (1990) and Bardhan (1993) in arguing that local institutions become more effective when arrangements in the external environment support them. Certainly, the intervention of external institutions may seem justified if, as Manig (1999) argues, local institutions, left to their own devices, may struggle to adapt to rapid socio-economic, environmental or political change.

This external–local institution nexus is implicitly addressed in this chapter, which examines the relationship between local wetland management institutions and the sustainability of wetland use in western Ethiopia. Critically, it explores how these local institutions have evolved in response to environmental, socioeconomic and political change during the last 150 years, and assesses the implications of external intervention in their operations, for their current and future sustainability.

People and Wetlands in Western Ethiopia

The people

The ethnic composition of Ethiopia's western highlands is diverse. In Illubabor and Western Wellega zones (see Fig. 8.1) the dominant ethnic group is the Oromo, who account for between 80 and 90% of the population. Some claim that the Oromo are not, however, indigenous to the area; a period of Oromo migration and expansion displaced and incorporated indigenous Omotic and Sudanic hunter-gatherer and agro-pastoralist groups, who subsequently migrated to lowland areas.

The second largest ethnic group is the Amharas (approximately 10%) who, during the late 19th century and under the leadership of Menelik II, expanded their empire to the south and the west of what was then the 'Abyssinian Kingdom'. Through this conquest and a continuous process of inward migration and resettlement, the lands of present-day Illubabor and Western Wellega were subsequently annexed into what became the modern Ethiopian state by the end of the 1890s. The remainder of the population is, as a consequence of migration, immigration and government resettlement schemes during the 1970s and 1980s, composed of Tigrayan and Gurages from the north and east of the study area, and Mocha and Keffa peoples from the immediate south, with small numbers of other ethnic groups from around the country.

Wetlands

Wetlands are a common feature of the landscape in the highlands of western Ethiopia, particularly Western Wellega and Illubabor. The warm, temperate climate, characterized by a

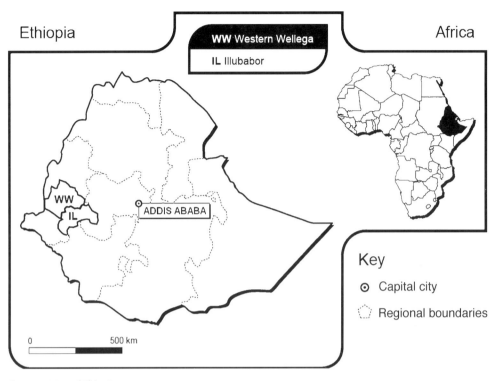

Fig. 8.1. Map of Ehiopia.

mean annual temperature of around 20°C and annual rainfall usually in excess of 1500 mm, together with the undulating to dissected topography, which ranges between 1000 and 2000 m above sea level, produce an environment characterized by steep-sided river valleys and flat, waterlogged valley bottoms. The accumulation of run-off, poor drainage and a high groundwater table in these valley bottoms promotes the formation of both permanent and seasonal swamp-like wetlands, ranging from < 10 to >300 ha (Dixon, 2003). It has been estimated that approximately 4% of Illubabor's land area is occupied by wetlands, and this proportion is likely to be similar for Western Wellega, to the north (Afework Hailu, 1998, unpublished).

The wetlands are vital natural resources, both in terms of their environmental functions and their products, which are used extensively by local communities. They represent a vital source of water throughout the year, in an area which receives half of its annual rainfall between June and August, and only 5% during the dry season

months of December, January and February. The storage and release of water from the wetlands and their peripheral springs ensure that local communities have access to clean drinking water throughout the year. The abundance of water in the wetlands also supports the growth of dense sedge vegetation known locally as *cheffe* (*Cyperus latifolius*) which, in addition to providing limited fodder for cattle, is traditionally harvested by local communities for use as a roofing and craft material. It is also used throughout the year in a range of ceremonies and celebrations and, as such, it is a marketable commodity. The wetlands are also a habitat for a variety of other plant communities, some of which are used for medicinal purposes by those living around the wetlands. For example, the plant known locally as *balawarante* (*Hygrophila auriculata*) is used as a treatment for various skin diseases (Zerihun Woldu, 1998, unpublished).

As reservoirs of soil moisture during dry periods, these wetlands are also valuable agricultural resources and many have traditionally been used, albeit on a small, informal scale, to cultivate maize much earlier in the agricultural calendar than on the uplands (Tafesse Asres, 1996; Wood, 1996). This practice, which includes the majority of the wetland maize crop being harvested before maturation, i.e. during its 'green' phase, facilitates the production of crops during a period of the year which is normally associated with food shortages.

Over the last century, however, it appears that wetland cultivation has extended beyond the use of wetland margins to include much larger areas, and nowadays the complete drainage and cultivation of wetlands are common phenomena throughout the region. Recent estimates have suggested that wetland cultivation provides somewhere between 10 and 20% of the annual food needs of the region's population (Ethiopia Network on Food Security, 2001) but, during the food shortage months, its contribution rises to 100% in some areas. In Western Wellega, in particular, the dependence on wetlands for food security is greater than that in Illubabor, and it appears that more intensive forms of wetland cultivation have existed over a longer period of time. This appears to be a product of population pressure in Western Wellega and the conspicuous degradation of upland farming areas. In contrast, Illubabor has less population pressure and the uplands have remained relatively fertile, in part because of the abundance of forests.

The current system of wetland drainage and cultivation is dominated by the cultivation of maize, although in some areas sugarcane, tef (*Eragrostis tef*) and vegetables, including cabbage and potato, are also grown in small quantities from year to year. The wetland agricultural calendar typically begins after the rains subside, usually in September or October, when some farmers may cultivate tef on the residual moisture. It is more common, however, for cultivation to begin with the sowing of a maize crop sometime between January and March. Prior to this, existing drainage channels are cleared of weeds or cheffe, or new channels are excavated – usually in a herringbone pattern – to ensure appropriate soil moisture conditions for the water-intolerant maize crop. The maintenance of drainage channels and the guarding of the crop against wild pests continue throughout the year until harvesting, which usually occurs during the start of the rains (between June and July).

Wetlands tend to be divided among numerous farmers from the local community. The allocation of wetland plots in some instances dates back to the early 20th century (see below), with the number of stakeholders depending upon the size of the wetland and the demand for agricultural land within each community. It is common for wetlands of 40 ha in size to have over 300 local cultivators (Afework Hailu et al., 2000, unpublished). Given the large number of stakeholders involved and the dynamic nature of the wetland environment, the whole system of wetland management requires a significant amount of coordination in terms of farming resources and manpower.

The origins of and changes in wetland cultivation

Although there are conflicting accounts of the origins of wetland cultivation in the area, it is generally agreed that more intensive forms of wetland cultivation were initiated in response to food shortages on the uplands caused by drought in the early years of the 20th century. During this period, land was effectively owned by a few feudal landlords, and installed by the expanding Amhara Empire in the late 19th century, who rented out their lands to peasant farmers. With the occurrence of drought and food shortages, many landlords either instructed their peasant farmers, or granted a request from farmers themselves, to cultivate wetlands in order to achieve food security. Following an initial period of trial and error, during which farmers experimented with different management practices with varying degrees of success, wetland cultivation became the mainstream agricultural activity in many areas.

With the overthrow of the Haile Selassie government by the Derg in 1974, the social dynamics of wetland use changed. In 1975, the Derg nationalized all rural land, with the result that wetland access was controlled by the newly established kebeles (peasant associations) that constitute the lowest administrative unit of the government. In most cases, however, those who previously had access to wetland plots were, on request, given custodianship over the same plots.

Wetland agriculture during the Derg period (1974–1991) was characterized by an increase throughout the region for several reasons. First, wetland cultivation was encouraged by the government in order to meet regional targets of food self-sufficiency. Failure to cultivate in accordance with this policy risked the reallocation of wetland plots to other farmers who were willing to expand into wetland cultivation (Afework Hailu, 1998, unpublished). Second, the expansion of coffee production in the area, and the wider commercialization of farming in particular, resulted in local shortages of upslope agricultural land and, hence, the cultivation of wetlands became the means of subsistence for some farmers. Finally, in response to the famine of 1984, Illubabor and Western Wellega were chosen by the government as resettlement areas for famine victims. This inward migration of approximately 100,000 people, often to localized areas, resulted in further agricultural land shortages (Alemneh Dejene, 1990) and, in many cases, wetland plots were allocated to settlers (who, unlike the Oromo population, had no experience of farming under such conditions).

The only significant change in wetland agriculture since the change in government in 1991 has been the pressure to cultivate wetlands more intensively. This has stemmed from a government initiative in 1999 which, in response to drought-induced food shortages, sought to establish a Wetlands Task Force in each kebele. Although precise details of the Wetlands Task Force policy are unclear, many kebeles throughout Illubabor and Western Wellega have formed committees specifically to oversee the complete cultivation of wetlands. Ironically, while the implementation of this policy may have an impact on food security in the short term, it arguably represents a major threat to the sustainability of wetland agriculture, since it threatens to override locally adapted management practices and the knowledge base and indigenous organizational activities within local wetland-using communities.

Recent research by the Ethiopian Wetlands Research Programme (EWRP) in Illubabor, undertaken in response to concerns about widespread unsustainable use and wetland degradation, have drawn attention to the important contribution of local knowledge to sustainable wetland management practices. This research reported that few wetlands showed signs of environmental degradation, mainly because of the application of farmers' knowledge and experience. Among other things, farmers have developed extensive knowledge of wetland eco-hydrological process, vegetation changes, different cropping scenarios, drainage layouts and, critically, mechanisms of ensuring sustainability.

One such mechanism involves the practice of retaining areas of cheffe vegetation at the head of each wetland. These areas act as a reservoir of water and ensure there is always enough moisture distributed throughout the wetland to facilitate crop production. Farmers also use various plants as indicators of the 'health' of their wetland. For example, the plant known locally as *kemete* (*leersia hexandra*) is considered an indicator of poor fertility, and on its colonization it is common practice to abandon the wetland plot until fertility is restored (indicated by a host of other plants).

Ongoing research in the area has also drawn attention to the importance of local institutions formed specifically to coordinate wetland management activities among the various stakeholders. These institutions have emerged as a key factor influencing wetland sustainability.

Wetland Management Institutions

Distribution and origins

Wetland management institutions (WMIs) exist throughout Illubabor and Western Wellega, although their development over time, organizational structure and functions is spatially and temporally variable. They are known locally by a variety of names, including *Abba Laga* (father/leader of the catchment), *Abba Adere* (father/leader of a group of villagers), *Cheffe Kore* (wetland committee) and *Garee Misooma* (development committee). The name Abba Laga is the most frequently used in conjunction with the WMIs, especially in Western Wellega where wetland cultivation has a longer history than that of Illubabor.

The extent (if any) to which Abba Laga and Abba Adere had played a role in the traditional Oromo Gadaa system of administration prior to the Amhara invasion in the late 19th century remains unclear. The Gadaa system of public administration was itself brought to Illubabor and Western Wellega during the Oromo invasion of the area during the 17th and 18th centuries and, although its form and application varied from place to place, it was essentially a traditional socio-political institution in which the male members of each community progressed through different life 'grades', each with its own associated rights and responsibilities.

Within the system, one grade ruled for 8 years, before being replaced by another and, within each 8-year period, an *Abba Gadaa* (father of power), *Abba Dula* (father of war) and *Abba Sera* (father of the law) were elected (Hassen, 1990; Watson, 2003; see Desalegn *et al.*, Chapter 9, this volume). Whilst there is no documented evidence of either Abba Laga or Abba Adere playing an essential role in the Gadaa administration, it is probable that Abba Laga was a title instituted when and where the need to coordinate land use occurred. In the current day Borana zone in southern Ethiopia, where remnants of the Gadaa system still exist, Watson (2003) reports that *Abba Konfi* (father of the well) regulates access to water, yet there is no indication that the title is intrinsically linked to the Gadaa life grades system.

Eventually, in western Ethiopia, the Gadaa system gradually eroded as a result of local warlords undermining its administrative officials, and the system appears to have disappeared some years before the Amhara conquest. The origins of Abba Laga and Abba Adere are, therefore, ambiguous, and require further investigation. Recent reports from farmers in the area suggest that the titles were simply modifications of previous Gadaa era roles, in response to the need for a new institution in the light of an increase in wetland use during the Haile Selassie era. What appears different to the traditional Gadaa administrative roles, however, is that the title (either Abba Laga or Abba Adere) is now used interchangeably to describe both the institution itself, that is made up of participating farmers, and the appointed head of the institution, rather than just the latter as during the Gadaa era. The *Cheffe Kore* and *Garee Missooma*, which are the names used more frequently in Illubabor, tend to be Derg or post-1991 reinventions or reproductions of

previous titles of Abba Laga and Abba Adere and, in some instances, there is evidence to suggest that these institutions were actually organized and initiated by the local agents of the Ministry of Agriculture Development.

Role and functions

Throughout Illubabor and Western Wellega, the role of the WMIs is similar. They are a mechanism through which wetland users, almost exclusively men,[1] coordinate and facilitate cooperation in all wetland management activities, particularly in the preparation of drainage ditches prior to cultivation. Given the often elaborate design of drainage networks and the large number of stakeholders involved in each wetland, there is a need to ensure that the correct depth and width of all the drains are adequately maintained at roughly the same time, so that the soil moisture conditions are optimal for cultivation. It is usual practice for the elected leader of the institution to decide the date when this and other activities will take place, and ensure that all farmers comply with the decision.

The WMIs are also involved in a range of other activities, including the coordination (or joint work) of farmers for guarding crops, hoeing, weeding, ploughing and sowing, although the general maintenance of crops and harvesting are undertaken on an individual basis. In some areas where water shortages are a recurrent problem during the dry season, the institutions coordinate farmers' access to wetland drain water (for irrigation) via a system of rotation, while they may also plan and coordinate the blocking of drainage ditches to maintain water levels in the wetland, especially at the time of sowing and seed germination. The institution is also involved in mobilizing labour for tasks such as building footpath bridges across wetlands.

In some respects, the WMIs play a role similar to that of other local institutions, particularly *Debo* and *Dado*, both indigenous work group organizations, formed whenever needed, for tasks such as forest clearance or harvesting. Dado differs from Debo in that the work arrangement is one of reciprocal labour, i.e. the person requesting the Dado for his land is obliged himself to work in Dado for another person. In Debo, the organizer provides food and drink rather than reciprocating work by himself. Both are often utilized for tasks such as ditch clearing and harvesting in the wetlands and, hence, they represent important organizational components of the WMI constitution. They are, however, voluntary organizations, and it is uncommon for all the cultivators of one wetland to belong exclusively to one Dado or Debo arrangement.

One key challenge for the WMI, therefore, is to coordinate the various groups so that an activity such as ditch clearing can be carried out at the same time. Whilst other indigenous institutions such as *Tula* and *Ider* also exist, these are associated with social welfare and funeral arrangements. In some cases, they are influential in mobilizing labour for wetland activities if one member of the WMI or wider community is sick, imprisoned or absent. They also provide a forum in which information on wetland management is communicated (Dixon, 2005) but, in general, they are regarded as having little to do with wetland management activities or the enforcement of the WMI constitution.

A key function of the WMIs, which differentiates them from these other local institutions, is their role in controlling potentially destructive agricultural practices, such as excessive cattle grazing, which leads to the compaction and erosion of soil in wetlands. Similarly, some institutions restrict wetland cultivation to only one crop per year, whilst also prohibiting the cultivation of tef, sugarcane or the increasingly ubiquitous eucalyptus that are damaging in terms of their soil moisture requirements:

> Some people like to plant potato and tef after the maize is harvested, but so far no one has planted eucalyptus trees on the wetland we are cultivating. If someone wants to, the committee will stop it. Usually following the maize harvest the wetland is fenced. Drainage ditches are blocked and the land is allowed to flood.
>
> (Farmer at the Hadesa Wetland, Illubabor, 6 March 2003)

WMIs make informed decisions on whether whole wetlands should be used for cultivation, reserved for cheffe production or whether the wetland is perceived as being degraded, abandoned and left to regenerate.

In order to function successfully, the institutions require all its members to cooperate and abide by a series of rules and regulations, which are either informally agreed upon or, in some cases, written in a constitution (see Fig. 8.2). If the latter is the case, each member is required to sign the constitution, which usually also states that the failure of a member to comply with the rules is punishable by either a fine or imprisonment. This, according to farmers, seldom happens in practice, since most conflicts are settled amicably by those involved. This process of conflict resolution, the stakeholders involved, constituent membership and the organizational structure of the WMIs appear to have evolved as the experience of wetland management has grown and the wetland environment itself has changed as a result of human interventions. Successive changes of governments have also played a key role in shaping these institutions, albeit indirectly through local adaptive responses to political and socioeconomic change.

Historical changes

During the Haile Selassie era (1930–1974), it was common practice for either the landlord to appoint an Abba Laga leader to set up a WMI (which would be known by the same name) and coordinate wetland cultivation in each wetland, or for farmers themselves to propose an Abba Laga. In those cases where landlords owned numerous wetlands, a *Teteri* (landlord's representative) was appointed to oversee the local Abba Laga (see Fig. 8.3). In compensation for their wetland management duties, the taxes of each appointed Abba Laga leader would be waived. Any problems arising in the wetland were first reported to the Abba Laga leader, then to a Teteri if present and, finally, to the landlord himself (if present, since many were absentees). The final decision on whether to cultivate the wetlands ultimately rested with the landlord.

Land nationalization during the Derg era (1974–1991) brought about the redistribution of wetland plots, largely on the basis of family size and, with no landlords present, the organization and election of WMIs – which occurred then for the first time in some communities –

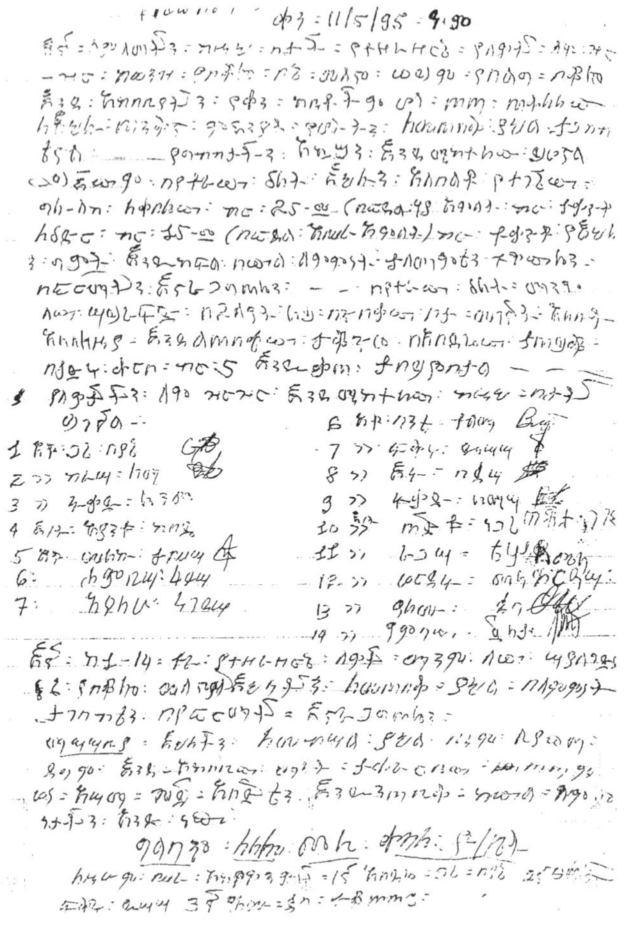

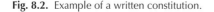

Date: 19th January, 2003

We, the users of Gafare Abba Mati wetland have agreed to cooperate with each other in wetland agriculture according to our local calendar.

The following are our terms of agreement:

- We have agreed to guard wetland crops against wildlife and livestock turn by turn.

- Anyone who fails to properly guard crops on his turn and as a result of which wetland crops may be damaged, will be penalized 25 Ethiopian birr to the kebele administration, 10 birr to the Ider, and pay the estimated amount of crop damaged to the owner of the crop.

- When the crop is mature, every neighbourhood should construct huts and is expected to guard the crop regularly against wildlife damage in the evening.

- Every individual on his turn is expected to visit the field without waiting for orders from coordinators, at 8:00 am, and he should properly guard the crop the whole day. If he fails to do so, his case will be investigated by the Ider and if found guilty he will be penalized 5 birr.

- The following three people are elected as the management committee
    Ambaw Jaleta - Abba Adere /Chairman
    Fikadu Disasa - Secretary
    Alemu Daba - Inspector

- The names of the signatories of this agreement are listed below:

1. Gari Beyene	5. Melku Tasisa	9. Benti Tesema	13. Tajitu Nagari
2. Birrisa Lemma	6. Hambisa Fayissa	10. Fikiru Disasa	14. Reggasa Tesso
3. Fikadu Lencho	7. Ayalew Rudsa	11. Ifa Bedasa	15. Hordofa Chala
4. Ayantu Kebede	8. Alemu Daba	12. Fikadu Lamessa	16. Ambaw Jaleta

We whose names are listed above have agreed to be governed by the rules mentioned and have approved it with our signature without any external pressure.

CC
To Lalo Chole kebele administration office

Fig. 8.2. Example of a written constitution.

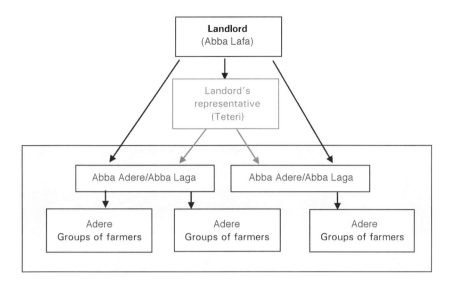

Fig. 8.3. A typical WMI structure during the Haile Selassie era.

were placed in the hands of the wetland users themselves. In many respects, these traditional labour and management associations complemented the new government's ethos of local-level cooperation and, indeed, evidence suggests that some of the WMIs formed in this era originated as farmer cooperatives:

> The wetland management committee was established here in 1982 when we first started to fully cultivate the wetland. The idea came from the *woreda* [district] agriculture office. They passed instructions to all kebeles to drain and cultivate wetlands, and they suggested that we organize ourselves to cultivate in a coordinated manner. Then we discussed it among ourselves, started draining the wetland and established the management committee.
> (Farmer at the Shenkora Wetland, West Wellega, 31 January 2003)

The newly formed kebeles (neighbourhood cooperatives and the smallest units of government), however, were assigned ultimate responsibility for overseeing the functioning of the WMIs, although the nature of their influence varied from location to location. Whilst the kebele played a key role as decision maker in wetland cultivation activities in some areas, in others its function was solely one of last resort for conflict resolution. In the opinion of most farmers, the kebele–wetland institution relationships formed during this era were the most productive in terms of efficient and successful wetland management, on account of the backstopping (largely enforcement-orientated) role played by the kebele.

The number of members in the leadership of the WMIs also increased during the Derg era and, in some areas up to seven persons, rather than one, are reported to have been elected to the WMI committee by the community of wetland cultivators (see Fig. 8.4). As in the Haile Selassie era, the committee of the WMI was responsible for the day-to-day operations of wetland management, namely coordinating activities and reporting problems. Farmers refusing to abide by the constitution of the WMI (which would have been discussed in an annual or biennial meeting of a Wetland Users Assembly) were then reported by the committee to the kebele administration, who may or may not have referred the matter to the kebele court.

According to farmers, however, conflict between those in the WMI was a rare occurrence during the Derg era on account of the strong punishments imposed:

> Even a farmer could be imprisoned for a month and his ownership right could be removed by the wetland committee without any approval from

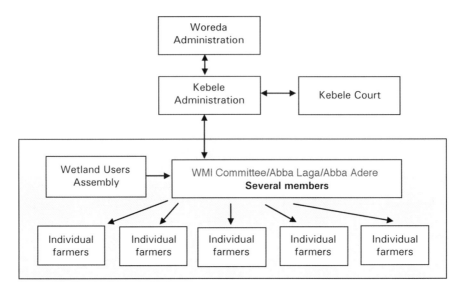

Fig. 8.4. A typical WMI structure during the Derg era.

the kebele administration. With one farmer, his legs and hands were tied with a rope and he was put next to termites for more than two hours to punish him … because of these practices coordination among wetland users was very strong and the wetland management committee was also very strong and powerful.

> (Farmer at the Minie Wetland, Illubabor,
> 8 February 2003)

The kebele administration was reportedly very active in supporting the WMI and although, like the landlord, it often intervened directly in wetland use issues in a top-down manner, it arguably provided a means of legitimizing the authority of the WMI committee.

The post-Derg era since 1991 has been characterized by further land redistribution and minor structural changes to WMIs (see Fig. 8.5). The federal government's focus on regionalization and decentralization appears to have had both positive and negative impacts. Most WMIs have undergone self-induced structural change to fit in with the new 'democratic' ideals promoted by the government: for example, increasing membership of the WMI committee to make it more transparent and accountable to its members. Many farmers, however, protest that the concept of democracy has been

misinterpreted by wetland users who wish to 'exercise their individual democratic right' not to participate in communal activities such as ditch digging or guarding against wild pests.

WMIs have also eroded further, say farmers, through the lack of interest shown by the kebele administration in addressing such problems, and in wetland management generally. The reasons offered by farmers for this lack of kebele support vary from corruption among the kebele committee to a simple lack of available time and resources, in the light of more pressing socio-economic and political concerns. The net result, however, is that most WMIs now exist in a more weakened state than ever before, since their mechanisms of enforcement have been removed. This, potentially, has major implications for the sustainable management of wetlands:

> Everyone says, 'it is my right to work or not'. It seems the current democracy given by the government is having a negative role in the area. The kebele administrations don't take any measures against those who refused to cultivate. In summary, the current situation is not conducive to using our wetland properly.
>
> (Farmer at the Korqa Wetland, West Wellega,
> 30 March 2003)

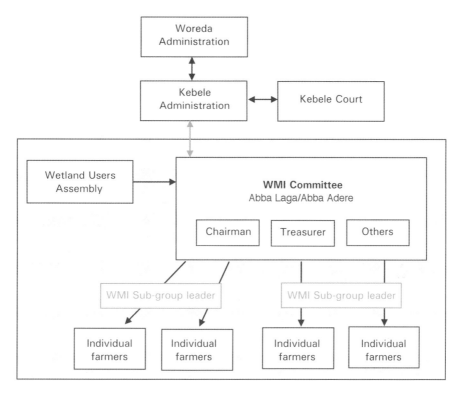

Fig. 8.5. A typical WMI structure in the post-Derg era.

Building Relations with the State: the Key to Wetland Institutional Sustainability?

What is clear is that a range of institutions external to WMIs have, to varying degrees, influenced their structure and functioning since wetland cultivation first began. In Illubabor, local government institutions such as the Ministry of Agriculture and the kebele and woreda administrations have played a particularly prominent role during the initial establishment of WMIs. Perhaps as a result of this, the WMIs in Illubabor now appear more dependent on the intervention of these external institutions for the enforcement of rules and regulations. Even in Western Wellega, where wetland cultivation has a longer history, and where external state institutions have been less influential in the actual formation of WMIs, farmers now insist that the kebele administration should play a fundamental role in conflict resolution.

Despite their informal linkages with external and indigenous institutions, the extent to which WMIs are recognized as functioning institutions in the wider context of wetland policy making is ambiguous. Some communities evidently work more closely with staff from the Ministry of Agriculture than others and, in some cases, members of the ministry staff assist in organizing meetings for the WMI committee. In such cases, local development agents undoubtedly have some sensitivity to the importance of wetlands and their local management arrangements.

At any level higher than that of the development agent, however, there appears to be no official recognition of WMIs in official agricultural or NRM policy, although members of the ministry staff do personally acknowledge their existence. Given that the majority of WMIs acknowledge the need for external intervention in their decision making, a key challenge exists in terms of raising awareness of the potential importance of WMIs (and arguably other local institutions) among external policy-making institutions, and transforming existing knowledge and awareness into real policies that support local-level wetland management.

Given that the functions and structure of the WMIs have much in common with those local organizations cited in much of the literature (Rasmussen and Meinzen-Dick, 1995; Blunt and Warren, 1996; Agrawal, 2001; Mazzucato and Niemeijer, 2002) a key question is: To what extent have they facilitated the sustainable management of wetlands in Illubabor, Western Wellega and possibly elsewhere in Ethiopia?

Whilst most wetland users would agree that crop yields have declined over the years as a result of falling fertility in the wetlands, there is little evidence of widespread wetland degradation characterized by falling water table levels and the inability of wetlands to support agriculture or cheffe production. As suggested earlier, the intimate knowledge and understanding of the wetland environment among wetland users in most cases inform management practices, such as fallowing and ditch-blocking for moisture management, that support sustainable utilization (Dixon, 2003). Furthermore, many wetland farmers are active in small-scale experimentation, seeking solutions to problems and adapting to change.

In examining the relatively few cases of serious wetland degradation, the Ethiopian Wetlands Research Programme drew attention to several potential causal factors. First, where local NGOs or government departments were particularly active in dictating drainage and cultivation regimes, local knowledge and management practices were effectively overridden, and the benefits of wetland agriculture under such scenarios were found to be short-lived (Wood, 1996; Afework Hailu et al., 2000, unpublished).

Secondly, the research highlighted the use of inappropriate practices, such as the excavation of deep drainage channels or the allowing of cattle unlimited access to wetlands, critically, despite recognition from farmers themselves that these practices were destructive. In examining why this might be the case, it was suggested that the lack of cooperation and communication between farmers, often from different kebeles, prevented the formation of institutional arrangements to govern equitable and sustainable wetland use. Similarly, in a recent study of the relationship

between communication networks and wetland sustainability (Dixon, 2005), the lack of social capital, in terms of cooperation, communication networks and common values among wetland-using communities, clearly manifested itself in the form of wetland degradation. In effect, a breakdown in communication, cooperation and mutual respect among wetland users, for reasons which require further investigation, was shown to lead to destructive practices such as overgrazing and double cropping.

It would appear, therefore, that in Illubabor and Western Wellega, the WMIs do make an important contribution to the environmental sustainability of wetlands via the regulation of management practices, their role in conflict resolution and, at the very least, bringing stakeholders together.

There remain, however, inherent problems with wetland management, suggesting that the current system, and indeed the institutions associated with it, lack the capacity to cope with elements of environmental, socio-economic and political change. Hence, there are concerns over the environmental, economic and social sustainability of wetland management, and the sustainability of the WMIs themselves.

The importance of the adaptive capacity of WMIs is illustrated by the concern among farmers that they are increasingly struggling to cope with unprecedented variability in the timing, duration and intensity of the rains, which affect soil moisture, weed growth, the prevalence of insect pests and, ultimately, the economic sustainability of wetland cultivation. Adapting to such changes in weather patterns means adjusting the wetland farming calendar and being flexible and responsive, a logistically difficult task given the communal nature of management and farmers' concurrent interests in the uplands. Clearly, the WMI has a critical role to play in facilitating adaptation to change in such circumstances; yet at the same time the institution appears to struggle with even the more mundane and predictable wetland management issues, such as coordinating farmers to guard against wild pests:

> We depend on wetland cultivation for food production but in recent years there have been problems with termites, wild pests and worms. The problem is getting worse because of a lack of coordination among each other. In the past, Abba Laga was powerful in coordinating farmers but

today it is powerless to take action against those who break the rules due to a misinterpretation of the new democracy, and lack of support from government. Some farmers abandon the land between plots, so the rodents and other wild animals hide there and attack the crops.

(Farmer at the Korqa Wetland, West Wellega,
30 March 2003)

The above quote reiterates the major problems influencing the effective functioning and sustainability of the WMIs, and the sustainability of wetland use itself at the present time: the loss of respect for the institutions among their membership, and the lack of support provided by the kebele administration. Both, according to farmers, have weakened the capacity of WMIs to coordinate wetland management and enforce what are ultimately reciprocally beneficial rules and regulations.

In response to these problems, farmers are seeking greater external intervention in the functioning of the WMIs. They regard the kebele administration and the associated kebele court as structures that legitimize and backstop the rules and decision-making process that are central to the effective functioning of the WMIs. Many would like to see the kebele playing a role in formulating written constitutions for wetland management, which include enforceable penalties for non-compliance. Others seek greater representation of wetland stakeholders on the kebele committee. If, however, more power were to be handed over to external institutions, this once again raises issues of local institutional legitimacy, effectiveness and sustainability (Ostrom, 1990; Richards, 1997; Serra, 2001, unpublished; Watson, 2003). Moreover, there is a danger that the kebele administration, as a government-biased institution, will encourage the use of wetlands in an unsustainable manner, in pursuit of politically important policies such as short-term regional food security.

With the local government intervening more in the day-to-day operations of the WMIs, one key concern is whether this would erode their flexibility and capacity to operate effectively outside the vagaries of bureaucratic government administrations. One could argue, however, that this erosion is unlikely to occur, since the WMIs of Illubabor and Western Wellega are atypical of the many 'indigenous' local institutions cited in the anthropological

literature. They have, in effect, always relied on external intervention in one way or another, whether this was via landlord edicts during the Haile Selassie era or the kebeles who helped establish many WMIs during the Derg era. Moreover, the need for external intervention emanates from the WMI members themselves and, hence, any linkages formed arguably represent 'bottom-up' adaptive responses to the current pressures facing the WMIs.

In addition to seeking increased participation from the kebeles, some WMI members have suggested that they should possess their own legal authority, thereby enabling them to implement penalties without having to refer cases to the kebele committee or court. Accordingly, most communities are in the process of increasing membership of the WMI leadership and establishing more subgroup leaders so that more land inspections can be carried out and conflict resolution dealt with in a more democratic fashion:

> To strengthen the committee we want to increase the committee members from five to seven. The addition of committee members is to strengthen the activities of team leaders and proper inspection of the whole wetland system. The wetland management committee also needs to establish a strong constitution which manages the whole wetland system including the catchment. The committee needs to get recognition by government bodies and considered a legal community organization.
>
> (Farmer at the Minie Wetland, Illubabor,
> 5 March 2003)

In those WMIs where problems exist, it seems most members are well aware of the nature of the problems and potential solutions to these problems. Whilst some have begun to make small-scale changes to their structure and functioning in response to new challenges, many appear powerless to implement changes at the present time, again because of the perceived withdrawal of local administrative support. If this support is provided, however, and WMI members are allowed scope to implement their suggested changes, then the adaptive potential of these institutions could dramatically increase. This would inevitably empower capacity for sustainable wetland management.

Conclusions

The local WMIs found throughout Illubabor and Western Wellega make a key contribution to the sustainable use of wetlands throughout the area. In those wetland-using communities where they are present and functional, empirical evidence suggests that wetland cultivation is not affecting the capacity of wetlands to continue to support agricultural activities, sedge production and natural functions, such as water storage. Although the repeated cultivation of wetlands has inevitably led to a decline in soil fertility and agricultural productivity in most areas, the complete degradation of wetlands to a dryland environment has largely been avoided, due mainly to locally developed management practices which are coordinated and regulated through WMIs.

Although most WMIs throughout the area are similar in terms of their objectives, structure and functioning, their evolution reflects site-specific experiences of wetland management and varying degrees of intervention from external institutions. In Western Wellega, where there are indications that wetlands have been used for agriculture for over a century, WMIs have developed directly from similar institutions within the indigenous Oromo Gadaa system. In Illubabor, many WMIs have been established more recently through consultation with external institutions such as the Ministry of Agriculture.

This chapter has drawn particular attention to the relationship between WMIs and external institutions, in the context of a wider debate in the literature that considers whether such a relationship is beneficial or detrimental to the functioning of local institutions and their sustainability. In this respect, it is difficult to draw firm conclusions, since closer inspection

of the history of many of the WMIs suggests that they have always operated in close contact with external institutions, whose influence has been spatially and temporally variable.

Moreover, the present situation is one where WMI members are actively pursuing stronger ties with external institutions for the enforcement of their own institutional arrangements. There is a danger, however, that in seeking support from external institutions, the WMIs will effectively hand over power and decision making to government structures renowned for their lack of sensitivity to local communities, and their deep suspicion of civil society groups. The main challenge for the future sustainability of the WMIs, and arguably wetland management itself, therefore, is achieving a level of external support which recognizes and values local knowledge, local decision making and social capital.

Acknowledgements

The research on which this chapter is based was achieved with the support of the UK Economic and Social Research Council (Grant Ref: RES-000-22-0112). The authors are also grateful to Ato Afework Hailu and Ato Legesse Taffa for their assistance during the implementation of the research, and for their comments on this chapter.

Endnote

[1] Although most women are actively involved in the collection of water and medicinal plants from wetlands, only the very few involved in wetland cultivation (those either widowed or divorced) are able to participate in the WMI.

References

Alemneh Dejene (1990) *Environment, Famine and Politics in Ethiopia: a View From the Village*. Lynne Reiner Publishers, Boulder, Colorado and London.

Agrawal, A. (2001) Common property institutions and sustainable governance of resources. *World Development* 29 (10), 1649–1672.

Bardhan, P. (1993) Symposium on management of local commons. *Journal of Economic Perspectives* 7 (4), 87–92.

Blunt, P. and Warren, D.M. (1996) *Indigenous Organizations and Development*. ITDG Publishing, London.

Boserup, E. (1965) *The Conditions of Agricultural Growth.* Allen and Unwin, London.

Dixon, A.B. (2003) *Indigenous Management of Wetlands: Experiences in Ethiopia.* Ashgate, Aldershot, UK.

Dixon, A.B. (2005) Wetland sustainability and the evolution of indigenous knowledge in Ethiopia. *The Geographical Journal* 171 (4), 306–323.

Dugan, P.J. (1990) *Wetland Conservation: a Review of Current Issues and Action.* IUCN, Gland, Switzerland.

Ethiopia Network on Food Security (2001) Monthly report, 12 February 2001. http://www.fews.net/centers/files/Ethiopia_200101en.pdf

Guri, B.Y. (2003) Indigenous institutions: potentials and questions. *Compas Magazine* 9, 16.

Hassen, M. (1990) *The Oromo of Ethiopia: a History 1570–1860.* Cambridge University Press, Cambridge, UK.

Hinchcliffe, F., Thompson, J., Pretty, J., Guijt, I. and Shah, P. (1999) *Fertile Ground: the Impacts of Participatory Watershed Management.* ITDG Publishing, London.

Howes, M. (1997) NGOs and development of local institutions: a Ugandan case-study. *The Journal of Modern Africa Studies* 35 (1), 17–35.

Hulme, D. and Woodhouse, P. (2000) Governance and the environment: policy and politics. In: Woodhouse, P., Bernstein, H. and Hulme, D. (eds) *African Enclosures? The Social Dynamics of Wetlands in Drylands.* James Currey, Oxford, UK, pp. 215–232.

Koku, J.E. and Gustafson, J.E. (2001) Local institutions and natural resource management in the South Tongu district of Ghana: a case study. *Sustainable Development* 11 (1), 17–35.

Leach, M., Mearns, R. and Scoones, I. (1999) Environmental entitlements: dynamics and institutions in community-based natural resource management. *World Development* 27 (2), 225–247.

Manig, W. (1999) Have societies developed indigenous institutions enabling sustainable resource utilization? *Journal of Sustainable Agriculture* 14 (4), 35–52.

Mazzucato, V. and Niemeijer, D. (2002) Population growth and environment in Africa: local informal institutions, the missing link. *Economic Geography* 78 (2), 171–193.

Ostrom, E. (1990) *Governing the Commons: the Evolution of Institutions for Collective Action.* Cambridge University Press, Cambridge, UK.

Pretty, J. and Ward, H. (2001) Social capital and the environment. *World Development* 29 (2), 209–227.

Rasmussen, L.N. and Meinzen-Dick, R. (1995) *Local Organisations for Natural Resource Management: Lessons from Theoretical and Empirical Literature.* EPTD Discussion Paper No. 11, International Food Policy Research Institute, Washington, DC.

Richards, M. (1997) Common property resource institutions and forest management in Latin America. *Development and Change* 28, 95–117.

Shivakumar, S.J. (2003) The place of indigenous institutions in constitutional order. *Constitutional Political Economy* 14, 3–21.

Silvius, M.J., Oneka, M. and Verhagen, A. (2000) Wetlands: lifeline for people at the edge. *Physical Chemistry of the Earth (B)* 25 (7–8), 645–652.

Tafesse, A. (1996) Agro-ecological zones of south-west Ethiopia. MSc thesis, University of Trier, Germany.

Tiffen, M., Mortimore, M. and Gichuki, F. (1994) *More People, Less Erosion: Environmental Recovery in Kenya.* John Wiley, Chichester, UK.

Uphoff, N. (1992) *Local Institutions and Participation for Sustainable Development.* Gatekeeper Series No. 31, IIED, London.

Wade, R. (1988) *Village Republics: Economic Conditions for Collective Action in South India.* ICS Press, Oakland, California.

Warren, D.M., Slikkerveer, L.J. and Brokensha, D. (1995) *The Cultural Dimension of Development: Indigenous Knowledge Systems.* ITDG Publishing, London.

Watson, E. (2003) Examining the potential of indigenous institutions for development: a perspective from Borana, Ethiopia. *Development and Change* 34 (2), 287–309.

Wood, A.P. (1996). Wetland drainage and management in south-west Ethiopia: some environmental experiences of an NGO. In: Reenburg, A., Marcusen, H.S. and Nielsen, I. (eds) *The Sahel Workshop 1996.* Institute of Geography, University of Copenhagen, Copenhagen, pp. 119–136.

9 Indigenous Systems of Conflict Resolution in Oromia, Ethiopia

Desalegn Chemeda Edossa,[1] Seleshi Bekele Awulachew,[2] Regassa Ensermu Namara,[3] Mukand Singh Babel[4*] and Ashim Das Gupta[4]**

[1]Haramaya University, Ethiopia; e-mail: dchemeda@yahoo.com; [2]International Water Management Institute (IWMI), ILRI-Ethiopia campus, Addis Ababa, Ethiopia; e-mail: s.bekele@cgiar.org; [3]International Water Management Institute (IWMI), Accra, Ghana; e-mail: r.namara@cgiar.org; [4]Asian Institute of Technology, School of Civil Engineering, Pathumthani, Thailand; *e-mail: msbabel@ait.ac.th; **e-mail: adg@ait.ac.th

Abstract

This chapter describes the role of the *gadaa* system, an institution developed for guiding the social, political, economic and religious life of the Oromo people in Ethiopia and for managing resources such as water, as well as its contribution in conflict resolution among individuals and communities. It discusses ways to overcome the difference between customary and statutory approaches in conflict resolution. A synthesis of customary and statutory systems of conflict resolution may facilitate a better understanding that will lead to improved management of resources, which are predominant variables for the socio-economic development of the country. It suggests that top-down imposition and enforcement of statutory laws that replace customary laws should be avoided. Instead, mechanisms should be sought to learn from the Lubas, elders who are knowledgeable in the gadaa system, about the customary mechanisms of conflict resolution so as to integrate them into the enactment or implementation of statutory laws.

Keywords: gadaa, indigenous institution, water management, Oromo, conflict resolution, Borana, Ethiopia.

Introduction

The water resources endowment of Ethiopia exhibits tremendous spatial and temporal variability and poses significant development and management challenges. The level of water supply in Ethiopia is among the lowest in Africa. A great majority of Ethiopians use unsafe and polluted water, and are at risk from a variety of water-borne diseases (Flintan and Imeru, 2002). The strong bias towards urban development means that the provision of water supplies in rural areas is particularly low. Its availability in the dry season is of great concern to the majority of rural populations across the arid and semiarid parts of the country, where villagers travel long distances to the nearest sources of water after local sources have become exhausted as a result of the prolonged dry season. For example, in the Awash river basin, spending 4–6 h on a daily basis for getting water is not uncommon for a rural household living far from a river course (Desalegn et al., 2004).

In the Dollo and Filtu districts of the Liban zone of the Somali regional state, there is hardly any perennial source of water between Genale and Dawa, the two main rivers in that state. Therefore, villagers in places like Filtu must rely on water tankers and boreholes, once the nearby local pond dries out (Ahrens and Farah, 1996). These situations apply equally to most pastoral lands in Oromia, which experience low annual precipitation, averaging between 400 and 700 mm. In many of pastoral areas drought occurs on a regular basis. As a result, pastoral land use depends on scarce water supply from the rivers and groundwater. Consequently, both intra- and inter-ethnic conflicts over the use of natural resources are commonplace in these areas.

Grimble and Wellard (1997) categorize conflicts in terms of whether they occur: (i) at the micro–micro or micro–macro levels, i.e. among community groups or between community groups and government; or (ii) within private or civil society organizations. Micro–micro conflicts can be further categorized as taking place either within the group directly involved in a particular resource management regime (e.g. a forest user group or ecotourism association), or between this group and those not directly involved (Conroy *et al.*, 1998). For instance, access to water and land resources is the major source of conflicts between clans and ethnic groups in the Awash river basin and the Borana zone, while territory is another important source of conflicts in the former. Consciousness of clan 'territory' is more intense nearer to the water source such as the Awash river, whereas exclusive rights to land are less important farther from the water source, indicating the significance of water resources to the pastoral communities' socio-economic survival.

An illustrative example of micro–macro conflicts is found in the Awash basin. A number of studies have attributed the cause of conflicts in the Awash river basin to the introduction of various large-scale irrigation schemes along river courses and the opening up of the Awash National Park on the land predominantly used by pastoralists for grazing during the dry season and during droughts, which also limited access to key dry-season springs (Flintan and Imeru, 2002). As a result, competition between pastoralist groups increased as they moved in search of pasture and water supplies. Many of the development projects

in the basin involve investment by international organizations with a top-down approach, bypassing the customary laws of the indigenous communities. Bassi (2003) presents the feelings of the local community Karrayu elders about the establishment of the Awash National Park in the year 1969 as follows:

> Haile Selassie [Ethiopian emperor] sent his ministers. They asked us whether we agree to the establishment of the park or not. Their question was not genuine, since they had already taken all the land without consulting us. It was intended to produce a pretext to arrest us as usual. We told them that we do not give all of our land since we have no other place, but part of it. We, then, agreed out of fear, obviously, to give the land east of Fantale Mountain for the park. They agreed to give us land west of the Fantale Mountain. We accepted since we could not do anymore. When they prepared a map of the park and began to protect the land, the thing was different. They reversed the agreement. The map of the park included areas west of Fantale Mountain, which they previously agreed to give us. They have begun to evict us. They built a camp in our settlement areas. We repeatedly asked the government and the park to respect our joint agreement but no one listened to us …
>
> (Karrayu elder, quoted in Buli Edjeta, 2001, p. 86 (cited in Bassi, 2003))

Some of the local people gained some economic benefits from developments in the Awash valley mainly through employment opportunities. However, such trends sowed the seeds of further conflict within Afar political structures as a growing Afar capitalist class undermined traditional clan elders. This was a factor in the violent conflict that was manifested in the *Derg*[1] period. The most common inter-clan and inter-ethnic conflicts are between the Karrayu and Ittu Oromo communities and the Afar and Issa communities, respectively.

During the Derg regime, peasant associations (PAs, or *kebeles*) were the powerful instrument of formal conflict resolution. They had their own judicial committee to oversee conflicts and had the power to impose decisions through fines and imprisonment. Under the current regime, kebele administrations (KAs) have been set up, bringing together two or three of the former PAs, with similar judicial powers to the latter. In addition, governmental teams have been established to represent a maximum of 50 households, thus bringing state institutions to an even more local level. Conflicts relating to natural resources management are nowadays often reported to the governmental teams and, through them, to the KAs.

There also exist various traditional institutions in the country that have their own customary methods of settling conflicts. In this regard, the *gadaa* system of conflict resolution is one that deserves attention. Although its powers have diminished over recent decades, this institution is well respected by the Oromo society at large in the country. If this indigenous knowledge can be harnessed, it can be a means through which sustainable development may be achieved (Watson, 2001). However, there exists a loose collaboration and, in some cases, even a contradiction between these statutory and customary institutions in the management of natural resources and conflict resolution.

This chapter presents the role of the gadaa system in conflict resolution, through better management of one of the scarce natural resources – water. Historical conflicts over the use of natural resources in Borana, the major

pastoral area of the regional state of Oromia, and local methods of resolving these conflicts are reviewed. The organizational structure of the gadaa system is explained and the current and potential interface between this institution and the statutory method of conflict resolution is discussed. Special emphasis is given to the gadaa system of Borana Oromo. In this area, the gadaa system of governance is still active as compared with that in other areas of the regional state. In addition, the area is facing various degrees of water scarcity and is the target of various water development projects in the country, and is therefore an area very susceptible to competitions and conflicts.

The Borana Zone of Oromia and the Nature of Conflicts over Water and Land Resources

The Regional State of Oromia comprises 13 administrative zones, including the Borana zone, which is located at the southern edge of Ethiopia bordering Kenya and Somalia (see

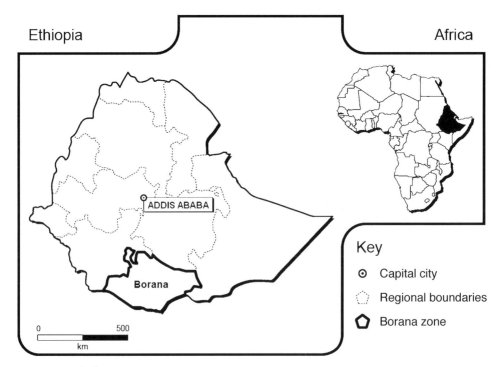

Fig. 9.1. Map of Ethiopia.

Fig. 9.1). The zone is made up of 13 districts called *Woreda*[2], divided between two agro-ecological zones – the semiarid lowlands to the south and the more humid lands at higher altitudes to the north (Tache and Irwin, 2003).

The mean annual rainfall across the districts varies between 500 and 700 mm, with an overall average of 648 mm. Surface evaporation is high. The mean maximum and minimum temperatures of the area vary from 25.26 to 28.79°C and 14.19 to 18.11°C, respectively (Luseno *et al.*, 1998). There are two rainy seasons: the main season, *ganna* (March–May) and the minor season, *hagayya* (September–October). The land is largely covered with light vegetation of predominantly pod-yielding Acacia species of low forage value. The ecological conditions favour pastoralism more than farming.

The Borana zone[3] is inhabited by different ethnic groups, including Oromo, Somali, Gedeo, Burji, Konso, Amhara, Worradube and Bonta. The most significant inhabitants (in terms of number) belong to the various Oromo and Somali clans. Among the Oromo clans, the Borana and Gabra reside mainly in the semiarid lowlands, while the Guji and Arsi Oromo clans are settled in humid lands at higher altitude. The various Somali[4] clans living in the zone include Garii,[5] Digodi, Merehan, Gurre, Duriane and Shabelle. Some of these groups, such as Garii, straddle land between Kenya, Ethiopia and Somalia due to their nomadic lifestyles. Amhara, Gedeo, Burji and Konso are settlers and live in and around towns and are engaged in trading and farming activities.

The farming activities that are usually practised on the hitherto pastoral lands through encroachment are the potential source of conflict. The Borana Oromo (Boran) are numerically the dominant group inhabiting the Borana zone. The area is endemic to conflicts between these rival pastoral groups over resources. During the 1990s, the frequency and magnitude of conflicts increased. For instance, in 2000, three major conflicts occurred between the major pastoral groups (Boran *versus* Garri, Merehan *versus* Digodi and Digodi *versus* Boran). These conflicts, in combination with severe drought, resulted in the death of hundreds of people and dislocations (Dejene and Abdurahman, 2002). There are serious tensions and sporadic violence between Garri returnees from Kenya, who currently claim

to be a Somali clan, and the Boran (Tache and Irwin, 2003).

According to a United Nations Development Programme (UNDP) report (Ahrens and Farah, 1996), while Borana and Liban zones in Ethiopia are prone to drought, adjacent areas in neighbouring Kenya and Somalia are even more likely to suffer from water scarcity. During times of complete failure of rainy seasons in northern Kenya and south-west Somalia, there are often apparent influxes of pastoralists from those countries into Ethiopia searching for water and pasture. These situations lead to conditions where local people and 'guests', often related by trans-border kinship and sharing common languages and cultures, have to compete for the use of the few perennial water resources. Similarly, Watson (2001) provides a thorough account of conflicts between Somali groups and the Boran over the use of natural resources. Coppock (2001) used results from 120 group interviews collected in 1998 to quantify how inhabitants across northern Kenya and southern Ethiopia perceived and ranked various risks to their livelihoods and found that reliable access to food, feed and water were the most common sources of risks in the area, these being related to drought, market inefficiencies or insecurity.

Institutions of Conflict Resolution and Natural Resources Management in the Borana Zone

The traditional mechanisms of resolving conflicts and managing natural resources (i.e. water, land and forests) in the Borana zone is derived from the Oromo institutions of gadaa, *aadaa*, *seera* and *safuu*, and the associated cultural administrative structure.

Gadaa is a system of social organization based on age-grade classes of the male population that succeed each other every 8 years in assuming economic, political, military and social responsibilities.[6] A complete gadaa cycle consists of five or six age-grades, excluding those stages following *luba* (see Table 9.1). The number of age-grades is cited differently in various literatures. For example, Gumii Bilisummaa Oromiyaa (2000) reports five grades in the whole cycle of the gadaa system whereas Constantinos (1999) and Workneh (2001)

Table 9.1. Different gadaa grades with their corresponding roles (adapted from Constantinos, 1999 and Workneh, 2001[a]).

Grades	Designation	Age limit	Remarks	Specific role in society
1.	*Dabballee*	0–8	Child is born; stage of childhood	None; immature, sons of the gadaa class or the luba; only symbolic role as mediator between God and humans
2.	*Foollee* (*Gaammee xixiqoo*)	9–16	Naming ceremony at home or *Nura Shrine* in *Liben* if *Ilmaan jaarsaa* or *Ilmaan korma,*[7] respectively	Some look after small stock around *ollaa*; perform light work
3.	*Qondaala* (*Gaammee gurguddoo*)	17–24	Intensification of the 2nd stage	Take livestock further away from *ollaa* and begin drawing water from *Eelaa*; can go long distances to hunt; perform heavy work
4.	*Kuusaa*	25–32	Politically significant	Nucleus of future gadaa leaders emerges through nominations by the current gadaa class (see grade 6, below); the nominated future luba are formally installed in office; however, they do not yet assume full authority
5.	*Raabaa doorii*	33–40	This and the kuusaa grade constitute a period of preparation for the assumption of full authority	Important military wing of the gadaa system; conduct raids; protect Boran territory and resources against enemies; men are allowed to marry
6.	Gadaa (Luba)	41–48	Politically the most active	Leadership grade – the most important of all stages; Luba assumes power/office; transition is marked by leadership ceremony; visit all Borana regions, settle serious disputes and convene assemblies
7.	Yuba I	49–56	Retirement stage	Advisory role in the society; receive a great deal of respect as wise, experienced authorities and repositories of law
8.	Yuba II	57–64	Retirement stage	
9.	Yuba III	65–72	Retirement stage	
10.	*Gadaamojjii*	73–80	Marked by rites at different sites	Senior advisor
11.	*Jaarsa*	> 80	Stage of old age	At a stage to be cared for

[a] Grade designation and age limits differ slightly between the two sources.

recognize six grades, regardless of the stages following luba.

The gadaa system organizes Oromo social life around a series of generation grades that assign obligations as well as rights to all the males in the society. Each man born to or adopted by Oromo parents is automatically placed for life into a ready-made pattern of positions and moved through it, performing various services for the public and also receiving certain privileges. Each man contributes his labour power in different capacities to the society as a whole.

The grades are also periods of initiation and training as well as periods of work and performance. The roles and rules attached to the age-grade system are the most important elements that regulate the gadaa system. Every Oromo man of specific age-grade is expected to perform a certain function according to specified rules and regulations. When an Oromo man passes from one stage to the next, his duties and way of life in society also change. For instance, during the grades of *qondaala*, *kuusaa* and *raabaa doorii*, individuals learn war tactics, Oromo history, politics, ritual, law and administration over a period of 24 years. When they enter the gadaa class or luba at the age of about 40 years, they have already acquired all the necessary knowledge to handle the responsibility of administering the society and the celebration of rituals.

This process ends with the partial retirement of the whole group of elders to an advisory and judicial capacity. Following luba, men automatically retire from gadaa and move into an advisory role known as *yuba*. By then they receive a great deal of respect, as wise, experienced authorities and repositories of law, but their decisions are no longer final, as they had previously been. They turn the bulk of their attention to private family businesses or religious activities while their sons enter gadaa, the public service.

Luba is the ruling grade. Its members hold all political authority, elect representatives to attend a national convention called *caffee*, where the laws of the land are amended by the vote of tens of thousands of lubas and where officials are elected to administer the society in a wide variety of capacities. The caffee elects nine gadaa officials. The following are the gadaa officials and their duties:

- *Abbaa bokkuu* or *abbaa gadaa* – president.
- *Abbaa bokkuu* I – vice-president.
- *Abbaa bokkuu* II – vice-president.
- *Abbaa caffee* – chairman of the assembly (caffee).
- *Abbaa dubbii* – speaker who presents the decision of the presidium to the assembly.
- *Abbaa seeraa* – memorizer of the laws and results of the assembly deliberations.
- *Abbaa alangaa* – judge who executes decisions.
- *Abbaa duulaa* – in charge of the army.
- *Abbaa sa'aa* – in charge of the economy.

These gadaa leaders are elected on the basis of wisdom, bravery, health and physical fitness (Workneh, 2001). Slight differences are observed among the Oromo communities across Oromia in the way they practice gadaa. The Boran have kept the system more intact than the Oromos in the other areas because of their relative isolation from external influences. In the case of the Boran, the entire gadaa presidium, consisting of nine

members, is called *Saglan Yaa'ii Boran* (nine of the Boran assembly). The current abbaa gadaa or Bokkuu (the president) is called *Guyyoo Goba;* he is the legitimate leader of the Boran. If the gadaa officials fail to carry out their duties, the caffee can replace them by another group from among the same gadaa class, which proves how accountability is entrenched in the governance system.

One major economic function of gadaa is the distribution of resources, by establishing who had to help whom, when and why, by settling conflicts between families over goods and by making laws. It is the system that governs the Boran's use of natural resources and enables the various groups to coordinate their use of important resources like water. According to gadaa, those people who have entered the luba grade (individuals in the expected age range of 40–48) are considered to be elders. Therefore, the lubas (elders) settle disputes among groups and individuals and apply the laws dealing with the distribution of resources, criminal fines and punishment, protection of property, theft, etc. Thus the elders in the community form a dominant component of the customary mechanisms of conflict management and natural resources management (Watson, 2001; Dejene, 2004; Desalegn *et al.*, 2004).

The authority held by the elders is derived from their position in the gadaa system. While the rules and regulations laid down by the gadaa tradition must be respected by all councils of elders, any problem regarding resources use which could not be solved by these elders would be handled by the higher gadaa leaders. Watson (2001) describes the role of abbaa gadaa in natural resources conflict resolution as follows:

> The abbaa gadaa is seen as the figurehead of the whole of Boran, and is often described as the president. As well as performing rituals, matters are referred to him and his council when a decision cannot be reached at a lower level. When conflict breaks out between *ollaas* or *araddaas*, or *maddaas*, then the abbaa gadaa will rule on the case. If there is conflict between ethnic groups, then he will be called in to help make peace. As the abbaa gadaa is responsible for dealing with matters of concern to the Boran, and as matters of concern are often related to access to the resources (water, land and forests), the abbaa gadaa is the highest level of institution of natural resources management in Borana.

Tache and Irwin (2003) also present evidence of how the diverse local communities, both Oromo and non-Oromo, in the Borana zone of Oromia coexist under the traditional negotiated systems of shared management of natural resources. Conflicts tend to be rapidly resolved through the traditional conflict resolution mechanisms.

The foundation of the gadaa system is rooted in the informal or customary Oromo institutions of aadaa (custom or tradition), seera (Boran laws), safuu (or the Oromo concept of Ethics) and *heera* (justice). These institutions form indigenous systems of knowledge and include the rules and regulations that determine access to natural resources. They define the access and the rights that a group has to natural resources. In the Borana zone, individuals, groups and organizations have different statuses regarding access to resource and use rights, and these institutions define their differentiated access and use. These indigenous institutions are rehearsed with both regularity and rigour and supported by networks of kin, and institutionalized in meetings and rituals. Natural resource access is governed by the combination of these different institutions, which are also conflict-resolution institutions and are uniquely placed to assist in tackling the interlinked problems of the environment, welfare, and conflict. The aadaa and seera are rehearsed at a meeting that is held every 8 years in Borana. Aadaa sanctions the different strategies that the Boran institutions at all levels adopt and restrict access to those parts of the pasture within their jurisdiction.

However, it is worth noting that gadaa is a male-oriented, socio-political and cultural system and excludes the Oromo women from its political and military structures. Taking the case of the Boran, Legesse (1973) states the following relationship between men and women:

> Men are in control of military and political activities. Only men can engage in warfare. Only men take part in the elections of leaders of camps or of age-sets and gadaa classes. Men lead and participate in ritual activities. However, ritual is not an exclusively masculine domain: there are several rituals performed by women. In these and a few other instances women do take an important part. Women are actively excluded

from age-sets. They are therefore heavily dependent on men for most political-ritual services and for all activities connected with the defence of Boran camps, wells, herds, and shrines.

However, there are parallel female-oriented institutions to gadaa known as *ateetee* and *siiqqee* institutions (Megerssa, 1993, unpublished PhD dissertation; Hussein, 2004). Oromo women used to practice ateetee as a way of strengthening their solidarity and as a tool to counter atrocities staged against them by men. Similarly, as a check and balance mechanism, siiqqee was institutionalized and women formed parallel organizations of their own that actively excluded men.

Another important informal institution with relevance to conflict resolution is the institution of *araara* (literal meaning, reconciliation) and *jaarsummaa* (literal meaning, the process of reconciliation between conflicting individuals or groups by a group of Jaarsaas). Dejene (2004) reported the effectiveness of the araara institution between the Karrayyu Oromo of the Upper Awash and its neighbouring ethnic groups like the Afar and Argoba. Araara is the process of conflict management involving individual clans within and outside the community. It is basically handled by the council of elders in the community and thus associated with the gadaa system, and called jaarsummaa in some localities. The term jaarsa is the Oromo version of elder, and thus jaarsummaa, is the process of reconciliation between conflicting individuals or groups by a group of jaarsaas (elders).

The Local Cultural Administrative Structures of Borana

In addition to the rules, laws, norms, customs and ethical values embedded in the gadaa system there are integrated sets of cultural administrative structures that regulate access to water, land and forests (see Table 9.2).

Water resources management in Borana

Management of water, as a common property, in Borana remains relatively intact to date (Tache and Irwin, 2003). Despite the collapse of

Table 9.2. Integrated sets of cultural administrative structures that regulate access to water, land and forests.

No.	Level of organization	Equivalent English terms or description
1.	*Warra*	The warra is the household; it is administered by the male head of the household, the abbaa warra, which literally means the father of the house.
2.	Ollaa	The ollaa is the smallest unit of settlement. It consists of between 30 and 100 warras. The head of the ollaa is called the abbaa ollaa (father of the ollaa), who is usually the first man to have founded that olla – or the senior descendant of the person who is considered to have done so.
3.	Araddaa	This is a small group of ollas – usually two or three only – who may cooperate together in their grazing patterns; they may jointly delineate and fence off grazing area for calves.
4.	*Madda*	This is the water point surrounded by a grazing land which is used by all those who use the water source. The abbaa madda, literally the father of the water source, is the authority at this scale of administration; he is the most senior male descendant of the man who originally found and excavated that water source; as he owns the water source and he has first right to it; he can decide who can and cannot use the water source. Related to this is the *abbaa konfi*, who owns ponds (developed through excavations), and the *abbaa herregaa* – an official responsible for the day-to-day supervision of watering procedures, including the maintenance and cleaning of wells, enclosures and environs; he is assigned at a meeting of the clan group council known as *kora eelaa*.
5.	*Dheeda*	This is the wider unit of grazing land used by different ollaas and araddaas. This spatial administrative unit is administered either by a council of elders (*jaarsa dheedaa*) or by an individual known as *abbaa dheedaa*. The size of the madda and dheedaa may vary and the boundaries may overlap.
6.	Borana-wide	The abbaa gadaa and his councillors are the governing body of Borana. The abbaa gadaa is the man who is elected to lead the Borana.

most of the indigenous institutions of Boran over the last 30 years, those concerned with the administration of water have sustained their importance (Homann *et al.*, 2004). It is important to note that access to water and grazing land is fundamental to the survival of Boran pastoralists because of the inherent nature of the ecological setting of the Borana zone. Thus, the water and land management functions of the gadaa system remained relatively robust.

Homann *et al.* (2004) gave detailed accounts of Boran's water management strategy under drought conditions, as follows:

- Wet season: after rainfall, open water sources are used and wells are closed.
- Dry season: herds are successively shifted to more distant ponds and traditional wells are re-opened to preserve water near the homestead.
- Progressing dry season (water scarcity): the drinking frequency of cattle is gradually and subsequently reduced to 1 day (*dhabsuu*),

2 days (*limaalima*) and finally 3 days (*sadeen*).

The coordination of access to water is also linked with tasks of cleaning, maintenance and rehabilitation. For example, cattle are restricted from entering the water sources by fencing off the sources and making them drink water hauled into troughs made from clay and cement (*naaniga*).

Traditionally, the Boran clearly define the rights to water for each of the various sources (wells, rivers and ponds). According to Watson (2001), the following are the most important sources of water (*madda*) that are highly regulated:

- Hand-dug shallow ponds (*haroo*): a pond is the property of an individual or his direct descendants who initially excavated it and the person is called *abbaa konfi*. Rights to use the pond are obtained by providing labour for the maintenance of the pond. Although the property of the abbaa konfi, the pond is administered by the local elders.

- Wells (*eelaa*): the wells are highly regulated in Borana. They are divided into two types, *adadi* (shallow wells) and *tulla* (deep wells). The tullas are famous because they can reach a depth of 30 m and water is drawn by a row of people standing one above the other and passing up containers of water. There are nine tullas throughout the Borana zone that contain water throughout the year and they are known as *tullan saglan* (the nine wells) (Helland, 1997).

Watson (2001) lists the following additional sources, where access is mainly opportunistic:

- Natural ponds containing water throughout the year, known as *bookee*.
- Rivers.
- Temporary ponds.
- Rainwater harvesting.

The opportunistic nature of access to these water sources implies that the right of access to the water depends, above all, on the reliability of the water supply (as they are either temporary or occasional sources) and landownership on the shoreline of the sources (the riparian rights doctrine). Watson (2001) reports that, in some cases, the rights to water from these sources have been privatized and are sold by individuals and groups. The access to these sources is characterized mainly by poor institutional development and little regulation. Tache and Irwin (2003) also maintain that occasional water sources (surface water from rain) have the most unreliable supply, and no restrictions whatsoever are imposed in accessing these.

By contrast, hand-dug ponds and wells are regulated and they are the most important sources of water as they are the most reliable and labour-intensive types. The wells are managed by a council of the clan group, which includes a retired *hayyuu* (special counsellors or individuals who hold ritual authority to judge (Watson, 2001)), the *jallaba* (a local lineage of clan elder or special messenger (Homann *et al.*, 2004)), the abbaa konfi (trustee of each well), the abbaa herregaa (the coordinator of water use and maintenance) and other members. Any violation of the customary rules of water use and maintenance is referred to and discussed by the kora eelaa in the presence of the culprit. Watson (2001) discusses a complex web of entitlements that enable an individual to gain access to water from any particular well and the turn that person is given in the rota for the watering of animals. It depends on the membership of the clan of the abbaa konfi and on the contribution to the labour of constructing the wells. Animals are given water according to a strict rotation: the abbaa konfi, the abbaa herregaa and then other clan members according to their seniority in the clan. In addition to these entitlements, the Boran aadaa and seera forbid the denying of anyone access to water or the request for its payment. In general, the ideology and social relations of Boran society are based on *nagaa* Boran (the peace of the Boran). Oromos define peace not as the absence of war but as a proper relationship within the localities and with God, *Waaqa*. The relationship between different clans, villages and households or any other social group is based on cooperation and mutual respect. Where a dispute arises, it is soon resolved through mediation by a council of elders (Constantinos, 1999).

Relationship between Statutory and Customary Institutions

Watson (2001) provides a thorough account of the professed interests of various NGOs in working with indigenous Boran institutions as a bridge to accessing and enabling the community in helping themselves. In general, it is underscored that the state and the NGOs show a strong commitment to working with indigenous institutions as a means of achieving development. However, no pragmatic collaboration is being realized between the statutory and the customary institutions. Bassi (2003) stated that the Boran political/judicial/governance system has never received any formal recognition from modern Ethiopia. It is still important in regulating interpersonal relations in the rural context and access to pastoral resources but it is, as a whole, losing relevance due to the overall, state-imposed allocation of land resources to the newcomers from other zones of Oromia and other regional states of Ethiopia. Consequently, the newcomers increase pressure on the water resources by claiming a substantial share of the existing water rights and often neglecting the local rules and agreements.

Similarly, some scholars shared their experiences of the prevailing relations between the formal government units for political administration, the KAs, and the gadaa institution in the Borana zone (Tache and Irwin, 2003; Homann *et al.*, 2004). The following excerpt is taken from Tache and Irwin (2003):

> A herder bringing his cattle to an area would traditionally negotiate grazing rights with the araddaa council. The decision would be made according to the number of cattle already grazing in the area and forage availability. If the area were already being used to its maximum potential, the herder would be asked to explore other areas to graze under the traditional grazing management system. However more recently, in the event of such a decision, herders who are 'refused' access may now go to the KA and gain legal permission to graze their animals in the area.

Tache and Irwin (2003) further argue that the KA officials – the youngest community members, alien to the indigenous system and inexperienced in rangeland management – are appointed and given powers of decision making at the local level. Today, the KA officials are linked to the territorial administration of the rangelands. They operate against the advice of the elders, who are delegated clan representatives and responsible for a more flexible organization of the rangelands. This has caused conflicts between generations and disagreements within and among the communities.

Conclusions and Recommendations

Both inter- and intra-ethnic (micro–micro) and macro–micro conflicts over the use of natural resources are common in Borana. Such conflicts are usually settled by the local elders using the principles of the gadaa system. According to the gadaa age-grade system, individuals in the age range of 40–48 are called luba and are considered to be elders, with a social responsibility for maintaining peace and stability within the local community. The relevance and application of this indigenous institution in dealing with conflicts that may arise over the use of natural resources have been assessed by many scholars.

There is only loose collaboration, if any, between this customary institution and the government in dealing with conflict resolution between individuals and communities. The government fails to appreciate, collaborate and complement the traditional methods of resource allocation and resolution of conflicts. Limited state understanding of the role played by the gadaa system has diminished the efficacy and relevance of this customary institution in conflict management in Oromia in general and in Borana in particular, which has contributed to the degradation of rangelands and weakened the resilience of pastoralists to droughts.

We propose that there should be an increased collaboration and networking between the statutory and customary institutions of governance. In particular, the state should recognize and support the customary courts and enforce their rulings. The customary laws are often more important than statutory laws and are relied upon in deciding access rights to natural resources and in resolving conflicts. Neglect of these norms and laws may have negative consequences for development policy of the nation in general and for the local community who rely on them in particular. A 'systematic combination' of customary and statutory institutions in the development and management of natural resources may facilitate cross-cultural understanding, thereby improving the socio-economic development of the country. However, enforcing the statutory rules on the local community without due consideration for their indigenous norms and values should be avoided on the side of the state. Access of what and by whom to the local communities should be established through customary institutions.

In the Boran tradition, natural resources management and conflict resolution are combined; and as a result of the great respect the customary institution receives from the local communities, it is the best institution to deal with the operation and management aspects of natural resources governance. Therefore, full authority should be given to the indigenous (gadaa) institution in making decisions regarding access rights to scarce natural resources. The involvement of government bodies (KA officials) in decision-making processes about natural resources (such as overruling the indigenous institution's decision) should be avoided. In general, the whole effort of the government should be directed at natural resources development, leaving the management and operational aspects to the traditional institution. Yet, the local community should be given a say in the development projects starting right from the planning stage. Furthermore, the role of local customary institutions in water resources management and conflict resolution should be spelled out clearly in the water resources policy of the country.

Acknowledgements

The authors of this paper would like to thank all individuals who contributed their ideas directly or indirectly during the preparation of the paper. Suggestions and comments given by Dr Messele Zewdie are invaluable.

Endnotes

[1] The word *Derg* (or *Dergue*) means a committee or a council in Amharic, one of the many languages spoken in Ethiopia. In the present context it refers to a coordinating committee of the armed forces, police and the Territorial Army, which ruled Ethiopia from 1974 to 1987 after ousting the government of Emperor Haile Selassie.

[2] A *Woreda* is one of the administrative divisions of Ethiopia immediately one step down the zonal administrative divisions.

[3] It is important to distinguish Borana or the Borana zone as an administrative unit from Borana Oromo or simply Boran, which is one of the Oromo clans inhabiting the Borana zone of Ethiopia and northern Kenya.

[4] It is also important to distinguish the ethnic Somali of Ethiopia residing in Somali Regional State from Somalia, which is one of the countries in the Horn of Africa.

[5] The ethnic identity of the Garii pastoralist group is usually a major source of contention. Some members of this group identify themselves with Oromo, while others claim to be Somali.

[6] For a detailed account of the gadaa system of the Oromo society see Asmerom Legesse (1973, 2000).

[7] The place for undertaking the naming ceremony of a child depends on the age of his father. The ceremony is usually conducted at a place called Nura Shrine, except when a child is born to a *jaarsaa* (stage of old age).

References

Ahrens, J.D. and Farah, A.Y. (1996) *Borana and Liban Affected by Drought: Situation Report on the Drought Affected Areas of Oromiya and Somali Regions, Ethiopia.* United Nations Development Programme, Emergencies Unit for Ethiopia, Addis Ababa, Ethiopia.

Bassi, M. (2003) Enhancing equity in the relationship between projected areas and local communities in the context of global change: Horn of Africa and Kenya, http://www.iucn.org/themes/ceesp/Publications/TILCEPA/CCA-MBassi.pdf (accessed 28 October 2004).

Conroy, C., Rai, A., Singh, N. and Chan, M.K. (1998) Conflicts affecting participatory forest management: some experiences from Orissa. Revised version of a paper presented at the *Workshop on Participatory Natural Resource Management in Developing Countries*, Oxford, UK, 6–7 April 1998.

Constantinos, B.T. (1999) Alternative natural resources management systems: processual and strategic dimensions. In: Okoth-Ogendo, H.W.O. and Tumushabe, G.W. (eds) *Governing the Environment: Political Change and Natural Resources Management in Eastern and Southern Africa.* African Centre for Technology Studies (ACTS), Nairobi.

Coppock, L. (2001) *Risk Mapping for Northern Kenya and Southern Ethiopia.* Global Livestock Collaborative Research Support Program, University of California, Davis, California.

Dejene, A. (2004) Fuzzy access rights in pastoral economies: case studies from Ethiopia. *The Tenth Biennial Conference of the International Association for the Study of Common Property, URL*, http://www.iascp2004.org.mx/downloads/paper_109a.pdf (accessed 20 October 2004).

Dejene, A. and Abdurahman, A. (2002) The root causes of conflict among the southern pastoral communities of Ethiopia: a case study of Borana and Degodia. Paper presented at the *Second Annual Workshop on 'Conflict in the Horn: Prevention and Resolution'*, Addis Ababa, Ethiopia.

Desalegn, Ch.E., Babel, M.S., Das Gupta, A., Seleshi, B.A. and Merrey, D. (2004) Farmers' perception about water management under drought conditions in the Awash River Basin, Ethiopia. *International Journal of Water Resources Development* 22 (4), 589–602.

Flintan, F. and Imeru, T. (2002) Spilling blood over water? The case of Ethiopia. In: Lind, J. and Sturman, K.

(eds) *Scarcity and Surfeit: the Ecology of Africa's Conflicts*. African Centre for Technology Studies (Kenya) and Institute for Security Studies (South Africa), Pretoria, South Africa.

Grimble, R. and Wellard, K. (1997) Stakeholder methodologies in natural resource management: a review of principles, contexts, experience and opportunities. *Agricultural Systems* 55, 173–193.

Gumii Bilisummaa Oromiyaa (2000) *Understanding the Gadaa System*. URL address, http://www.gumii.org/gada/understd.html (accessed 1 November 2004).

Helland, J. (1997) Development interventions and pastoral dynamics in Southern Ethiopia. In: Hogg, R. (ed.) *Pastoralists, Ethnicity and the State in Ethiopia*. Haan Publishing, London.

Homann, S., Dalle, G. and Rischkowsky, B. (2004) *Potentials and Constraints of Indigenous Knowledge for Sustainable Range and Water Development in Pastoral Land Use Systems of Africa: a Case Study in the Borana Lowlands of Southern Ethiopia*. GTZ, Tropical Ecology Support Programme (TOEB), Germany.

Hussein, J.W. (2004) A cultural representation of women in Oromo society. *African Study Monograph* 25 (3), 103–147.

Legesse, A. (1973) *Gadaa: Three Approaches to the Study of African Society*. Free Press, New York.

Legesse, A. (2000). *Oromo Democracy: an Indigenous African Political System*. The Red Sea Press, Inc., Asmara, Eritrea.

Luseno, W.K., Kamara, A.B., Swallow, B.M., McCarthy, N. and Kirk, M. (1998) *Community Natural Resource Management in Southern Ethiopia*. SR/GL-CRSP Pastoral Risk Management Project Technical Report No. 03/98, Utah State University, Logan, Utah.

Tache, B. and Irwin, B. (2003) Traditional institutions, multiple stakeholders and modern perspectives in common properties: accompanying change within Borana pastoral systems. *Securing the Commons* 4.

Watson, E. (2001) *Inter Institutional Alliances and Conflicts in Natural Resources Management: Preliminary Research Findings from Borana, Oromiya region, Ethiopia*. Marena Research Project. Working Paper No. 4, Department for International Development, UK.

Workneh, K. (2001) *Traditional Oromo Attitudes towards the Environment: an Argument for Environmentally Sound Development*. Social Science Research Report Series, No. 19, OSSREA, Addis Ababa, Ethiopia.

10 Kenya's New Water Law: an Analysis of the Implications of Kenya's Water Act, 2002, for the Rural Poor

Albert Mumma

Faculty of Law, University of Nairobi, Nairobi, Kenya; e-mail: cepla@nbnet.co.ke

Abstract

This chapter analyses the implications of Kenya's Water Act, 2002 for the rural poor in the management of water resources and delivery of water services. It is premised on the belief that recognizing pluralistic legal frameworks is necessary for the effective management of water resources and delivery of water services to the rural poor. The chapter argues that, to the extent the Water Act, 2002 depends on state-based legal frameworks, its effectiveness in meeting the needs of the rural poor will be limited, particularly given the limitations of technical and financial resources the Kenyan state is facing. Consequently, it is necessary that a conscious policy of pursuing the use of the limited opportunities the law presents be adopted in order to maximize the law's potential in meeting the needs of the rural poor.

Keywords: Kenya, water law, rural water supply, water services, water resources management, rural poor, legal pluralism.

Background

The present institutional arrangements for the management of the water sector in Kenya can be traced to the launch in 1974 of the National Water Master Plan, the primary aim of which was to ensure availability of potable water, at reasonable distances, to all households by the year 2000 (Sessional Paper No. 1 of 1999). The Plan aimed to achieve this objective by actively developing water supply systems, which required the government to directly provide water services to consumers, in addition to its other roles of making policy, regulating the use of water resources and financing activities in the water sector. The legal framework for carrying out these functions was found in the law then prevailing, the Water Act, Chapter 372 of the Laws of Kenya, which had been enacted as law in the colonial era.

In line with the Master Plan, the government upgraded the Department of Water Development (DWD) of the Ministry of Agriculture into a full Ministry of Water. The DWD, which continued to exist as a department in the newly created Ministry, embarked on an ambitious water supply development programme. By the year 2000 it had developed, and was managing, 73 piped urban water supply systems serving a population of about 1.4 million and 555 piped rural water supply systems serving a population of 4.7 million. Typically, in rural areas, the consumers used the water supplied for both domestic and small-scale irrigation, a practice that continues to date. Indeed, the rules used in implementing the Water Act,

Chapter 372 allowed irrigation of up to 2 acres as part of domestic use of water.

As a consequence of this practice and the rules applied, the use of water for small-scale irrigation (informally referred to as 'kitchen gardening') is hardly ever separately accounted for. Consequently, no distinction is drawn in documents relating to the permits granted for water abstraction between the water to be used for drinking, cooking and washing and the water to be used for kitchen gardening, and no clear records for such use are maintained by the Registrar of Water Rights.

In 1988, the government established the National Water Conservation and Pipeline Corporation (NWCPC) as a state corporation under the State Corporations Act, Chapter 446 of the Laws of Kenya, to take over the management of government-operated water supply systems that could be run on a commercial basis. By 2000, the NWCPC was operating piped water supply systems in 21 urban centres serving a population of 2.3 million and 14 large water supply systems in rural areas serving a population of 1.5 million.

Alongside the DWD and the NWCPC the large municipalities were appointed as 'water undertakers'. A water undertakership was the term given to the licence issued under the Water Act, Chapter 372 to supply water within an area. By the year 2000, ten municipalities supplied 3.9 million urban dwellers under an undertakership granted to them by the Minister.

Additionally, about 2.3 million people were receiving some level of service from systems operated by self-help (community) groups that had built the systems, often with funding from donor organizations and technical support from the district officials of the DWD (Government of Kenya, 1999).

Persons not served under any of the above arrangements did not have a systematic water service, and had to rely on such supply as they were able to provide for themselves, typically by directly collecting water from a watercourse or from some other water source on a daily basis. Indeed, despite the government's ambitious water supply development programme, by 2000 less than half the rural population had access to potable water and, in urban areas, only two-thirds of the population had access to potable and reliable water supplies.

Supplying water by commercial and other large-scale irrigation schemes was carried out under the Irrigation Act, Chapter 347, first enacted in 1967. The Irrigation Act established the National Irrigation Board as being responsible for the development, control and improvement of national irrigation schemes in Kenya. Further, the Act gave the Minister powers to designate any area of land as a national irrigation scheme. Once an area was designated as such a scheme, the National Irrigation Board would be responsible for settling people on it and for administering it, including making arrangements for the supply of irrigation water to the scheme.

Apart from irrigation carried out through designated irrigation schemes, private individuals engaged in irrigated agriculture were required to apply for, and obtain, a permit for water abstraction, following the permit application procedures that applied to abstraction for any other use. The Water Act, Chapter 372 stipulated, however, that the use of water for domestic purposes took priority over the use of water for any other purposes, including irrigation purposes.

In the 1980s, the government began experiencing budgetary constraints and it became clear that, on its own, it could not deliver water to all Kenyans by the year 2000. Attention therefore turned to finding ways of involving others in the provision of water services in place of the government, a process that came to be known popularly as 'handing over'.

There was general agreement over the need to hand over government water supply systems, but much less agreement over what it meant for the government to hand over public water supply systems to others. In 1997, the government published a manual giving guidelines on handing over of rural water supply systems to communities (Ministry of Land Reclamation, Regional and Water Development, 1997).

The manual indicated that: '... at the moment the Ministry is only transferring the *management* of the water supply schemes. The communities will act as custodians of the water supply schemes, including the assets, when they take over the responsibility for operating and maintaining them.' However, the goal of community management should be *ownership* of the water supplies, including the associated assets.

The manual stated the criteria for handing over to be: (i) the capacity of the community to take over; (ii) the ability to pay; (iii) the capacity to operate and maintain the system; (iv) the involvement of women in management; and (v) the ability and willingness to form a community-based group with legal status. By 2002, ten schemes serving about 85,000 people had been handed over to community groups under these guidelines, focusing on management and revenue collection, but not on full asset transfer.

Building on this experience, the government developed a fully fledged policy, The National Water Policy, which was adopted by Parliament as Sessional Paper No. 1 of 1999. The development of the National Water Policy was largely funded by donor organizations whose predominant interest was with regard to domestic water supply, and not with irrigated agriculture or even with water resources management. Key among these donor organizations were GTZ – interested primarily in urban water supply, SIDA – interested largely in rural domestic water supply and the World Bank.

The National Water Policy stated that the government's role would be redefined away from direct service provision to regulatory functions: service provision would be left to municipalities, the private sector and communities. The Policy also stated that the Water Act, Chapter 372 would be reviewed and updated, attention being paid to the transfer of water facilities. Regulations would be introduced to give other institutions the legal mandate to provide both water services and mechanisms for regulation.

The Policy justified the handing over, arguing that ownership of a water facility encourages proper operation and maintenance: facilities should therefore be handed over to those responsible for their operation and maintenance. The Policy stated that the government would hand over urban water systems to autonomous departments within local authorities and rural water supplies to communities.

While developing the National Water Policy, the government also established a National Task Force to review the Water Act, Chapter 372 and draft a bill to replace the Water Act, Chapter 372. The Water Bill 2002 was published on 15 March 2002 and passed by Parliament on 18 July 2002. It was gazetted in October 2002 as the Water Act, 2002 and came into effect in 2003, when effective implementation of its provisions commenced.

The Reforms of the Water Act, 2002

The Water Act, 2002 has introduced comprehensive and, in many instances, radical changes to the legal framework for the management of the water sector in Kenya. These reforms revolve around the following four themes: (i) the separation of the management of water resources from the provision of water services, which is explained further below; (ii) the separation of policy making from day-to-day administration and regulation; (iii) decentralization of functions to lower-level state organs; and (iv) the involvement of non-government entities in both the management of water resources and the provision of water services. The institutional framework resulting from these reforms is represented diagrammatically in Fig. 10.1.

Separation of functions

Under the Water Act, Chapter 372, the DWD carried out all the functions in the water sector. It developed and supplied water for consumption and for productive use in irrigated agriculture, among other uses; it regulated the sector by issuing permits and carrying out policing; it

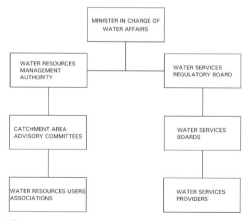

Fig. 10.1. Diagrammatic representation of the new institutional structure for the management of water affairs in Kenya.

was responsible for conserving and managing water resources and for determining funding allocations between water resources management and water supplies. Over the years, it became clear that priority was being given by the DWD to its role as a water supplier. The financial resources and the attention that the DWD gave to water resources management declined markedly in the 1970s and 1980s. This led to a dramatic deterioration in the effectiveness of the systems and arrangements that were in place for managing water resources. Given the water scarcity in Kenya generally, inattention to water resources management did not augur well for the sustainability of the resource.

The Water Act, 2002 separates water resources management from the delivery of water services. Part III of the Act is devoted to water resources management, while Part IV is devoted to the provision of water and sewerage services. It establishes two autonomous public agencies: one to regulate the management of water resources and the other to regulate the provision of water and sewerage services.

The Act divests the Minister in charge of water affairs of regulatory functions over the management of water resources. This becomes the mandate of a new institution, the Water Resources Management Authority (the Authority), established in Section 7 of the Act. The Authority is responsible for, among other things, the allocation of water resources through a permit system. The framework for the exercise of the water resources allocation function comprises the development of national and regional water resources and management strategies, which are intended to outline the principles, objectives and procedures for the management of water resources.

Similarly, the Act divests the Minister in charge of water affairs of regulatory functions over the provision of water and sewerage services and vests this function in another public body, the Water Services Regulatory Board (the Regulatory Board), which is created in Section 46. The Regulatory Board is mandated to license all providers of water and sewerage services that supply water services to more than 20 households. Community-managed water systems therefore need to obtain a licence from the Regulatory Board to continue providing water to their members. This is a departure from the practice previously prevailing under which community water systems, unlike the other systems, operated without a licence.

Decentralization of functions

The Water Act, 2002 decentralizes functions to lower-level public institutions. It does not, however, go as far as to devolve these functions to the lower-level entities. Ultimate decision making remains centralized.

With regard to water resources management, Section 14 of the Act provides that the Authority may designate catchment areas as areas from which rainwater flows into a watercourse, as they are so defined. The Authority shall formulate for each catchment area 'a catchment area management strategy', which shall be consistent with the national water resources management strategy. Section 10 states that the Authority shall establish regional offices in, or near, each catchment area. Section 16 provides that the Authority shall appoint a committee of up to 15 persons in respect of each catchment area to advise its officials at the appropriate regional office on matters concerning water resources management, including the grant and revocation of permits. The regulatory functions over water resources management, currently performed by the district offices of the Ministry in charge of water affairs are, supposedly, under the new legal framework, to be transferred to the catchment area offices of the Authority.

The development of large-scale infrastructure for harnessing water resources, including the building of dams and other infrastructure for flood control and water conservation, has been made the responsibility of the NWCPC. In order to facilitate infrastructural projects, the Water Act, 2002 stipulates that the NWCPC shall receive funding from Parliament. These projects are therefore seen as 'state schemes', because they will comprise assets and facilities developed under public funding. It is for this reason that this role has been vested in a state corporation. The NWCPC shall therefore supply water 'in bulk' for downstream use by others.

With regard to the provision of water and sewerage services, Section 51 of the Act establishes water services boards (WSBs), whose area of service may encompass the area of jurisdiction of one or more local authorities. A WSB is responsible for the provision of water and sewerage services within its area of coverage and, for this purpose, it must obtain a licence from the Regulatory Board. The WSB is prohibited by the Act from engaging in direct service provision. The Board must identify another entity, a water services provider, to provide water services, as its agent. The law allows WSBs, however, to provide water services directly in situations where it has not been possible to identify a water services provider who is able and willing to provide the water services. WSBs are regional institutions. Their service areas have been demarcated to coincide largely with the boundaries of catchment areas.

The role of non-governmental entities

The Water Act, 2002 has continued – and even enhanced – a long-standing tradition in Kenya of involving non-governmental entities and individuals in the management of water resources, as well as in the provision of water services. The Act envisages the appointment of private individuals to the boards of both the Authority and the Regulatory Board. Rule 2 of the First Schedule to the Act, which deals with the qualification of members for appointment to the boards of the two public bodies, states that, in making appointments, regard shall be had to, among other factors, the degree to which water users are represented on the board. More specifically, subsection 3 of section 16 states that the members of the catchment advisory committee shall be chosen from among, inter alia, representatives of farmers, pastoralists, the business community, non-governmental organizations as well as other competent persons. Similarly, membership on the board of the WSBs may include private persons.

Most significantly however, the Act provides a role for community groups, organized as water resources user associations (WRUAs), in the management of water resources. WRUAs

constitute a concept that builds on associations (previously known as 'water user associations') under which local community members who wished to develop water projects for domestic use (including small kitchen gardening) tended to organize themselves. The Water Act, 2002 opted to rely on voluntary membership associations rather than on other institutional mechanisms such as local authorities. The reason for this is the belief that, being voluntary in nature, these associations can draw on the commitment of the members as social capital, as opposed to attempting to rely on more formal statutory structures, which might not necessarily be able to call on that social capital.

Section 15(5) of the Act thus states that these associations will act as forums for conflict resolution and cooperative management of water resources. Consequently, water user associations, where they exist, will have to reconstitute themselves to take on board water resources management issues. Where such associations do not exist, which is the case in most parts of the country, new associations will need to be formed to carry out the role, which the new law has given to WRUAs. Inevitably, there will be financial cost and time involved in setting up new institutions. However, being institutions that depend for their success on the initiative of the members and the belief by the members in the usefulness of the association in meeting their water resources management needs, the investment of time and resources in setting up an association is likely to strengthen the commitment of the members to sustain the association.

With regard to water services, Section 53(2) stipulates that water services shall be provided only by a water services provider, which is defined as 'a company, non-governmental organization or other person providing water services under and in accordance with an agreement with a licensee [the WSB]'. Community self-help groups providing water services may therefore qualify as water services providers. In the rural areas where private-sector water services providers are likely to be few, the role of community self-help groups in the provision of water services is likely to remain significant, despite the new legal framework.

The role of non-governmental entities in both the management of water resources and the provision of water services is thus clearly recognized. However, given the state-centric premise of the Water Act, 2002, the role assigned to non-governmental entities, particularly self-help community groups, is rather marginal.

The Water Act, 2002 and State Centrism

In my view, the Water Act, 2002 is based on a notion of law that is unitary and state-centred. Its design and operation are premised on the centrality (indeed monopoly) of central state organs and state systems in the management of water resources as well as in the provision of water and sewerage services. It makes only limited provision for reliance on non-state-based systems, institutions and mechanisms. More fundamentally, the new Law continues the tradition of the Law it replaces of not recognizing the existence in Kenya of a pluralistic legal framework. It assumes that the legal framework in Kenya comprises a monolithic and uniform legal system, which is essentially state-centric in nature.

The continued denial of the existence in Kenya of a pluralistic legal framework is, in my view, inimical to the success of the new Law in meeting the needs of the rural poor who, more than urban-based Kenyans, live within a legally pluralistic environment. For this purpose, legal pluralism is understood as referring to a situation characterized by the coexistence of multiple normative systems all experiencing validity (see, for instance, von Benda-Beckman *et al.*, 1997). Kenya's rural poor, typically, live within normative frameworks in which state-based law is no more applicable and effective than customary and traditional norms. The new water law, however, ignores this reality.

The long title of the Water Act, 2002 states that it is: 'an Act of Parliament to provide for the management, conservation, use and control of water resources and for the acquisition and regulation of rights to use water; to provide for the regulation and management of water supply and sewerage services … and for related purposes.'

Part II of the Act deals with ownership and control of water. Section 3 vests ownership of 'every water resource' in the state. The term

'water resource' is defined to mean 'any lake, pond, swamp, marsh, stream, watercourse, estuary, aquifer, artesian basin or other body of flowing or standing water, whether above or below ground'. The effect of this provision, therefore, is to vest ownership of all water resources in Kenya in the state. Previously, the Water Act, Chapter 372 vested ownership of water 'in the government'. The replacement of the word 'government' with the word 'state' does not, in reality, represent a significant departure in the legal status of water resources.

The right to use water from any water resource is also vested in the Minister. Accordingly, Section 6 states that:

> [N]o conveyance, lease or other instrument shall be effectual to convey, assure, demise, transfer, or vest in any person any property or right or any interest or privilege in respect of any water resource, and no such property, right, interest or privilege shall be acquired otherwise than under this Act.

The right to use water is acquired through a permit, provision for which is made later in the Act. Indeed, the Act states that it is an offence to use water from a water resource without a permit.

Section 4 of the Act deals with control of water resources. It states that the Minister shall have, and may exercise, control over every water resource. In that respect, the Minister has the duty to promote the investigation, conservation and proper use of water resources throughout Kenya. It is also the Minister's duty to ensure the effective exercise and performance by authorities or persons under the control of the Minister of their powers and duties in relation to water.

The state centrism of the Water Act, 2002 – and its predecessor, the Water Act, Chapter 372 – is self-evident. Like its predecessor, the Water Act, Chapter 372, it has vested all water resources throughout the country in the state, centralized control of water resources in the Minister and subjected the right to use water to a permit requirement. This has far-reaching implications for the management of water resources and provision of water services to the rural poor who have only limited access to state-based systems. Matters are compounded by the administrative, financial and technical constraints inhibiting the ability of the Kenyan state to implement the Water Act, 2002 and to enable rural households to derive full benefits from its provisions.

The acquisition and exercise of water rights

As indicated, the Act imposes a permit requirement on any person wishing to acquire a right to use water from a water resource. Section 27 makes it an offence to construct or use works to abstract water without a permit. There are however three exceptions to the permit requirement. These relate to: (i) minor uses of water resources for domestic purposes (representing uses of water for domestic purposes abstracted without the assistance of equipment. Equipment is defined to mean any device for the abstraction of water, including a hand-held mobile pump); (ii) uses of underground water in areas not considered to face groundwater stress and therefore not declared to be groundwater-conservation areas; and (iii) uses of water drawn from artificial dams or channels, which – being artificial rather than natural – are not considered to be water resources of the country.

Application for the permit is made to the Authority. Section 32 stipulates the factors to be taken into account in considering an application for a permit. These include:

• The existing lawful uses of the water. As noted below, under the Registered Land Act, Chapter 300, discussed further below, customary rights of access to water are recognized as 'overriding interests', which remain valid and lawful even if they are not registered against the land.
• Efficient and beneficial use of the water in the public interest.
• The likely effect of the proposed water use on the water resources and on other water users.
• The strategic importance of the proposed water use.
• The probable duration of the activity for which the water use is required.
• Any applicable catchment management strategy.
• The quality of water in the water resources that may be required for the reserve.

These considerations are designed not only to enable the Authority to balance the demands of competing users, but also to take into account the need to protect the general public interest in the use of water resources as well as the imperative to conserve water resources.

Further guidance is given to the Authority in deciding on allocation of the water resources as follows:

• That the use of water for domestic purposes shall take precedence over the use of water for any other purpose and, in granting a permit, the Authority may reserve such part of the quantity of water in a water resource as is required for domestic purposes. It is to be recalled that, in rural settings, the use of water for domestic purposes typically includes the use for minor irrigation ('kitchen gardening') purposes.
• That the nature and degree of water use authorized by a permit shall be reasonable and beneficial in relation to others who use the same sources of supply.

Permits are given for a specified period of time. Additionally, unlike under the previous Act, the Authority is given power to impose a charge for the use of water. The charge may comprise both an element of the cost of processing the permit application and a premium for the economic value of the water resources being used. Charging a premium for the use of water resources represents the use of charging as a mechanism for regulating the use of water. It is made possible by the fact that ownership of water is vested in the state, which is entitled to grant and administer the right to use water resources. Details of the charges to be imposed, including the amounts to be charged, and the uses for which a charge may be imposed will be spelt out in subsidiary rules that have not yet been made.

As stated earlier, the permit system is state-centric in orientation. In operation, it privatizes water rights to a small section of the community, essentially property owners who are able to acquire and use water resource permits. By the same token, poor rural communities that are unable to meet the requirements for obtaining a permit – principally landownership – are marginalized from the formal statutory framework by the permit system.

Permits run with the land so that, where the land is transferred or otherwise disposed of, the permit also passes to the new owner of the land. Section 34 requires a permit to specify the particular portion of any land to which it is to be an appurtenant. Where the land on which the water is to be used does not abut on the watercourse, the permit holder must acquire an easement over the lands on which the works are to be situated. It is thus not possible, under the law, to obtain a permit in gross (i.e. that is not linked to a particular land).

This provision reinforces the predominance of landowners – or those with a property interest in land – with regard to the use of water resources. It is premised on a land tenure system, which prioritizes documented individual or corporate ownership of land over communal systems of access to land and land use, and which does not require documented title, such as extant in most parts of rural Kenya. The Act therefore marginalizes collectivities – such as poor rural community groups – in the acquisition and exercise of the right to use water resources. This could potentially undermine the ability of poor rural communities in Kenya to effectively utilize water resources in economically productive activities, such as irrigation and commercial livestock rearing. Given the pluralistic land tenure system prevailing in Kenya, this issue will influence the effectiveness of the implementation of the new water law.

Kenya's land tenure systems

In Kenya, three land tenure systems apply: government lands, trust lands and private lands. These land tenure systems are provided for in a series of statutes dating back to the early colonial days.

In traditional Kenyan society, before the advent of colonial rule, land was owned on a communal basis by small community groups. Individuals and families acquired use rights and rights of access to land by virtue of membership to a social unit, such as a clan. Rights of access and use operated for all practical purposes as title to land, even though there was no documented title.

Following the declaration of a protectorate status over Kenya in 1895, the British Colonial

Government passed the Crown Lands Ordinance to provide a legal basis for alienation of land to white settlers. The Ordinance declared 'all waste and unoccupied land' to be 'Crown Land'. By a 1915 amendment of the Crown Lands Ordinance, Crown lands were redefined to include land that had hitherto been occupied and owned by the natives. Further, in 1938, the Crown Lands (Amendment) Ordinance excised native reserves, which became vested in the Native Lands Trust Board. A Native Lands Trust Ordinance was passed to provide for this and for the control and management of 'trust lands'. After independence these lands became vested in county councils.

In the 1930s and 1940s, the Colonial Government adopted the policy of enabling Africans to obtain documented title to land as a way of promoting better agricultural productivity. The Swynnerton Plan of 1955 recommended the consolidation and registration of fragmented pieces of land held by Africans into single holdings that could be economically farmed.

The Native Lands Registration Ordinance was passed in 1959, under which Native Land Tenure Rules were made. These authorized the alienation of trust lands to individual members of the native communities. This required the ascertainment of the entitlements of the individuals to the portions of land to which they laid a claim, the registration of the entitlements in the names of the individuals and the issuance of title documents. To facilitate this, the Land Adjudication Act was enacted. Lands within the native areas (trust lands) that were not alienated remained trust lands, while lands outside of trust lands that had not been alienated to private individuals and entities remained 'crown land' and later became known as government lands. Three land tenure systems thus arose: government land, trust land and private land.

The government as a landowner can obtain a water resources permit with respect to its land, but the Water Act, 2002 exempts state schemes from the requirement for a permit.

Under the Constitution and Trust Lands Act, Chapter 288 of 1962, trust lands are held by county councils for the benefit of the ordinary residents of the county council. Currently, trust lands comprise what remains of lands that were designated as native reserves. These lands, which are predominantly in the arid and semi-arid areas of Kenya, are occupied by semi-nomadic pastoralist communities. The Constitution stipulates that County Councils 'shall give effect to the rights, interests and other benefits in respect of trust land as may, under the African customary law for the time being in force and applicable thereto be vested in any tribe, group, family, or individual'.

In effect, therefore, the trust land tenure system contemplates the continued operation of customs and traditions of granting land use rights and access systems without the necessity for formal documents of title. This means that occupiers of trust land – who comprise largely the rural poor – would not be able to demonstrate ownership of land for purposes of an application for a water permit as required by the Water Act, 2002. Consequently, the effective implementation of the Water Act, 2002 is dependent on the implicit recognition in practice of a legally pluralistic land tenure regime, which the Water Act, 2002 has not expressly done.

Private land is registered under either the Land Titles Act, Chapter 281 or the Registration of Land Act (RLA), Chapter 300. The RLA provides for the issuance to landowners of a title deed and, in cases of leasehold interests a certificate of lease, which shall be the only *prima facie* evidence of ownership of the land. The RLA provides that the registration of a person as the proprietor of land vests in that person the absolute ownership of that land 'together with all rights and privileges belonging or appurtenant thereto and free from all other interests and claims whatsoever'.

Land registration, which grants private ownership, has been completed in those regions of the country with high agricultural potential whereas, in the areas in which pastoralism is predominant, communal tenure is recognized by the law. Despite the registration of land in the names of private individuals, empirical evidence suggests that, even in areas of high agricultural potential, among rural communities land use and access rights continue to be based largely on customary and traditional systems, notwithstanding statutory law. Indeed, studies have revealed what has

been described as 'a surprising recalcitrance of indigenous institutions and land use practices' (Migot-Adhola *et al.*, 1990, unpublished).

The widespread application of traditional and customary rights over even registered land can therefore be explained on the basis of the existence of a pluralistic legal framework with respect to land tenure. Indeed, rural communities tend to assume that the individuals registered as owning the land hold it in trust for other family or clan members, in line with customary practices. The discovery that, following registration, the registered landowner holds the land absolutely, and free from the claims of other family members, has led to a great deal of social upheaval, insecurity of title and access rights and to much litigation. To date, local beliefs and practices have not changed significantly.

The absolute nature of private ownership is qualified under Section 30 of the RLA, which states that all registered land shall be subject to such priority interests as may for the time being subsist and affect it, even if not recorded on the register, including:

- Rights of way, rights of water and profits subsisting at the time of first registration under the Act.
- Natural rights of light, air, water and support.

Consequently, collective rights of access to water under traditional and customary laws subsist despite the registration of a private individual as an absolute owner of land. Such rights need therefore to be taken cognizance of in allocating water rights under the permit system established by the Water Act, 2002, even if this Act makes no reference to them.

The implication of the existence of a pluralistic land tenure regime for the administration and the Water Act, 2002 and the management of water resources is that the sections of rural communities who have title documents to their land will be able to meet the requirements of the Water Act, 2002 for purposes of acquiring a water rights through a permit. Rural communities practising communal land tenure systems are unlikely to be able to operate within the straitjacket of the Water Act, 2002. It is likely that these communities comprise predominantly the rural poor. Consequently, in considering revision of the

Water Act, 2002, it will be important to examine and provide mechanisms for granting water rights to community members who do not have land titles.

The acquisition and operation of a water supply licence

The right to provide water services is also subject to licensing requirements. Section 56 states that no person shall provide water services to more than 20 households or supply more than 25,000 l of water/day for domestic purposes – or more than 100,000 l of water/day for any purpose – except under the authority of a licence. Indeed, Subsection (2) stipulates that it is an offence to provide water services in contravention of the licence requirement.

Consequently, community groups must obtain a licence in order to be able to continue or commence supplying water to their members. This is likely to have far-reaching implications for member-based rural water supplies, given the requirement for technical and financial competence that is a precondition to obtaining a licence. Many such groups will probably have great difficulty demonstrating such competence, and this may result in water service agreements being granted only to well-established community groups and other organizations having access to technical and financial resources, to the detriment of the self-help initiatives of the local community.

Section 57 provides that an application for a licence may be made only by a WSB, which therefore has a monopoly over the provision of water services within its area of supply. As earlier indicated however, the WSB can only provide the licensed services through an agent known as a water services provider, which can be a community group, a private company or a state corporation that is in the business of providing water services.

In order to qualify for the licence the applicant must satisfy the Board that:

- Either the applicant or the water services provider by whom the services are to be provided has the requisite technical and financial competence to provide the services.

- The applicant has presented a sound plan for the provision of an efficient, affordable and sustainable service.

- The applicant has proposed satisfactory performance targets and planned improvements and an acceptable tariff structure.

- The applicant or any water services provider by whom the functions authorized by the licence are to be performed will provide the water services on a commercial basis and in accordance with sound business principles.

- Where the water services authorized by the licence are to be provided by a water services provider that conducts some other business or performs other functions not authorized by the licence, the supply of those services will be undertaken, managed and accounted for as a separate business enterprise.

Unlike that with respect to a permit for the use of water resources, there is no property involved in a water services provision licence and, as stipulated in Section 58(2), the licence shall not be capable of being sold, leased, mortgaged, transferred, attached or otherwise assigned, demised or encumbered.

Ownership of the assets for the provision of water services is vested in the WSB, which is a state corporation. Section 113 provides for the transfer of assets and facilities for providing water services to the WSBs. Where the assets and facilities belong to the government they are required to be transferred outright to the WSBs. Where, on the other hand, they belong to others, including local authorities and community groups, only use rights may be acquired by the WSBs. A WSB may require the use of assets and facilities presently used by community groups in order to integrate them into a bigger and more cost-effective water service. In arranging to use the assets and facilities belonging to communities for its purposes, the WSB would be required to pay compensation to the community group.

The likely effect of this provision is that WSBs will be inclined to reach agreements with those community groups having their own assets. Those community groups without assets – mostly, the most marginalized rural communities – are likely to find that their ability to develop water services facilities will diminish over time as funding for infrastructural development is channelled increasingly to WSBs directly, rather than to communities. Furthermore, in order to be able to enter into contracts for the provision of water services as an agent of the WSB, the entity concerned needs to be a legal person, which – as we shall show below – many poor community self-help groups are not.

Local community water systems

As already indicated, by the year 2000, less than half the rural population had access to potable water, and even in urban areas only two-thirds of the population had access to potable and reliable water supplies. Typically, the people without access to reliable water services often represent the poorest and most marginalized of the Kenyan people. This chapter is premised on the belief that these are the people least likely to take advantage of, and benefit from, the legal framework in the Water Act, 2002 for the provision of water services, and the ones likely to suffer most from inadequate management of water resources.

The ability of rural communities to provide water services through community groups is demonstrated by the fact that presently a population of no less than 2.3 million get water services from systems operated by self-help (community) groups – traditionally known as 'water user associations' (WUAs). These systems are diverse in nature and capacity, ranging from fairly sophisticated systems with well-structured tariffs to simple gravity schemes operated without any formal processes (Njonjo, 1997).

The history of community provision of water services in Kenya is long. Most of the systems are small in scale, serving perhaps one constituency and serving between 500 and 1000 families. Even in the areas served, the systems rarely serve everyone, tending to be restricted to those who qualify as members according to criteria stipulated for the system by its initiators.

The phrase 'self-help' – which is often used to describe these systems – is an apt one. Many such systems arose out of the initiative of a small group of visionary and energetic community members who sought to redress the lack of water services in their local community whether for domestic water consumption strictly speaking or for irrigation or both. Typically, these individuals or groups of individuals would have approached some donor organization, church group or even community members living abroad and successfully negotiated funding support.

Also typically, it was a condition of donor support that the community make a contribution of up to 15% of the cost of the project in labour and cash. The organizers of the project would then have had to raise funds from community members and other well-wishers through a system commonly described in Kenya as a *harambee*, in which people get together once – or, more commonly, repeatedly – to raise funds from members of the public for a community development – or other – project. Additionally, members of the community in which the project was to be constructed would have contributed to the cost of the project 'in-kind', that is by providing direct manual labour at the site in digging trenches, carrying and laying pipes, backfilling and doing other non-skilled tasks.

Another important element of the community's contribution to the project has often taken the form of a donation of land for the physical facilities, such as the storage tanks and reservoirs, the treatment facilities and even the standpipes. Donations of land are often a contribution by one of the initiators of the project, as a gesture of support for the project. It is not unusual to find that the title to the land – if one exists – remains in the name of the person donating the land, even though for all practical purposes the person ceases to be the owner of the land in question, and the land is perceived as being communal in ownership. The common reason for the failure to transfer the land formally to the community often relates to the lack of a corporate entity into whose name to transfer the land, the cumbersome nature of the paperwork and the expense involved in effecting the transfer, as well as to the belief by the community members and the landowner that the transfer is as good as complete with the oral donation of the land by its owner.

Typically, technical input into the design and supervision of the project will have been provided by the water engineers stationed at the local district office of the Ministry in charge of water affairs. Indeed, the Ministry's policy over the years has been to encourage its officials, as part of their official duties, to provide technical and backstopping support to community projects, at no cost to the communities. The actual construction of the water system, however, is often carried out by private constructors paid for by the donor organization and the community group.

Given these origins, the formal ownership of these community systems under formal statutory frameworks is far from clear. They are truly 'community systems' in the sense that many have contributed to their development one way or another, but no one contributor can lawfully claim formal ownership of the system. Legal disputes over ownership are rarely, if ever, heard of and, in the experience of the writer, those involved in the development and management of these systems do not perceive this as being of significance. That the question of ownership is not perceived as being an issue in Kenya can only be explained on the basis of the existence and active operation of a parallel concept of ownership of these community-developed and -managed water systems.

The registration of community water systems

Many organizations operating community self-help water systems are registered under an administrative registration system operated by the Ministry in charge of community development. The registration is carried out at the district office of the Ministry, where there is a community development officer. To be registered, the community members must choose a name for the project, form a committee of officials – including a chairman, a secretary and a treasurer – and draft a constitution setting out their objectives and the rules that will govern the affairs of the group. Following approval, the community development officer will issue a certificate of registration.

The registration of a self-help group by the community development officer is relatively easy and inexpensive. It is, however, a purely administrative exercise as the statutory laws do not provide for it. Registration under this administrative system does not provide the group with any legal personality; neither does the group acquire corporate identity under the statutory laws. The group cannot, for instance, own land in its own name under the prevailing land laws of Kenya. Lack of legal and corporate personality notwithstanding, most of the community projects operated by such self-help groups work quite well. This is so particularly among rural communities in which concepts such as legal personality and corporate identity in terms of statutory law have relatively little relevance. It is an example of the existence of a parallel normative framework governing the existence and operation of community self-help groups in Kenya based, in this instance, on a normative framework established purely on the basis of administrative arrangements.

Statutory law, on the other hand, provides for various systems for registration of organizations that could be adopted by communities. These can be categorized broadly into membership-based organizations and non-membership-based organizations. Membership-based organizations are typified by the *society*, also known as the *association*. The Societies Act, Chapter 108 of the Laws of Kenya provides for the registration and control of societies. It defines a society as an association of 12 or more persons. Registration of the association as a society grants the association legal personality under the laws of Kenya.

Unlike self-help groups, societies are registered by the Registrar of Societies, who is an official based in Nairobi. This makes it difficult – and expensive – for the marginalized rural communities to register a society, as they would have to travel to Nairobi or engage an agent – often a lawyer – in Nairobi to carry out the registration on their behalf. Strictly speaking, a society is unincorporated in law, but this fact is rarely appreciated and rarely does it give rise to any legal issues in the administration of the affairs of the society.

The Cooperative Societies Act, Chapter 490 of the Laws of Kenya, provides for a form of association known as the 'cooperative society', which is regulated by the Commissioner of Cooperatives, but not by the Registrar of Societies. The key difference between this and societies registered under the Societies Act is that the objective of a cooperative society is the promotion of the economic interest of its members. Cooperative societies have therefore not been commonly used for rural community-based water projects, but have been used often by farmers' organizations in rural areas.

Rural communities have rarely perceived rural-community water projects as existing to advance the economic interests of the members. Typically, they have perceived such projects as existing largely to advance the social welfare of the members of the community. This is despite the very real link between the availability of water supplies and the economic benefit to the consumers arising from the use of the available water for productive economic activities such as irrigation and livestock rearing. This factor partly explains the difficulty many self-help groups experience in enforcing tariff payments for water consumption, as there is rarely the will to cut off supplies to community members who fail to make payments.

The failure to make the link between the provision of water services and economic benefit to particular community members, together with the assumption that water services are a social service, is further evidence of the existence of pluralistic normative frameworks among poor rural communities. Such communities will face real difficulty in making the transition to the new legal framework, which is premised on the belief that water services must be operated *on a commercial basis and in accordance with sound business principles*.

Non-member-based organizations are the second type of organization that could be adopted by communities. The existing types of non-member-based organizations used for community water projects are non-governmental organizations (NGOs), trusts and companies limited by shares. It is rare to find a community project registered as either a trust or a company limited by shares, particularly in rural areas. The main form of non-member-based organization found implementing community rural water projects tends therefore to be the NGO.

Non-governmental organizationss are set up under the Non-Governmental Organization Registration Act of 1990. This provides for the registration of an organization whose objective is the advancement of economic development. It requires three directors, an identified project and a source of funding. NGOs have been favoured mostly by persons external to the community who have received funding for a community project and wish to implement the project themselves, rather than through the community members. It is also commonly the case that the NGO will be an urban-based organization.

The Water Act, 2002 has provided for the provision of water services by water services providers, described as 'a company, a non-governmental organization or other person or body providing water services under and in accordance with an agreement with a [WSB]'. Under the Interpretation and General Provisions Act, Chapter 2 of the Laws of Kenya, the word 'person' refers to a legal or natural person. As the self-help group is not a legal person, it would not qualify to be a water services provider. Consequently, it will be necessary for these community organizations to acquire legal personality by registering themselves as societies if they are to continue providing water services. The considerable advantages of the system provided by the present system for registering self-help groups at district level will therefore be lost under the new regime.

Conclusions and Recommendations

This review of the Water Act, 2002 has highlighted significant implications for poor rural communities arising out of the provisions of the Water Act, 2002. These must be seen in the context of the existence in Kenya of a pluralistic legal framework, which has not been recognized or provided for in the new Law. To the extent that the new Law is premised exclusively upon a formal statutory legal system, it is likely to prove inappropriate to the needs and circumstance of the Kenyan rural poor.

The reasons, which have already been explained, are that Kenya's rural poor have not been integrated into the private land tenure and other formal regimes upon which the Water Act, 2002 is premised. They depend largely on land rights arising from customary practices that however have been systematically undermined over the years by the statutory provisions governing land rights and which are not recognized by the Water Act, 2002.

It is unlikely, therefore, that the new Law will be able to facilitate Kenya's achievement of the Millennium Development Goals with respect to the provision of water and sanitation by 2015, particularly for the poor rural communities. This chapter argues that, in order to address the circumstances of the rural poor, there is a compelling case for continued reliance – in the management of water resources and in the provision of water services – on alternative and complementary frameworks drawn from community practices.

This chapter argues further that there is little benefit to be gained, in the foreseeable future, by attempting to incorporate community self-help water systems into formal legal frameworks through, for instance, formalization of ownership arrangements. There is even a risk that disputes will be engendered in the process, as community mechanisms are undermined, as was experienced in the land registration process. Giving community systems due recognition and legitimacy calls for the recognition of existing pluralistic legal frameworks. In this respect, the implementation of the provisions of Section 113, which deals with mechanisms for giving use rights over community assets to the WSBs, requires considerable legal innovation. But it is precisely through such innovative interpretation of the provisions of the new law that the potential of the new law to address the needs and circumstances of the rural poor can be enhanced.

With respect to the management of water resources, one possibility for enhancing the role of local communities in water resources management is to utilize WUAs as an institutional mechanism for allocating water resources to a community-based entity as opposed to an individual landowner. This recommendation is to the effect that, in appropriate circumstances, a water resources use permit could be allocated to a WUA on behalf of all the members of the association. The association would then, in turn, allocate the water resources to its members according to internally agreed rules. The association would also enforce its rules with respect to the use of the water resource in question.

The above proposal would enhance the role and authority of the WUA. It would also utilize community compliance mechanisms as a supplement to the enforcement efforts of the Authority. Its success however would depend on the cultivation of strong and effective WUAs. It is recommended that the government support the nurturing of WUAs as institutional mechanisms for community management of water resources.

The WUAs can build on the local associations that have already been formed in areas with significant water scarcity, such as in the Nanyuki district in which the necessity for water users to cooperate in sharing the resource, brought on by water scarcity, has fostered the growth of community groups. These associations have proved that they can provide a viable community-based mechanism for conflict resolution and cooperative management of a scarce resource. Additionally, being voluntary entities, their formation does not require funding from the government, but they are funded by contributions from the members. Ordinarily, the costs are met from the membership subscriptions.

With respect to the provision of water services, the government should reinforce the capacity and role of district community development officials as a means of providing support to community self-help organizations. Furthermore, the rules governing water services providers should take account of the need to foster and promote community self-help schemes as systems for meeting the water supply needs of the rural poor who are unlikely to receive attention from private operators or from financially hard-pressed public systems.

Looking further ahead, the Water Act, 2002 will need to be amended to take on board legal pluralism as the basis for the design and operation of a water law. This would require that rights of access to land – which arise from customary rights of use – be recognized as a legitimate basis for the provision of a water permit.

References

Government of Kenya (1999) *National Policy on Water Resources Management and Development*, Sessional Paper No. 1 of 1999. Government Printer, Nairobi.

Ministry of Land Reclamation, Regional and Water Development (MLRRWD) (1997) *Community Management of Water Supplies Projects: Guidelines, Modalities, and Selection Criteria for Handing Over Water Supply Schemes*. MLRRWD, Nairobi.

Njonjo, A. (1997) Study of community managed water supplies – final report on case studies and experience exchange. In: World Bank (ed.) *World Bank RWSG-EA, 1994: Survey of Community Water Supply Schemes*. JICA, Republic of Kenya.

von Benda-Beckman, F., von Benda-Beckman, K. and Spiertz, H.L.J. (1997) Water rights and policy. In: Spiertz, J. and Wiber, M.G. (eds) *The Role of Law in Natural Resources Management*. VUGA, Amsterdam.

11 Coping with History and Hydrology: how Kenya's Settlement and Land Tenure Patterns Shape Contemporary Water Rights and Gender Relations in Water

Leah Onyango,[1] Brent Swallow,[2] Jessica L. Roy and
Ruth Meinzen-Dick[3]

[1]*Department of Urban and Regional Planning, Maseno University, Maseno, Kenya;
e-mail: leahonyango@yahoo.com;* [2]*Environmental Services, World Agroforestry
Centre, Nairobi, Kenya; e-mail: B.Swallow@cgiar.org;* [3]*International Food Policy
Research Institute (IFPRI), Washington, DC, USA; e-mail: r.meinzen-dick@cgiar.org*

Abstract

Like many other African countries described in this volume, Kenya has recently enacted several new policies and public-sector reforms that affect its water sector. This chapter considers those reforms in the context of the country's particular history of land tenure and settlement, a history that continues to have a profound influence on contemporary patterns of land and water management as well as on gender relations in water. The chapter focuses on the particular case of a river basin in Western Kenya, the Nyando river basin (3517 km^2), that has its outlet in Lake Victoria. Over the last century, the Nyando river basin has experienced a history that has shaped spatial patterns of land tenure, settlement and water management. The plural land management systems that exist in the basin today are the product of three distinct periods of historical change: (i) the pre-colonial era that was dominated by customary landholding and land rights systems; (ii) the colonial era in which large areas of land were alienated for specific users and the majority of the Kenyan population confined to native reserve areas; and (iii) the post-colonial era that has encouraged large-scale private ownership of land by men and a small public-sector ownership of irrigation land, all against the backdrop of customary norms and the colonial pattern of settlement and land use. Both colonial and post-colonial institutions have largely disregarded women's rights to land and water resources. Although customary norms are consistent in ensuring access to water for all members of particular ethnic groups, in practice access and management of water points vary across the basin depending upon the historically defined pattern of landownership and settlement. Customary norms that secure the rights of women to water resources tend to have most impact in former native reserve areas and least impact in ethnically heterogeneous resettlement areas held under leasehold tenure. Recommendations are made on how new policies, legislation and government institutions could be more effective in promoting the water needs of rural communities in Kenya.

Keywords: legal pluralism, land tenure, water tenure, gender roles, integrated natural resources management, property rights, policy framework, community participation.

Introduction

Like many other African countries, Kenya has recently enacted new policies and legislation regarding its water resources sector. Reforms of the institutions that govern water supply and water resources management were still in progress as of 2005, with new boards and authorities coming to grips with their responsibilities. The Water Act, 2002 was founded on modern principles of integrated water resources management, empowering water user associations and basin authorities with responsibility for managing water resources and regulating water service providers for efficient, equitable and sustainable use of water.

The Water Act appears to be based on the following propositions: (i) that land and water management are quite distinct areas of administration and governance; (ii) that customary institutions have little influence over contemporary patterns of governance; (iii) that formal administrative structures for water management will be able to have a large influence over land management that affects water resources; and (iv) that private sector and large-scale non-governmental organizations (NGOs) will replace the government as the main supplier of water services. This chapter considers these propositions for the particular case of the Nyando basin in Western Kenya. While representing relatively small portions of Kenya's land and water resources, the Nyando basin displays a surprising level of diversity. Historical processes of settlement and land tenure change have resulted in contemporary differences in land and water management.

Three strands of literature influence the approach taken in this research. The first is the historical and evolutionary approach to property rights and institutional change promoted by institutional economists such as North (1990). By that theory, property rights change is a continual, path-dependent process influenced by a confluence of external and internal forces – forces based on political, social and economic power. The second strand of literature is the theory of legal pluralism promoted by legal anthropologists. In a nutshell, legal pluralism proposes that de facto property rights are always affected by multiple sources of legal, social and political authority, including customary and religious law, local norms and even project regulations. All of these frameworks can be the basis for claims over land, water and trees. Access to, and control over, water and other resources are thus the outcome of the interplay between these different types of claims, the negotiation processes that take place and the relative bargaining power of different claimants.

The third strand that influences this research, often associated with political ecology, is generally based on the premise that contemporary patterns of resource use and management are embedded in historical processes involving competing and cooperating social actors. Cline-Cole (2000) describes a landscape as a 'produced, lived, and represented space constructed out of the struggles, compromises and temporary settled relations of competing and cooperating social actors'. This strand of literature is firmly grounded in historical studies of the African landscape, such as those undertaken by Fairhead and Leach (1996), Cline-Cole (2000) and Ashley (2005).

This chapter draws information from the Safeguard project, 'Safeguarding the rights of the poor to critical water, land and tree resources in the Nyando River basin in Western Kenya'.[1] This chapter also synthesizes information from the Safeguard project regarding the distinct histories of land management and settlement that have unfolded across the Nyando basin and how these histories shape contemporary water rights and gender relations to water. Conclusions are drawn on the ways in which formal water management authorities can take better account of those local realities to improve water quality and water access in rural Kenya.

The chapter begins by introducing the history and hydrology of the Nyando basin, followed by a brief discussion of the methods used in the Safeguard project. It then provides an examination of the evolution of land tenure in the Nyando basin from the pre-colonial Kenya period to the present day, highlighting the changing property rights over time and space. Gender relations over water are examined at both household and community levels in the context of contemporary property rights and legal pluralism. The chapter concludes by proposing ways of integrating natural resources management at all levels as a way of improving community participation in water management within the existing legal and institutional framework.

Overview of the Nyando River Basin Study Site

The Nyando river drains into one of the largest lakes of the world, Lake Victoria (*see* Fig. 11.1). The Nyando river basin covers an area of approximately 3517 km^2 and had a population of approximately 746,000 as of 1999 (Mungai *et al.*, 2004). At that time, the average population density was 212 persons/km^2 across the basin, with large areas supporting up to 750 persons/km^2 and other large areas with as few as 50 persons/km^2.

As of 1997, the incidence of poverty, as measured by food purchasing power in Kenya's poverty mapping study, was generally high in the Nyando basin, with an average poverty incidence of 58% in the Kericho district, 63% in the Nandi district and 66% in the Nyando district, compared with the national average of 53% (CBS, 2003; World Bank, 2005). Poverty

incidence is variable across space, with an estimated incidence ranging from 36 to 71% across the administrative locations of the Nyando district (*see* Fig. 11.2). HIV/AIDS prevalence is 28% in the Nyando district, 7% in the Nandi district and 12% in the Kericho district (Swallow, 2004). The basin is primarily inhabited by two ethnic groups – the Luo, who occupy the lowlands and part of the midlands and the Kalenjin, who occupy the highlands. Small numbers of a third ethnic group, the Ogiek, occupy parts of the forest margin at the uppermost parts of the basin. Almost the whole basin falls within the three administrative districts of Nyando, Nandi and Kericho, with small portions of the basin falling within other neighbouring districts.

The upper reaches of the basin rise as high 3000 m above sea level (m asl) and receive an annual average rainfall of between 1200 and 1600 mm/year. The highest parts of the basin are within gazetted forests of the Mau forest complex – known as one of Kenya's five 'water towers'. The uplands also support some large-scale tea plantations, smallholder subsistence and commercial agriculture. The Kano plains form the

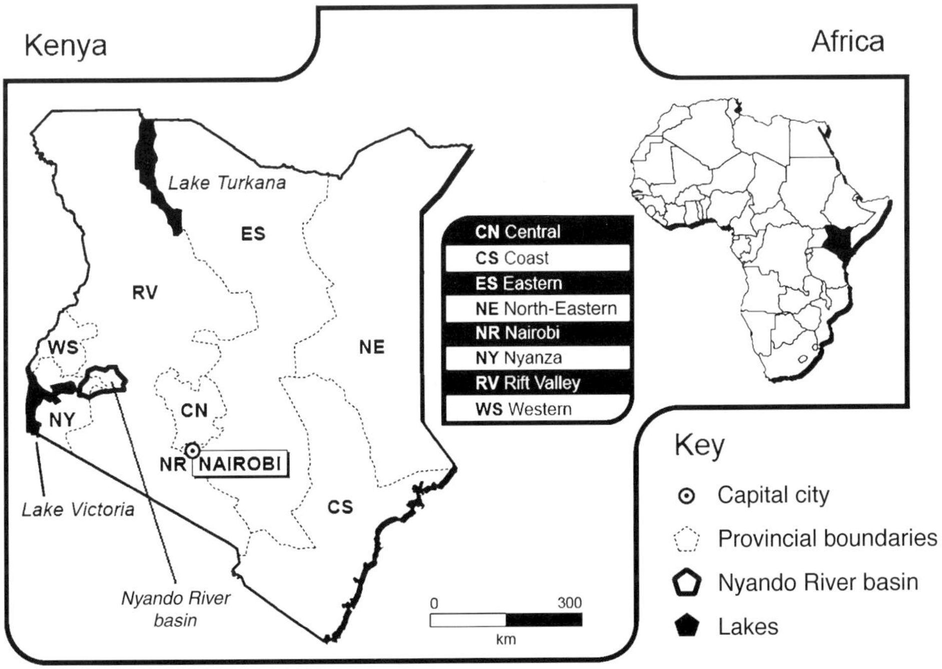

Fig. 11.1. Map of Kenya with Nyando basin.

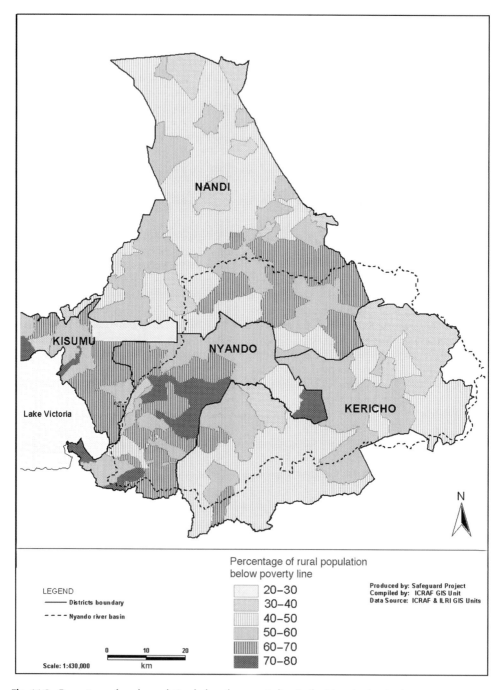

Fig. 11.2. Percentage of rural population below the poverty line in the Nyando river basin.

lowest parts of the basin, adjacent to Lake Victoria where the Nyando river discharges. The Kano plains lie between 1100 and 1300 m asl and receive an annual average rainfall of between 800 and 1200 mm/year. The Kano plains are prone to both floods and droughts. Economic activity in the lowland and midland areas ranges from small-scale subsistence agriculture to large-scale commercial production of sugarcane. A part of the Kano plains has been developed for large-scale irrigation.

The Nyando basin is endowed with unevenly distributed surface water and ground-water. The highlands have many streams and springs, which are the main sources of water for human and livestock use. The lowlands experience less rainfall, fewer rivers and streams, and saline groundwater (Okungu, 2004). The Nyando basin experiences serious land degradation. Walsh *et al.* (2004) estimate that about 61% of the basin suffers moderate to high erosion (soil loss of 40–70 t/ha/year), while the other 39% accumulates sediment (soil accretion of 38–61 t/ha/year). Although upland areas are being eroded, the most severe consequences are felt in the lowlands, which experience serious flooding and siltation. The net erosion rate for the entire basin is estimated at 8.8 t/ha/year. The Nyando river carries very high levels of sediment and is a major source of pollution and sediment loading of Lake Victoria (Walsh *et al.*, 2004).

Research Methods

The main research method used in the Safeguard project was a set of 14 village-level studies of poverty and property rights dynamics. A multi-stage sampling procedure was used to select villages. The first stage involved the characterization of the basin into strata based on altitude, hydrology, land tenure, ethnicity and agricultural production system. The sample sites were drawn from the strata to represent the range of situations that exist in the basin. Table 12.1 of Swallow *et al.* (Chapter 12, this volume) provides additional information about each of the 14 villages. In each selected sub-location, one village containing between 50 and 100 households was selected for a participatory analysis of poverty and livelihood dynamics. The foundation of that analysis was the 'Stages of Progress' methodology developed by Anirudh Krishna and applied in India (Krishna, 2004; Krishna *et al.*, 2005), Kenya (Krishna *et al.*, 2003) and Uganda (Krishna *et al.*, 2004).

By systematically prompting and guiding discussions among a village representative group, the method generates a Stage of Progress ladder for the village, which ranges from absolute poverty to relative prosperity, poverty and prosperity lines defined by those stages, and measures of the level of poverty:prosperity and poverty dynamics for each household in the village. In the Safeguard project, we added in a stronger focus on livelihood strategies pursued by households in the village, as well as the assets required for those livelihood strategies. A village resource map and calendar of historical events concerning land, water and tree assets in the village were also generated for each village.

Information on land and water use and management was collected through both focus groups and household surveys. First, an inventory of water points was compiled as part of a resource-mapping exercise that was conducted with the village representative group. Secondly, a list of 18 questions on land and water management in the village was posed to the village representative groups during a half-day discussion period. Questions focused on access, equity, control of land and water as well as on women's rights to land, water and trees. Additional questions about irrigation management were posed, where relevant. Thirdly, a household survey was conducted with a sample of about 30 households in each village, largely focusing on land and water management. The household sample was stratified by poverty level and poverty dynamics.

Besides data from the 14 Safeguard villages, complementary information was compiled from a wide range of secondary information sources. Registry Index Maps (RIMs) were combined for the three districts to generate a set of land tenure maps for the basin. Maps were generated to depict the situation as it was in 1964, at the eve of independence, and how it was in 2004, 40 years after independence. The history of land tenure and institutional change was compiled from a variety of literature sources and government documents.

Evolution of Land Tenure Systems in the Nyando River Basin and Access to Water under Each Type of Tenure

Land in the Nyando river basin is held under different tenure systems in different parts of the basin, with each system changing over time. In pre-colonial Kenya, all natural resources were owned communally and claims were determined by clans. In the colonial era, the Crown Lands Ordinance of 1902 gave authority to the Crown to alienate land. Any land not physically occupied by local people was considered wasteland (free land) and free for alienation to the European settlers. Local people's rights to land were defined by occupancy, while settlers were given freehold titles by the Crown.

Two parallel landholding systems thus developed. When settlers wanted to gain control over land that was occupied by locals they had to negotiate the right of occupancy with local people. The settlers advocated for grouping the Africans in defined reserves far removed from any lands deemed to be suitable for European settlement. The Crown Lands Ordinance of 1915 allowed the Governor to create Native reserves and provided for the settlers to be given agricultural leases of 999 years. Following the Kenya Land Commission (Carter Commission) of 1934, the Native Lands Trust Ordinance of 1938 re-designated Native Reserves as Native Land and removed them from the Crown Lands Ordinance. This created a set of laws to govern native lands and another set to govern crown land. Even after independence both sets of laws were still in force, which in part explains the current state of confusion in land administration in Kenya.

The Native Land Trust Board under the Chief Native Commissioner held native land in trust for the communities. Local people lost all their rights to lands outside of the native lands. The Crown Lands Ordinance was amended to define the highlands, which were administered by a separate Highland Board. Both boards and their boundaries were set up by 1939 and remained the same up to the time of independence in 1964 (Juma and Ojwang, 1996). The highlands are commonly referred to as the white highlands. They were often the most productive parts of the country and developed cash economies, whereas the native lands were often the less productive and developed subsistence economies. These patterns persist to date. Figure 11.3 is an attempt to illustrate the evolution of land tenure in the Nyando river basin.

This study identified seven ways in which land is currently held in the Nyando basin: (i) trust land – not titled; (ii) government land – not

titled; (iii) adjudicated land – freehold titles on completion of adjudication; (iv) settlement schemes – freehold titles on discharge from the Settlement Fund Trustee (SFT); (v) large-scale farms with leasehold titles; (vi) land-buying companies – freehold title on subdivision to small units; and (vii) forest land – reserved on gazettement. The landholding types in (iii), (iv), (v), (vi) and (vii) all fall under the category, labelled as 'private land' by Mumma (Chapter 10, this volume). This study has generated a map of land tenure for 1964 (see Fig. 11.4) – when the country achieved independence – and one of land tenure for 2004 (see Fig. 11.5) for purposes of analysing changes in land tenure. Using the two maps it is possible to examine the changes that have occurred over the last 40 years and how these changes explain contemporary water rights and gender relations. The remainder of this section describes land and water management under each land tenure type currently existing in the Nyando basin.

Management of land and water in trust lands

Public land in the *native lands* is held in trust for the people by the local authorities and is referred to as trust land. Before adjudication, all land in the native areas was trust land. In the Nyando basin there are three county councils (Nandi, Kipsigis and Nyando), as well as several municipal and town councils. All trust land that is not identified and gazetted for a specific use is held in trust by these local authorities. However, a survey of the three county councils established that most trust land in the basin had already been alienated. What remains under the jurisdiction of the county councils today are schools, cattle dips, dispensaries and some wetlands in the floodplains adjacent to Lake Victoria.

Water and other resources found on trust lands are mostly open to all people who live within the local community. People from outwith the local community may be allowed access to those resources, although locals are given priority, especially if the commodity is scarce. Access to natural water sources such as streams and springs is generally more open than access to constructed water facilities.

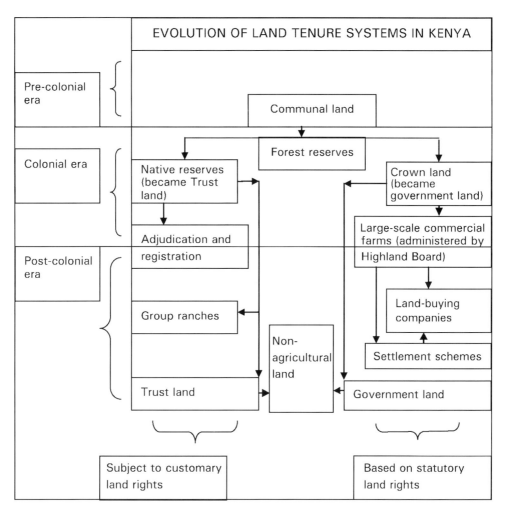

Fig. 11.3. Evolution of land tenure in the Nyando river basin (from authors' conceptualization based on extensive literature reviews).

Management of land and water in government lands

At independence, all crown land was converted to government land and was administered by the Commissioner of Lands on behalf of the President. All government land that is not alienated is still held in the same way. No one has any right to use or occupy such land unless granted a lease by the government, although it is common to find unofficial users of these lands. In the Nyando basin, government land is found only in the urban centres such as Kericho and Muhoroni and in the riparian reserves that abut streams, rivers and wetlands. Access to water resources on government land tends to be poorly regulated. Local authorities would ideally be the custodian of such land, but they rarely take up that responsibility, so riparian reserves remain designated as government land. The lack of enforcement of regulation of riparian areas on government land means that these are the places where high-density slums tend to be located all across Kenya. In the Nyando basin, similar trends are already becoming evident in towns like Muhoroni and Ahero. A very negative impact of the de facto open access to riparian reserves is that the

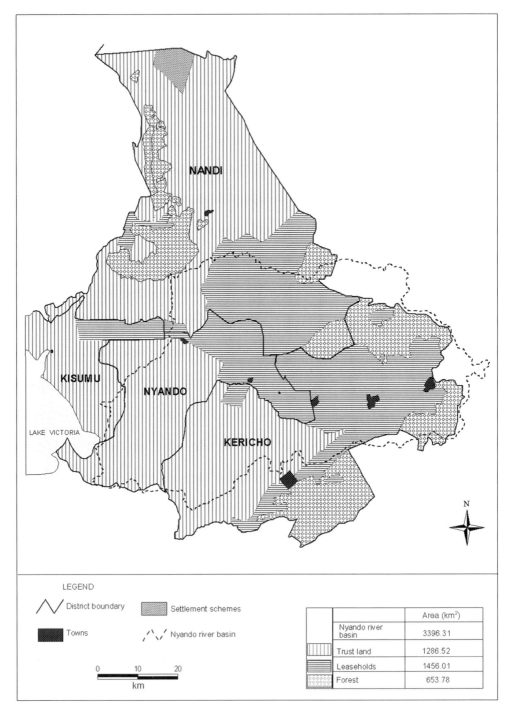

Fig. 11.4. Land tenure in the Nyando river basin in 1964.

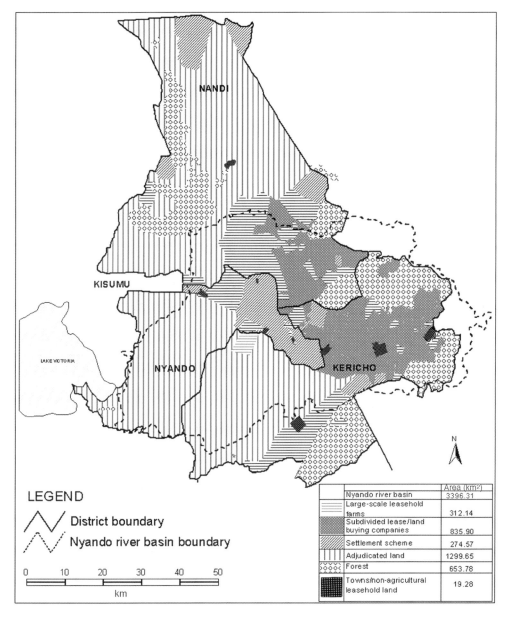

Fig. 11.5. Land tenure in the Nyando river basin in 2004.

	Area (km²)
Nyando river basin	3396.31
Large-scale leasehold farms	312.14
Subdivided lease/land buying companies	835.90
Settlement scheme	274.57
Adjudicated land	1299.65
Forest	653.78
Towns/non-agricultural leasehold land	19.28

rivers in towns are recipients of refuse from both people and industries. In Nairobi, it is reported that 34% of urban vegetable producers divert untreated sewerage from trunk sewers on to their riverine gardens (Cornish and Kielen, 2004).

Management of land and water in adjudicated land

Land adjudication is the process through which land in the native reserves is surveyed and registered as freehold. This process started in

1956 in some parts of central Kenya, but was widely implemented only after independence in 1964. The process of adjudication was slow because it had an inbuilt mechanism for hearing and determining disputes and, in many instances, included land consolidation. On completion of adjudication a freehold interest is registered and a title deed issued. The process of adjudication was prompted by the Swynnerton Plan of 1955, as the colonial government looked for ways of improving agricultural production in native lands. The Swynnerton Plan recommended that agricultural production could be enhanced if titles were issued to the Africans for the land they cultivated. The government was to provide loans for improving agriculture, using the titles as security.

Adjudication sections are carved out along ethnic lines and thus tend to be homogeneous in terms of ethnicity. As a result, property rights to adjudicated lands are heavily influenced by culture. In the two dominant ethnic groups found in the basin, land is controlled by men and managed through male-dominated councils of elders. Both the Luo and Kipsigis communities practice polygamy. Land is inherited from fathers through the mothers. Each married woman is allowed to cultivate specific pieces of land which are referred to as her land, although the women are rarely if ever registered as the title-holders of that land. When her sons grow up, the woman gives each of them a piece of land from the portion she has been cultivating. Sons are allowed to transfer the land and can acquire a title deed. Most women will only get land registered in their names if their husbands die while their children are still legal minors. An examination of the adjudication registers in the Ketitui sub-location (Village 3), which is Kalenjin-speaking, showed that 8% of the land was registered in the names of women while 92% was registered in the names of men. In the Agoro East sub-location (Village 12), a Luo-speaking community, the register indicated that of the registered parcels of land 12 and 88% were in the names of women and men, respectively.

Land adjudication is followed by a survey to establish the boundaries and the area of a parcel of land for the purposes of registration. The land adjudication process in the Nyando basin did not make any allowances for riparian reserves, but instead used the rivers as boundaries between individual plots. This had the effect of privatizing riparian reserves. Anyone whose land did not reach the river did not gain access to the riparian reserve. Since the river was drawn as a thin, straight line it was part of the two pieces of land on either side. Public access to the river was at the bridge where the road and the river meet. All adjudicated land is former ancestral land and is subject to customary norms. The customary norms of both the Luo and Kipsigis dictate that no one should be denied water. A Kalenjin proverb summarizes this by saying: 'Even the hyena [the least respected of the animals] has a right to water'. Because of this customary norm, people will let others pass through their private property to access river water even where there is no demarcated road. The lack of fencing, which is characteristic of adjudicated land in the Nyando river basin, makes it possible to create and use short cuts across individual lands.

When the river water is harnessed for a piped water supply, two methods can be used to secure passage through private land. The safest and most secure is to obtain an easement or a way leave, which will allow the pipes to pass officially through private property without interference from the registered owner. This is provided for in the Way Leaves Act, Cap. 292. The other alternative, which is more often used, is to seek verbal permission from the owners of the land through which the pipes will pass. Because water projects serve many people and because the customary laws dictate that no one should be denied water, this approach works, but it is not secure. In the event of any of the landowners falling out with the rest of the group, then he or she can cause a lot of trouble to other members.

Springs as water sources were also not accounted for in the adjudication process. As a result, all springs in the adjudication areas fall on private land. There are rarely public roads leading to the springs, so people use the roads passing closest to the springs and, where the roads end, they create trails passing through private land. By Luo and Kipsigis custom, no one denies other members of the community water from the springs. However, access to the springs is in fact becoming more restricted over time as land is subdivided and more fences are erected. Many water projects around springs have not yet entered into any legal or written agreement with the landowners on which the springs are found. This study established that most water projects around springs in the basin rely on customary laws to secure rights to the springs, sometimes backed up by a 'No objection' form signed by the landowner (Were *et al.*, 2006).

Management of land and water in government settlement schemes

At the time of independence, the new Government of Kenya set up settlement schemes as a way of transferring land in the white highlands from European settlers to African farmers. This was done in several ways. One of these was through the Settlement Trust Fund (STF) paying off the white farmer, planning and subdividing the land and then settling African farmers on it. The STF allocated the land on loan and registered a charge with the Permanent Secretary in the Ministry of Lands. When a farmer paid off the cost of the land to the Settlement Trust Fund, he obtained a certificate of discharge from the Permanent Secretary in the Ministry of Lands and registered a freehold interest in his favour. The STF also provided farmers with loans for working capital. The five settlement schemes (Koru, Oduwo, Muhoroni, Songhor and Tamu) in the Nyando district fall in the mid-altitude part of the basin and were set up to promote rain-fed sugarcane farming. Three sugar factories – Miwani, Chemelil and Muhoroni – were constructed to process the sugarcane produced in these settlement schemes. The cash economy that had been started by the white farmers was continued. In the Kericho and Nandi districts there are fewer settlement schemes and they promoted mixed farming (dairy, tea and food crops).

Settlement schemes were a creation of the government and, although a lot of planning was carried out in other aspects of land use (e.g. steep hillside areas), they did not take care of the riparian reserve. This oversight can be blamed in part on the legislation under which the land was registered, which did not state clearly the width of the riparian reserve. As a result, the river was used as a boundary between farms, which again had the effect of allocating the riparian reserve as private land. The government involved professional land use planners who took care of springs, dams and swamps as sources of water. They were identified, surveyed and reserved as Special Plots to be held in trust by the local authority for the community. The land reserved was substantial and allowed for catchment protection and conservation. However, due to lack of a focused land policy and enforcement, some of these special plots have recently been allocated to individuals. Other special plots have become de facto open access plots.

The people who settled in the schemes usually did not have their origin in the same community, so they lack the cohesion that comes with a common heritage. Most settlers moved to the area with the hope of making a better life for themselves, so that economic factors are foremost in their dealings. Statutory rights protect individual rights that promote accumulation, as opposed to customary rights that advocate the communal use of resources. There is better enforcement of statutory rights than in the native lands where people choose not to prosecute their close kin even when the livestock of these kin destroys their crops. As a result, statutory rights are more powerful but customary rights still exist.

Management of land and water in large-scale leasehold commercial farms

Large-scale farms are found only in the former white highlands and are operated as commercial enterprises. All large-scale farms hold 999-year leases from the government. Most large-scale farms in the higher altitudes are tea plantations, while in the mid-altitude areas they are sugarcane plantations. Multinational companies such as Unilever operate most of the tea plantations, but most of the sugarcane plantations are locally owned. There are a number of factories located within the region to process both sugarcane and tea. The large-scale farms employ large numbers of labourers, many of whom are provided with housing within the farms. There are people who have lived on the plantations all their lives and have come to feel entitled to land on the plantations. Several large-scale farms, including eight tea estates, have excised portions of their land to settle these long-term farm workers, popularly known as squatters. The land claims of long-term farm workers pose a serious challenge to large-scale farms. Where these claims have been ignored, the squatter populations have been known to take the law into their hands and invaded the farms. This was the case in one of our sample communities, i.e. Kapkuong (Village 8).

The operations of the large-scale farms are strictly guided by the statutory laws. Water resources on the large-scale farms are accessed only by persons authorized by the farm owners. Environmental management on most large-scale commercial farms is exemplary, and their water sources are well protected. Riparian areas are conserved and the natural vegetation left intact. Most have employed environmental officials in response to increasing concerns about environmental protection and the long-term sustainability of their operations. Large-scale commercial farms in the Kericho district have assisted the District Administrative Office in detecting illegal use of forest resources.

Land and water management in subdivided, large-scale farms purchased by land-buying companies

Land-buying companies emerged as an important phenomenon after Kenya's independence, as a way of transferring landownership from white settlers to interested Africans. Commercial land-buying companies emerged in part because the government was unable to purchase all the land from the white settlers who wanted to sell. The government therefore allowed the white settlers to negotiate sales agreements with anyone who was willing and able to make an outright purchase. Very few Africans were in a position to do this, so they came together to form land-buying companies or cooperatives. The members contributed money for the purchase of land and were allocated land in proportion to their share of the contribution.

There were no rules restricting membership in the land-buying companies and this led to problems. Some companies had so many

members that they were unable to be accountable to all the members, and these members lost their money. At other times, they were allocated very small parcels of land. They also did not pay much attention to topography, land use suitability or the provision of public utilities such as water points, roads, schools and clinics. As a result, people were allocated land on very steep slopes, swamps, river banks and hilltops. Most of the land in the upper reaches of the Nyando river basin was bought by land-buying companies (see Fig. 11.4). On subdivision, the land was converted to freehold and each member of the land-buying company was issued with a freehold title.

Yet, many land-buying companies have not issued their members with their title deeds, e.g. the Kotetni farm in Chilchila division of the Kericho district, purchased in 1968. People were allowed to settle in the land before they completed the process of subdivision and issuance of title. Meantime, the members were issued with share certificates as evidence that they had a right to a share of the land. These certificates were inadequate because they indicated only that a member owned shares but did not specify the location of the land allocated to him or her. Companies have taken a very long time to process the documents, and sometimes the final survey results did not tally with the actual position of the plots where members had already settled. Corruption and lack of accountability were rife in the workings of the land-buying companies. In some instances, the President had to intervene in the issuance of the title deeds to be issued. Such appeals are common and are reported in the daily newspapers.

The processes of subdividing the large-scale farms were spearheaded by the private sector. The private firms wanted to allocate as much land as possible to their members, so they did not spare any land in the riparian reserves. All springs are also located on private land, with no public access routes. Shareholders in land-buying companies often come from different places and thus have different cultural norms regarding land and water management. People have therefore tended to rely more on written laws than on cultural norms. Most plots of cultivated land are fenced, making it more difficult for people to access water sources. Private property rights are very strong, and barbed-wire fences were put up to keep away tres-passers and discourage free ranging of live-stock. Springs are all located on private land, with no provision for public access.

Trust relations among people living in the subdivided large-scale farms are generally low. People who live in the region have moved in only since independence in 1964, and have come from all parts of Kenya. Lingering tensions between the ethnic groups occupying this area have been heightened by political manipulations, resulting in the well-known tribal clashes of 1992, 1994 and 1997. In a community with such diverse origins and a history of distrust, statutory laws are stronger than customary laws.

The region has experienced dramatic land use changes in the last 40 years as the land has been converted from large-scale farming to intense smallholder cultivation. Over the same time period, the population of the area has increased as people move in to occupy the subdivided farms. The Nyaribari 'A' village in the Bartera sub-location (Village 2) was formerly the Lelu farm (LR.1442/2), which was previously owned and managed by one farmer. The Lelu farm was purchased by the Nyagacho land-buying company that subdivided it and settled its members. Today, the Lelu farm makes up the Bartera sub-location with a popu-lation of 2810 people, 526 households and a density of 273 persons/km^2 (Republic of Kenya, 2000b). The impacts of these changes on the environment are seen in the emerging environ-mental problems, such as deforestation and landslides. The area has also experienced high rates of erosion (Walsh *et al.*, 2004).

Management of land and water in forest reserves

The Crown Lands Ordinance that established the native reserves was the same measure that made provision for the establishment of forest reserves through gazette notices. Once land has been gazetted as a Forest Reserve it cannot be put to any other use unless it is de-gazetted through another gazette notice. The forests in the Nyando basin include the Tinderet Forest, the North Tinderet Forest, the Londiani Forests and the West Mau Forest. The gazetting of forest reserves displaced the forest dwellers such as people of the Ogiek ethnic group. Many Ogiek people remain landless or illegally reside on the fringes of the forest land. Such was the case of our sample village in Ng'atipkong sub-location (Village 5). The government prohibits entry into the forest, yet the Ogiek way of life is to use the forest resources for their subsistence. Chronic tension between the Forest Department, the local administration and the Ogiek communities becomes more heated when the government chooses to implement tighter restrictions on forest use.

Water sources in the forest are not easily accessible, due to the government policy that aims to keep people away from the forests. Forests are guarded by forest guards, who often harass the local people whom they suspect of encroaching upon the forest. However, the forests are the source of many permanent springs. The Kaminjeiwa village in Kedowa sub-location (Village 1) is a forest frontier commu-nity, and here the most permanent and the cleanest sources of water are within the forest. The people use these sources although they suffer constant harassment from the forest guards. The Ngendui village in Ngatipkong sub-location (Village 5) sits on the edge of a gazetted forest. In this village, the livestock and people draw water from the same point. Crops are cultivated up to the eye of the spring and, despite being on forestland, almost all the trees have been cut down. People living in that area are not provided with government agricultural extension services because they are considered to be squatters. The new Forest Bill passed in 2005 holds some prospect for more effective co-management of forest resources by the government and local community groups.

Gender Roles in Community Water Relations

Household water relations

In rural Kenya, water interactions occur at both the household and community level. At the household level, water interactions concern water demand, supply and allocation. How much water is needed? Who fetches it? What is it used for? This study confirmed that, in

Nyando, as in many parts of Africa, it is women who fetch water for the whole household. Of the household respondents in the Safeguard study 77% indicated that women are the most important collectors of water. Of the households surveyed, 18% had water within their homesteads, 70% obtained their water from other sources within their villages and 12% had to go beyond the village to fetch water. Men herd and take livestock to collective water points, while women carry water to livestock (especially dairy cows) kept within the compounds. In the lower basin, drinking water is collected from rivers in the early morning before water is contaminated by people and livestock upstream.

It is common to find most members of the family going to bathe in the river to save women from having to carry water to the homestead. There are separate designated bathing spots for men and women along the rivers. Men tend to bathe downstream and women upstream, since men usually go along with their livestock, which disturb water for downstream users. Women carry laundry to the river to save them from carrying water back to their homesteads and, when they go to bathe in the river, they carry water back. The 1999 population census (see Fig. 11.6) indicates that most people in the basin obtain their domestic water supply from rivers.

Women have primary responsibility for providing water for domestic needs in the Nyando basin. Table 11.1 lists the first, second and third most important collectors of domestic water: clearly, wives and children are the main collectors of water for the 150 households involved in the Safeguard household survey. Fewer than 10% of respondents indicated that husbands ever collected water.

Community organizations involving water

Community water interactions go beyond individual households. Organizations involving water are often spearheaded by men, although the impetus for organization is often provided by women. Why? Although women are given responsibility for providing their households with water, they are handicapped when it comes to organizing water supplies, because water is found on land, or passes over land,

controlled by men. Therefore, it is the men who can make decisions about what can or cannot happen on the land. At a glance, the men appear to take leadership of the projects while the women just enjoy the benefits of improved water supply. However, further probing reveals that the women work behind the scenes and make many contributions to the instigation and implementation of water projects through providing labour, food and even money (Were et al., 2006). Communities in the lower Nyando organize around water in irrigation areas and the irrigation committees are male-dominated.

Piped water supplies

A water supply refers to the supply of water from a system that has been improved and involves reticulation and improved water quality. Water supplies tap water from rivers, springs and bore holes. Water supplies in the Nyando basin have been implemented by the Ministry of Water and Irrigation, local community groups, the National Water Corporation and Pipeline Company (NWCPC), a variety of NGOs, private companies and public schools. An unpublished assessment undertaken for the Lake Victoria South Service Board indicates that, in the Kericho district, the Ministry of Water and Irrigation operates 48% of the water supplies. These are mostly fixed pump-water supplies, with high costs of implementation and maintenance.

Community groups own 28% of the water supplies, most of which are low-cost, gravity-fed projects. The public schools own 14%, the private companies 7% and the NWCPC 3%. In the Nyando district, the community groups manage 39% of piped water supplies, private companies 26%, the Ministry of Water and Irrigation 19%, public schools 13% and the NWCPC 3%. In the Nandi district, the community groups manage 20% of water supplies, private companies 33%, public schools 21%, the Ministry of Water and Irrigation 11% and the NWCPC 3%. Some households can afford to have piped water in the homesteads, while most of them purchase water from communal stand points or community kiosks, where they are charged per unit of water used. Community kiosks are managed by women's groups or other private vendors.

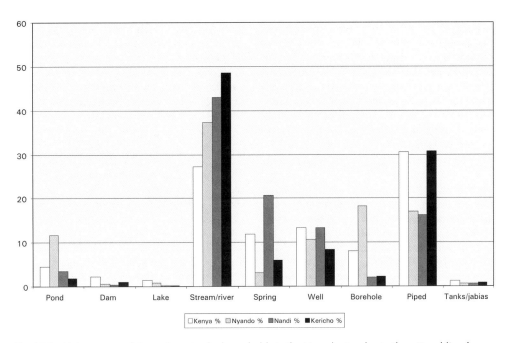

Fig. 11.6. Main sources of domestic water for households in the Nyando river basin (from Republic of Kenya, 2000b).

Table 11.1. Most important collectors of household water (from authors' analysis of data from Safeguard project (see text)).

		Most important		Second most important		Third most important	
		Frequency	%	Frequency	%	Frequency	%
Valid							
	Wife	107	71.3	14	9.3	3	2.0
	Husband	4	2.7	6	4.0	4	2.7
	All children	20	13.3	53	35.3	7	4.7
	Male children	7	4.7	23	15.3	20	13.3
	Female children	7	4.7	12	8.0	8	5.3
	Relatives	4	2.7	8	5.3	1	0.7
	Total	149	99.3	116	77.3	43	28.7
Missing							
from system		1	0.7	34	22.7	107	71.3
Total		150	100.0	150	100.0	150	100.0

Boreholes

There are no community-initiated water supplies tapping water from boreholes in the basin because of the high costs incurred in drilling and pumping the water. On average, the cost of sinking a borehole is about US$10,000 (A. Adongo, Kisumu, 2005, personal communication). Community-initiated water supplies tap water from springs and rivers.

Rivers

Communities organize around river water in areas where there is irrigation. In the irrigation area, water is used for production of rice and horticultural crops. The irrigation schemes are organized into scheme committees. There are a few women in the committees, but the majority of committee members are men. In the five Safeguard villages where irrigation is practised, only the rice-growing communities have water management committees: Ahero (Village 9), Jimo middle (Village 13) and Achego (Village 14).

Springs

Springs are common in the upper and middle reaches of the basin. Spring protection is one area where communities organize around water. In some instances the spring is just protected and the people come and fetch water, but there are also instances in which springs are protected and piped. When the spring water is piped, only members who have contributed towards the effort benefit by getting individual connections to the water supply. There are no communal standpipes or water kiosks from which such water is sold.

A spring census carried out in the part of Kericho district that is within the Nyando basin yielded an inventory of 143 springs, 17% of which were protected but not piped and 17% of which were protected and piped. Community groups took the lead for about half of the protected and protected-and-piped projects (D. Buntdotich, unpublished data, 2005). In only one of these community-based schemes was it evident that women took the lead in instigating and implementing the project.

Water pans

Water pans are a common source of water in the lower Nyando basin. Many water pans have been constructed through community initiatives. Construction involves digging up land to enable harvesting of surface run-off. A few water pans have also been constructed by individuals. Community water pans tend to be managed by the village elders and other village leaders, with maintenance done on a voluntary basis by the community. Some water pans were constructed by government agencies under donor-aided projects, such as the Winam Gulf Project, or through food-for-work projects. Water pans are used by the community mainly to water livestock and irrigate vegetable

gardens. During water-scarce times, water from the pans is also used for cooking and drinking, although they tend to be heavily polluted by livestock. This can be a potential source of conflict between men and women, with men responsible for watering livestock and women responsible for household water supplies.

Shallow wells

Shallow wells are a common source of water in the lower Nyando basin. Shallow wells are usually hand-dug to a depth of 70–100 feet (21.5–30.8 m), and are fitted with a hand-pump. Some are financed by donors, while others are financed and constructed by individuals. Only a few are initiated by the community. The cost of a 70-foot-deep, hand-dug, shallow well ranges from Ksh25,000 to 30,000 (approximately US$345–415), excluding the cost of the pump (M. Vardhan, Kisumu, 2005, personal communication).

In the case of donor-supported shallow wells, there is always an element of cost-sharing between the community and the donor. The community contributes local materials and food for the masons, while all other materials and payments are borne by the donor. The day-to-day management and operation are taken over by the community group on completion of construction. In most cases, women groups are responsible for management of the shallow wells through a committee. Members of the women's group pay a monthly flat rate to draw water from the well, whereas the non-members pay per unit of water collected. Water from shallow wells is used for drinking, cooking, washing, watering livestock and vegetable plots. A common problem with donor-assisted shallow wells is that the wells are located on private land, without the support of signed land-easement agreements with the landowners.

Rainwater

Rainwater harvesting in the basin is carried out at both the individual and group level. Several NGOs provide technical assistance to roof catchment projects. However, the household seeking such a project has to contribute financial and material inputs. The cost of erecting storage facilities for the water is too high for the majority of the people in the basin. A larger number of the people engage in a simpler method where rainwater is collected into pots and pans. This has a lower cost, but it provides less storage and does not make optimal use of the roof catchments.

Summary of Results

Property rights and legal pluralism in the Nyando basin

The analysis presented in this chapter has established that there are multiple sources of authority governing access to water in the Nyando basin, most of which are in fact related to the governance of the land where water points are located. Ambiguity and institutional overlap of land and water management lead to solutions that are context-specific, subject to re-negotiation and latent conflict.

Property rights in the Nyando basin change across the landscape. Large parts of the upper part of the catchment are in large-scale farms or have been subdivided by land-buying companies to people who are socially disconnected. Where there are no common customary norms among the residents of an area, statutory sources of authority tend to be more important. The statutory land laws tend to privilege individual use, thereby weakening the access of others to key water sources located on private land. In gazetted forests and areas that are designated as riparian reserves under state authority, lack of effective enforcement means that these are effectively open-access areas but, rather than providing a better water service, this often leads to degradation of the water sources through pollution or depletion, because no one takes responsibility to protect the water sources.

Most of the lower parts of the basin are former native reserve lands where people are more tightly connected by customary norms, and statutory laws are applied less frequently. Sharing is strongly encouraged, weakening the potential for wealth accumulation and infrastructural investment. No one is expected always to be rich, which is summed up in a Luo saying: *inind diere inind tung*. Loosely translated, this means that, today you sleep between the others and you are kept warm and safe, but tomorrow you will sleep on the edge where no one will shield you from the vagaries of life.

The institution of the chief and the village elder is a point where customary law and statutory law merge. The chief in the rural setting is usually a local person who is well versed in the customs of local people. He works with village elders who know the customs as well as the situations of individual families. Many issues are resolved at the level of the chief and village elders. Where it has to go beyond them to the courts of law, then cultural rights are represented by the chief. In this way non-statutory laws blend with statutory laws. This approach is commonly applied for the settlement of land cases. It could play a greater role in resolving water management cases.

Constraints to community water development

One of the Millennium Development Goals is 'reducing by half the proportion of people without sustainable access to safe drinking water'. Water development in rural Kenya faces constraints in meeting this goal, especially amongst the poorest in the community because of the 'user pays principle'. The poorest are so poor that they are unable to pay for water. The history of the basin has created geo-spatial patterns of wealth and poverty, with pockets of very poor communities. In areas with high incidence of poverty, such as the former native lands in the lower Nyando (see Fig. 11.5), the communities have less capacity to contribute financially to community water development: they rely heavily on donor-funded water development.

Landownership and settlement patterns influence community management of water. People find it difficult to access water pans or rivers whose banks are owned as private property. They have to negotiate for the use of land to locate water facilities because in many cases no land was set aside. Within some communities there is little social cohesion. This is especially so where people have moved into an area from diverse origins, e.g. land-buying companies and settlement schemes. Where there is community cohesion, development of community projects is easier, but it may be useful for such groups to obtain written agreements from those whose land is affected, so that statutory law can reinforce the agreements in case of local conflicts.

Although women have primary responsibility for water management, they are constrained in instigating infrastructural investments because these investments will invariably affect some areas of land which tend to be controlled by men. Gender relations therefore have a profound effect on water management. At the minimum, men need to be convinced that the water management activities of their wives will provide benefits in terms of increased water availability and improved household income.

Implications for Kenya's Water Sector Reforms

This study has established that Kenya's historical processes of settlement and land tenure differentiation have created a plurality of land and water property rights across the Nyando basin. There is therefore a need to explore ways through which policy formation processes can be made more meaningful to local communities affected by different combinations of land and water rights. One way of doing this is by improving community involvement in water resources management within the existing legal and policy framework. An integrated approach to water resources management would link community water development to other natural resource policies, strategies and legislation and enhance the performance of community organizations (see also Mumma, Chapter, 10 this volume).

The Water Act, 2002 (Republic of Kenya, 2002) creates the Water Resources Management Authority (WRMA) for management of water resources and the Water Service Regulatory Board (WSRB) to regulate the provision of water and sewerage service. These institutions implement and inform policy. A diagrammatic representation of the institutions created by the Water Act, 2002 is found in Mumma, Chapter 10, this volume. The regional offices created by the Water Act, 2002 should link their work with relevant government institutions that also affect water management. Coordinated action by government agencies on a common local platform would create a forum for effectively linking up with local communities

to ensure that their interests and issues are addressed and their capacities enhanced. Table 11.2 indicates some key government ministries involved in natural resources management with direct impact on water resources management. Integration of institutions found in the last column can be an ideal point for creating a forum where the government agencies meet local communities.

Resource conservation and protection by community organizations

The Environmental Management and Coordination Act (EMCA) of 1999 (Republic of Kenya, 2000a) established the National Environment Management Authority (NEMA) as the principal instrument of the government for implementation of policies related to the environment. Every 5 years, NEMA must produce a National Environmental Action Plan (NEAP), Provincial Environment Action Plans (PEAP) and District Environment Action Plans (DEAPs). The preparation of a DEAP is designed to be a participatory process that can be a forum for the governmental agencies and the local communities to come together to analyse natural resources (including water) and develop a workable plan of action for sustainable use, protection and conservation. One of the concerns of downstream communities in the Nyando basin is that rivers are polluted by upstream users, particularly by industries such as the sugarcane factories. The Environmental Management and Coordination Act (Section 42) addresses the protection and conservation of the environment, with specific reference to rivers, lakes and wetlands.

Environmental restoration orders (Section 108), Environmental easements and Environmental conservation orders (Section 112) can be used to prevent pollution of rivers and hold polluters accountable for the damages they generate. The Agriculture Act supports the Environmental Management and Coordination Act by providing for land preservation orders. Local community groups can be made more aware of these tools and of their rights by the government agencies within the participatory processes.

Table 11.2. A framework for integrating Kenya's statutory natural resources management institutions (from authors' conceptualization based on literature review and field study).

Statutes	Implementing ministry	National institution	Regional representatives	National policy tools	Regional policy tools
Water Act, 2002	Ministry of Water and Irrigation	Water Resource Management Authority Water Service Regulatory Board	Water Service Boards Catchment Area Advisory Committees	National Water Services Strategy National Water Resources Strategy	Catchment Management Strategy
Environmental Management and Coordination Act of 1999	Ministry of Environment and Natural Resources	National Environmental Management Authority	District Environment Coordinators	National State of the Environment Report National Environment Action Plan	District State of Environment Report District Environment Action Plan Environmental Easements
Agriculture Act Cap 318	Ministry of Agriculture	Forest Department	District Agricultural Officers Agricultural Extension Workers		Conservation Orders
Local Government Act Cap 265	Ministry of Local Government		Municipal Councils County Councils	Council Committees	Local Development Plans Environment Committees
Physical Planning Act of 1996	Ministry of Lands and Housing	Department of Physical Planning	Physical Planning Officers	National Liaison Committees	Physical Development Plans (Land Use Plans) District Liaison Committees
	Ministry of Planning and National Development		District Development Officers	National Development Plans (Economic Plans) District Focus for Rural Development Strategy	District Development Plans (Economic Plans) District Development Committees (DDCs)

Improving physical access to water resources and water facilities

Kenya's current policies and organizational set-up mandate many organizations to be involved in improving physical assess to water resources and water facilities. Under the Water Act, 2002, the Water Resource Management Authority is mandated to mobilize communities to identify their needs related to water resources management – including infrastructural requirements, areas that should be set aside for conservation of water sources and access roads. Those needs should be integrated into participatory processes of preparing Physical Development Plans by the Department of Physical Planning (Republic of Kenya, 1996) which, on approval, become the official documents that guide development in a particular area. The Land Acquisition Act can be used to provide for compulsory acquisition of land needed for public utilities, and the Water Service Trust Fund should compensate the displaced owners of that land. The National Environment Management Authority should make sure that water management plans are integrated into the District Environmental Action Plans, which are then integrated into the National Environmental Action Plans.

In order to be in line with the national budgeting requirements, it is necessary to transmit these needs to the national offices. The next step would then involve the government line ministries mandated to implement the specifics included in the plan. The Ministry of Roads is empowered to begin the process of compulsory land acquisition for the proposed roads; the Water Service Board or Water Service Trust Fund would avail the money for the compensation; the Ministry of Water and Irrigation will institute the process of compulsory acquisition of the land around the proposed water points; while the Ministry of Environment would institute the process of registering environmental easement on behalf of the Water Resources Management Authority and the community.

Such concerted effort by multiple government agencies is a possible objective for the government agencies implicated in water development. However, current experience also shows that many of these state agencies lack the resources to effectively reach most rural communities. They also lack incentive to work effectively across agencies. Government involvement must therefore be seen as a supplement to, rather than a substitute for, community involvement over the long term.

Financial empowerment of community organizations

The financial constraints of rural communities have been addressed by the Water Act, 2002 through the establishment of the Water Trust Fund (WTF). Its mandate is to give financial assistance to the rural communities in improv-

ing community capacity to participate in the development of water supplies. Groups that want to supply water must register with the Water Service Board (WSB) as Water Service Providers (WSP). Any group that needs to be eligible to obtain funding from the WTF must have a legal identity. This is obtained through registration with the Registrar of Societies or the Registrar of Companies at the Attorney General Chambers.

Studies carried out in the Nyando river basin have found out that there are many community groups involved in water supply and management, but none of them is registered with the Attorney General. Most of these are registered with the District Social Development Officer (DSDO) as Community-based Organizations (CBOs). This situation disqualifies all the CBOs from being funded. The Water Service Boards must find a way of ensuring that they do not lock out all community groups. A simplified system of registration, preferably at the district level, is a high priority.

Ultimately, there is considerable scope for government agencies to facilitate improved access to water, but they are unlikely to be able to fully replace customary authorities and community organizations in managing water. It is both more realistic and more effective to recognize the need to coordinate with both land management and the local social organization to build viable institutions for continued water services.

Conclusions

The introduction to this chapter suggested that Kenya's Water Act, 2002 was based on four implicit assumptions. The analysis presented in this chapter for the Nyando river basin implies the need for some rethinking of those assumptions. The first assumption was that land and water management are quite distinct areas of administration and governance. Contrary to that assumption, we find that use, access and governance of water points, riparian areas and fragile parts of the catchment depend largely on land tenure. These interactions between land and water need to be given explicit consideration in policies and strategies affecting land, environment and water.

The second implicit assumption of the Water Act, 2002 was that customary institutions have little influence over contemporary patterns of governance. Like several other case studies presented in this volume, and echoing Mumma's analysis of the Kenya context (Mumma, Chapter 10, this volume), we find that the customary institutions of the Luo and Kipsigis people continue to hold sway in the Nyando basin, guaranteeing community access to water on the one hand, while reducing private investment in water point infrastructure on the other.

The third implicit assumption of the Water Act, 2002 was that formal administrative structures for water management would be able to have considerable influence over land management affecting water resources. The analysis presented in this chapter shows that formal administrative structures may have more sway in the resettlement areas than in former native reserve areas. At the local level the main face of the government is the chief – administrative structures for water management will need to garner the support and active involvement of the chief in order to have any real effect.

Finally, the Water Act, 2002 was based on the implicit assumption that private and large-scale NGOs would replace the government as the main supplier of water services. Data and analysis presented in this chapter suggest that informal community groups rather than the government were the main suppliers of water services in rural Kenya before the policy change; they are likely to play even more central roles with the policy change. More explicit attention should be paid to the needs and constraints of such groups.

In addition to providing a useful analysis of water and land management in Western Kenya, this chapter has also demonstrated the merits of the analytical approach that was taken. The underlying approach drew upon three strands of literature: the legal anthropology approach to legal pluralism, the evolution approach to property rights change and the political ecology approach to interpretation of landscape dynamics.

Acknowledgements

We would like to thank the staff of Survey of Kenya and Lands Department in Nandi, Kericho and Kisumu offices for their help in obtaining information that enabled us to put together the land tenure maps for the Nyando basin. We also acknowledge funding from the Comprehensive Assessment of Water Management in Agriculture and the European Union for providing funds to carry out the Safeguard study, from which most of this information is drawn. We would also like to acknowledge the many useful comments provided by Barbara van Koppen and Mark Giordano.

Endnote

[1] The Safeguard project, which was a component of a larger project on 'Breaking the vicious cycle: managing the water resources of East and Southern Africa for poverty reduction and productivity enhancement', was funded through the Comprehensive Assessment of Water Management in Agriculture. The main objective of 'Breaking the vicious cycle' was to bring together and synthesize information on the links between water resources, poverty and productivity in Africa. Other components of 'Breaking the vicious cycle' were the African Water Laws Symposium, South Africa, January 2005 and a regional review of the dynamics of water resources and poverty in the Lake Victoria basin, by the Africa Centre for Technology Studies. The Safeguard project has been a collaborative project between Maseno University, the World Agroforestry Centre (ICRAF) and the International Food Policy Research Institute (IFPRI), and is entitled 'Safeguarding the rights of the poor to critical water land and tree resources in the Nyando River basin'. Additional financial support for the Safeguard project was also provided by the European Union, through the World Agroforestry Centre.

References

Ashley, R. (2005) *Colonial Solutions, Contemporary Problems: Digging to the Root of Environmental Degradation in Kabale, Uganda.* Agroforestry in Landscape Mosaics Working Paper Series, World Agroforestry Center, Tropical Resources Institute of Yale University and the University of Georgia, New Haven, Connecticut.

CBS (Central Bureau of Statistics) (2003) *Geographic Dimensions of Well-Being, Kenya: Where are the Poor?* Central Bureau of Statistics, Ministry of Planning and National Development, Nairobi.

Cline-Cole, R. (2000) Knowledge claims, landscape, and the fuelwood-degradation nexus in dryland Nigeria. In: Broch-Due, V. and Schroeder, R. (eds) *Producing Nature and Poverty in Africa.* Nordiska Africikainstitutet, Stockholm, pp. 109–147.

Cornish, G.A. and Kielen, N.C. (2004) Wastewater irrigation – hazard or lifeline? Empirical results from Nairobi, Kenya and Kumasi, Ghana. In: Scott, C.N., Faruqui, I. and Raschid, L. (eds) *Wastewater Use in Irrigated Agriculture: Confronting the Livelihood and Environmental Realities.* CABI/IWMI/IDRC, Wallingford, UK.

Fairhead, J. and Leach, M. (1996) *Misreading the African Landscape: Society and Ecology in a Forest–Savanna Mosaic.* Cambridge University Press, Cambridge, UK.

Juma, C. and Ojwang, J.B. (eds) (1996) *In Land we Trust. Environment, Private Property and Constitutional Change Acts.* ACTS, Nairobi.

Krishna, A. (2004) Escaping poverty and becoming poor: who gains, who loses and why? *World Development* 32 (1), 121–126.

Krishna, A., Lumonye, D., Markiewicz, M., Kafuko, A., Wegoye, J. and Mugumya, F. (2003) *Escaping Poverty and Becoming Poor in 36 Villages in Central and Western Uganda.* http://www.pubpol.duke.edu/Krishna/documents/Uganda_Nov9.pdf

Krishna, A., Kristjanson, P., Radeny, M. and Nindo, W. (2004) Escaping poverty and becoming poor in 20 Kenya villages. *Journal of Human Development* 5.

Krishna, A., Kapita, M., Porwal, M. and Singh, V. (2005) Why growth is not enough: household poverty dynamics in northeast Gujarat, India. *Journal of Development Studies* 41 (7).

Mungai, D., Swallow, B., Mburu, J., Onyango, L. and Njui, A. (eds) (2004) *Proceedings of a Workshop on Reversing Environmental and Agricultural Decline in the Nyando River Basin.* World Agroforestry Centre (ICRAF), the National Environment Management Authority of Kenya (NEMA), the Water Quality Component of the Lake Victoria Environment Management Programme (LVEMP) and the Ministry of Agriculture and Rural Development, Nairobi.

North, D.C. (1990) *Institutions, Institutional Change and Economic Performance.* Cambridge University Press, New York.

Okungu, J. (2004) Water resource management. In: Mungai, D., Swallow, B., Mburu, J., Onyango, L. and Njui, A. (eds) *Proceedings of a Workshop on Reversing Environmental and Agricultural Decline in the Nyando River Basin.* World Agroforestry Centre (ICRAF), the National Environment Management Authority of Kenya (NEMA), the Water Quality Component of the Lake Victoria Environment Management Programme (LVEMP) and the Ministry of Agriculture and Rural Development, Nairobi.

Republic of Kenya (1996) *Physical Planning Act of 1996.* Government Printers, Nairobi.

Republic of Kenya (2000a) *The Environmental Management and Coordination Act of 1999.* Government Printers, Nairobi.

Republic of Kenya (2000b) *1999 Population and Housing Census. Counting Our People for Development.* Vol. 1, Population Distribution by Administrative Areas and Urban Centers. Central Bureau of Statistics, Nairobi.

Republic of Kenya (2002) *The Water Act, 2002.* Government Printers, Nairobi.

Swallow, B. (2004) Kenya's Nyando basin: problems of poverty agriculture and environment. In: Mungai, D., Swallow, B., Mburu, J., Onyango, L. and Njui, A. (eds) *Proceedings of a Workshop on Reversing Environmental and Agricultural Decline in the Nyando River Basin.* World Agroforestry Centre (ICRAF), the National Environment Management Authority of Kenya (NEMA), the Water Quality Component of the Lake Victoria Environment Management Programme (LVEMP), and the Ministry of Agriculture and Rural Development, Nairobi, Kenya, pp. 22–26.

von Benda-Beckmann, F. (1995) Anthropological approaches to property law and economics. *European Journal of Law and Economics* 2, 309–333.

Walsh, M., Shepherd, K. and Verchot, L. (2004) Identification of sediment sources and sinks in the Nyando river basin. In: Mungai, D., Swallow, B., Mburu, J., Onyango, L. and Njui, A. (eds) 2004 *Proceedings of a Workshop on Reversing Environmental and Agricultural Decline in the Nyando River Basin.* World Agroforestry Centre (ICRAF), the National Environment Management Authority of Kenya (NEMA), the Water Quality Component of the Lake Victoria Environment Management Programme (LVEMP), and the Ministry of Agriculture and Rural Development, Nairobi, Kenya, pp. 27–30.

Were, E., Swallow, B. and Roy, J. (2006) *Water, Women and Local Social Organization in the Western Kenya Highlands.* ICRAF Working Paper No.12, World Agroforestry Centre, Nairobi, 36 pp.

World Bank (2005) *World Development Indicators.* http://devdata.worldbank.org/wdi2005/index2.htm

12 Irrigation Management and Poverty Dynamics: Case Study of the Nyando Basin in Western Kenya

Brent Swallow,[1] Leah Onyango[2] and Ruth Meinzen-Dick[3]

[1] *World Agroforestry Centre (ICRAF), Nairobi, Kenya; e-mail: b.swallow@cgiar.org;*
[2] *Maseno University, Maseno, Kenya; e-mail: leahonyango@yahoo.com;*
[3] *International Food Policy Research Institute (IFPRI), Washington, DC, USA;*
e-mail: r.meinzen-dick@cgiar.org

Abstract

Three distinct pathways of irrigation development have been pursued in Kenya over the last 20 years: a top-down planning approach, a centralized service approach and an unregulated smallholder approach. All three pathways have simultaneously unfolded in the Nyando basin flood plain in Western Kenya. Data from a participatory analysis of poverty and livelihood dynamics from villages around the Nyando basin indicate that the incidence of poverty is higher in the flood plain than in the other parts of the basin. Within the flood plain, there are distinct patterns of poverty and livelihood dynamics in areas associated with different approaches to landownership and irrigation management. Over the last 10 years, poverty has risen rapidly to over 40% in the area following the top-down planning approach, increased slowly in smallholder mixed farming areas and remained relatively stable in areas supported by the centralized service agency. Recent changes in Kenya's water policy offer new opportunities for reforming and reviving the irrigation sector.

Keywords: poverty, livelihood strategies, irrigation, Kenya, property rights, land tenure.

Introduction

The history of irrigation development in Africa parallels that of agriculture on the continent. Between the mid-1960s and the mid-1980s, parastatal irrigation agencies were established and irrigation infrastructure was installed in significant tracts of land. Besides installing infrastructure and providing support services, many agencies took on responsibility for purchasing inputs, selling outputs and organizing production processes, in fact taking on the character of 'command-and-control' operations, with smallholder farmers largely treated as labourers.

Over time, it has become clear that this approach has not been financially sustainable, that the high level of government involvement served to 'crowd out' private investment and initiative by individual farmers, and that farmers had become highly dependent upon state subsidies and direction. Efforts to reform these systems have generally proved to be problematic and, in many cases, government and project support has ended abruptly, leaving farmers with insufficient capacity to self-manage their systems. The downsizing and withdrawal of government support have led to the contraction or collapse of smallholder irrigation systems across Africa, from Sudan to South Africa, to Senegal and to Kenya.

In contrast to the failure of state-sponsored smallholder irrigation, commercial irrigation operations, in which commercial farmers pay for efficient irrigation services, have generally remained operational and profitable (Shah *et al.*, 2002). Given that experience, Shah *et al.* (2002) call for a shift in government approach to irrigation management in Africa, with private sector firms or professional farmers' associations supporting irrigation farmers with a range of high-quality services that farmers pay for.

While state-controlled irrigation systems with smallholder farmers have been declining over the last 10 to 20 years, there has been a quiet and consistent increase in 'tiny' irrigation, which has largely been unregulated and given scant support from irrigation or agricultural extension systems. Across Africa, smallholder farmers and women's groups have begun irrigating very small plots of land, producing a variety of vegetables for the expanding urban markets. There are limitations to these systems, however. Many rely on

untreated sewage, thus producing vegetables with potential negative health consequences for both farmers and consumers. Most of the very small irrigation farms have been established in riparian areas, reducing the value of these areas for biodiversity conservation and filtering of pollutants and nutrients that otherwise enter waterways. Furthermore, most small-scale farmers lack access to the capital that would be necessary to achieve significant economies of scale and income streams.

Irrigation development in the Nyando basin of Western Kenya (see Fig. 12.1) mirrors the situation that has unfolded across the rest of Africa. In the mid-1960s the National Irrigation Board (NIB) converted 1700 ha of wetlands into irrigated agriculture through two pilot irrigation schemes (Ahero and West Kano irrigation schemes). These schemes were located on wetlands within former native reserve lands that had not been subjected to the process of adjudication.[1] Following the apparent success of the first two pilot schemes, the Provincial Irrigation Unit (PIU) of the Ministry of Agriculture,

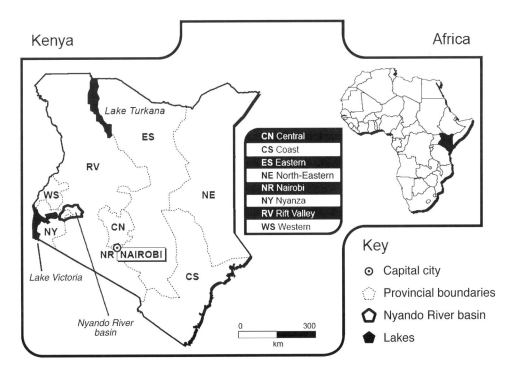

Fig. 12.1. Map of Kenya with Nyando river basin.

Livestock Development and Marketing supported the development of additional irrigation schemes on 4000 ha of the remaining 7000 ha of wetlands. Beginning in the mid-1990s, however, many of the irrigation schemes began to encounter severe problems, with some having become inoperable since 1997. At the same time, there is anecdotal evidence that an increasing number of farmers outside the formal irrigation schemes have been engaging in very small-scale irrigation in the riparian areas (Ong and Orengo, 2002).

Laws and regulations governing the irrigation sector in Kenya are embodied in the Irrigation Act (Cap 347), with the NIB created to be the main government organization involved in irrigation development. The irrigation sector can be divided into three main subsectors: (i) the public or national schemes managed by the NIB; (ii) smallholder schemes managed by the farmers with support from the Irrigation and Drainage Branch through the PIU; and (iii) private schemes. The Irrigation Act is now considered to be outdated and in need of revision. Its main weakness is that it does not allow for farmer participation in irrigation development or management.

According to the National Irrigation Board Cooperate Plan for 2003–2007, a draft legislative framework on irrigation and drainage in Kenya has been finalized and presented to parliament for approval. It proposes reforms similar to those in the water sector reforms, with the formation of the National Irrigation and Drainage Development Authority and the Irrigation and Drainage Regulatory Board. The draft legislative framework proposes the following: (i) reduced government involvement in non-core functions; (ii) increased stakeholder involvement; (iii) development of farmer capacity through Participatory Irrigation Management (PIM); and (iv) gradual transfer of management to farmer-based organizations through Irrigation Management Transfer (IMT). The Water Act, 2002 does not govern the irrigation sector per se. Rather, it provides for the creation of water user associations (WUAs) at the community level. These are some of the farmer-based organizations that will be used by the NIB for IMT.

Towards the end of 2004, the Government of Kenya revived the Ahero irrigation scheme based on the proposals contained in the draft legislative legal framework. Farmer participation has increased and they have formed a WUA. The NIB handles water extraction, then hands over the scheme to the WUA for management and distribution to the farmers. The NIB is currently the Water Service Provider supplying the scheme with water and is licensed by the Water Service Board. It is envisaged that this role will eventually be taken over by the WUA.[2] More information on the new structure of Kenya's formal sector water management institutions is given by Mumma and Onyango *et al.* (Chapters 10 and 11, this volume, respectively).

Ong and Orengo (2002) have focused on the links between irrigation development in the Nyando flood plain and the broader ecosystem. Their analysis of three irrigation schemes in the Nyando flood plain indicates that sedimentation of the intakes and irrigation canals was one of the major causes of the failure of the systems. Overall, the Nyando basin is an area of high erosion, with 60% of the basin – characterized by Walsh *et al.* (2004) – as having moderate to high rates of erosion. High rates of sediment carried through the Nyando river system have resulted in the need for very frequent desiltation operations, the costs of which could not be justified by the modest returns generated from irrigated rice production. Ong and Orengo (2002) also note that the conversion of the Nyando wetlands into agriculture has reduced the filter function of the wetlands, leading to higher rates of sediment deposition in Lake Victoria. Walsh *et al.* (2004) have documented an increasing rate of sediment deposition over the last 100 years, which was punctuated during El Niño events of the mid-1960s, 1986 and 1997.

This chapter focuses on another aspect of irrigation development in the Nyando basin: the links between irrigation and poverty. The chapter draws upon a study undertaken by the World Agroforestry Centre, Maseno University and the International Food Policy Research Institute (IFPRI), known as the Safeguard study. Safeguard is short for *Safeguarding the rights of poor and vulnerable people to critical land, water and tree resources in the Nyando basin of Western Kenya.* This chapter reports results from Safeguard pertaining to poverty and property rights dynamics in the lower flood plain area of the Nyando river adjacent to Lake Victoria. The results demonstrate how three different types of irrigation development have shaped poverty and livelihood dynamics in the area.

Methods

The Safeguard project employs a package of research methods grounded on the following principles:

- Addressing nest scales – collection of data at multiple, nested scales in recognition of the 'fractal' nature of poverty processes (see Barrett and Swallow, 2006, for a formal treatment of fractal poverty traps).
- Representing the range of circumstances in the basin, and then sampling villages to represent that range.
- Understanding intergenerational dynamics – focusing on processes that have had effects over the last 10–25 years (following the intergenerational approach to poverty dynamics proposed by Krishna (2004)).
- Recognizing diverse livelihood strategies – it is important to recognize and explicitly collect data on the full range of options that people employ to earn a livelihood (e.g. Ellis, 2000).
- Addressing multiple facets of poverty – explicitly considering the consumption, vulnerability and agency aspects of poverty (Narayan *et al.*, 2000).
- Adopting an inclusive and participatory research approach – the population under consideration should provide their own definitions of poverty, livelihood strategies and their own assessment of poverty and livelihood trends (Krishna, 2004; Krishna *et al.*, 2004).
- Adopting a legal pluralism approach to property rights – recognizing that there often are multiple and overlapping sources of sanction for property rights (Meinzen-Dick and Pradhan, 2002; Meinzen-Dick, Chapter 2, this volume).

In order to meet these criteria, the basin was characterized according to its hydrologic and land-tenure zones (for results, see Onyango *et al.*, Chapter 11, this volume). Based on this characterization, villages were chosen to represent 12 distinct zones in the basin. Altogether 14 villages were selected, one village for each of ten zones and two villages for each of the two

zones in the flood plain. These results therefore represent the variation found across the basin, including variations in elevation, production system, ethnicity, as well as land and water rights, but results cannot be simply aggregated to represent the whole basin (see Fig. 12.2).

Within each village, the Stages of Progress method (SPM) developed by Anirudh Krishna was used to study factors affecting intergenerational poverty dynamics. The method has been applied in India (Krishna, 2004) and Kenya (Krishna *et al.*, 2004). By systematically prompting and guiding discussions among a village representative group, the method generates a Stage of Progress ladder for the village, which ranges from absolute poverty to relative prosperity, poverty and prosperity lines defined by those stages, and measures of the level of poverty/prosperity and poverty dynamics for each household in the village. In the Safeguard project, we added a stronger focus on livelihood strategies pursued by households in the village, as well as the assets required for those livelihood strategies.

In addition to the village-representative group, interviews that generated the SPM data, other key informant and separate group interviews with men and women were conducted, with a focus on how people access and manage land, water, trees and other natural resources. This analysis of property rights used a legal pluralism approach centred on people's own experience, with access and control of resources and their personal strategies for claiming and obtaining resources. Additional discussion probed for the role of statutory and customary institutions as sources of land and water rights, and the implications for gender relations.

For information on the household scale, a stratified random total of 30 households was selected and interviewed with a structured survey in each of the 14 villages. Because of the study's focus on poverty dynamics, the whole village was stratified according to households that remained poor, became poor, became non-poor and stayed non-poor, based on the findings from the SPM. The household survey focused on rights and access to land, water and trees and livelihood strategies.

Table 12.1 presents descriptive information on the 12 zones and 14 villages included in the study. Note that the area represents a wide range of conditions. Land tenure varies from adjudi-

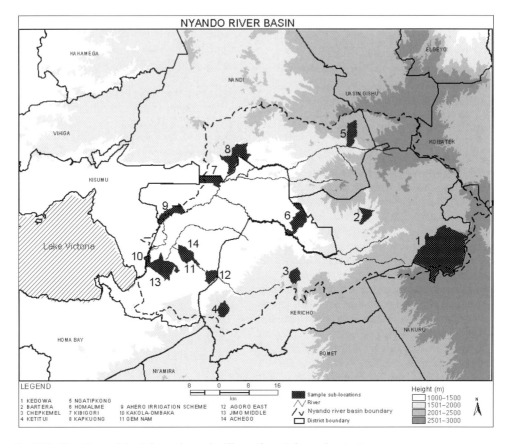

Fig. 12.2. Elevation and the Safeguard sample villages (from Safeguard project).

cated areas, large-scale leaseholds, subdivided leaseholds, settlement schemes, squatting in the forest reserve, and contested property rights in an irrigation area. Average income poverty rates vary from 40% to 70%. Population density varies from less than 100 to more than 1000 persons/km^2. Elevation varies from 1100 m asl near Lake Victoria to over 2500 m asl in the headwaters (see Fig. 12.1). The majority ethnic group in the lower part of the basin is Luo; the Kipsigis and Nandi Kalenjin are the majority in the upper part of the basin. The study also covered minority populations of Ogiek and Kisii in the uppermost parts of the basin.

Results

Villages 9, 10, 11, 13 and 14 are the foci of the current study. Village 9, known as Nakuru, was part of the national irrigation scheme owned and operated by the NIB. Villages 10 and 11, named Kasinrindwa and Karabok, respectively, are smallholder farming communities located outwith the public irrigation areas, where some farmers have developed private micro-scale irrigation farming of vegetables. Villages 13 and 14, named Kasiwindhi and Awach Scheme, respectively, are smallholder irrigation schemes supported by the PIU.

Land and water governance

The village representative groups and women-only focus groups were asked questions about access to, and control over, water. Follow-up questions were also included in the household survey. The results are remarkably similar from village to village, except for the NIB village and

Table 12.1. Characterization of the Safeguard study (from unpublished data compiled by the Safeguard project; poverty and population data from the Central Bureau of Statistics).

Zone, elevation (m asl¹)	Land tenure status	Irrigation development	Safeguard village number, name	District(s)	Population density (persons/km²)	Main ethnic group	Production system	Below poverty line in location (%)
Flood plain (1100)	Adjudicated	Smallholder mixed farming; some ad hoc irrigation	10, Kasirindwa; 11, Karabok	Nyando, Kisumu	224–1000	Luo	Smallholder mixed farming, some private irrigation	37 (Village 10), 55 (Village 11)
Flood plain (1100)	Adjudicated	Irrigation development supported by PIU, operational in 14, not operational in 13	13, Kasiwindhi; 14, Awach scheme	Nyando, Kisumu	224–1000	Luo	Smallholder commercial irrigation and dryland agriculture	68 (Village 13), 72 (Village 14)
Flood plain (1100)	Contested; formally owned by NIB but promised to local residents	Irrigation by NIB	9, Nakuru	Nyando, Kisumu	224–1000	Luo	Designed for irrigated rice; more diversification since NIB collapse in 1998	63
Lower Awach catchment (1250)	Adjudicated	None	12, Miolo	Nyando	224–527	Luo	Mixed subsistence, NR extraction	65
Upper Awach catchment (1700)	Adjudicated	None	4, Chepkemel	Kericho	88–149	Kipsigi/Kalenjin	Mixed cash/subsistence, coffee, dairy, maize, banana, smallholder tea	49
Mid-altitude part of Kapchorean basin	Undivided leasehold	None	6, Ongalo	Nyando	< 88	Luo	Commercial sugarcane	47
Lower Nyando basin	Resettlement scheme	None	7, Kimiria Aora	Nyando	150–303	Luo	Commercial sugarcane	48
Mid-altitude (1500)	Large-scale leasehold	None	8, Poto poto	Nyando	88–149	Nandi Kalenjin	Commercial sugarcane and mixed farming	48
High altitude (2000)	Adjudicated	Some home garden irrigation from springs	3, Kiptagen	Kericho	224–500	Kalenjin	Small-scale tea, some coffee, sugarcane, maize	49
High altitude (2100)	Subdivided leasehold	None	1, Kaminjeiwa; 2, Nyaribari A	Kericho	224–400	Mixture of Kalenjin, Kisii and others	Smallholder mixed farming	41
High altitude (2200)	Indigenous forest dwellers on forest land	None	5, Ngendui	Nandi	87–400 mixed	Ogiek/Nandi Kalenjin	Small-scale mixed farming	60

¹ Metres above sea level.

Village 14, which still has an operational irrigation system supported by the PIU.

All villages are predominately Luo and all except Village 9 have been adjudicated, so that individuals hold secure title to their land. Luo custom holds that water access should be freely available, particularly for basic household uses. In Village 10, for example, it was reported that: *Everybody has access to all community water points. No one is allowed to block the recognized community water points.* Luo custom also supports public access to private land resources for grazing, collecting firewood and passing through. With few physical or social fences, access to water resources is relatively free. It appears that it is only in irrigation areas that have had strong involvement of external agencies that the Luo customs have not held sway.

One possible drawback of the Luo custom for land and water governance is that there is relatively little incentive for private individuals or small groups to invest in protecting existing water sources or creating new water sources. This has particular impacts on women, who are responsible for provisioning the household with water and for providing health care within the household.

Land tenure security is much more restricted for farmers in Village 9. When the NIB built the irrigation system, they appropriated all land in the area. Standardized plots (50 × 50 m for homesteads and 4 acres of irrigated fields) were then allocated to farmers, who remain 'tenants' of the system. The farmers are forbidden to plant trees or own livestock, or even to bury the dead on this land. To add to the insecurity of tenure, farmers can be evicted for 'laziness' or failure to cultivate their land. The plots cannot be subdivided, which violates Luo customary norms that all sons are entitled to inherit land from their fathers. Because land rentals are also restricted on NIB land, landless sons have more difficulty in obtaining any land to cultivate.

A detailed investigation of water governance in Village 9 found not only that the NIB influences irrigation water management but that it also has some spillover effects onto the management of other water resources in the village (see Table 12.2). Customary norms play more of a role in granting authority over water sources used primarily for domestic uses. The Nyando river, which is used for irrigation, falls more under statutory law and government agency management. This contrasts with other villages in the area, where the management of river water is primarily governed by customary norms.

Results from the household survey indicate a fairly high level of social organization around water management in Village 14, the only village that had a functional collective irrigation system at the time of the survey. As indicated in Table 12.3, 29 out of the 30 surveyed households in Village 14 pay water fees, compared with only 3 out of 21 households in Village 13 and 17 out of 27 households in Village 9. The irrigation system in Village 14 is not without conflict. Table 12.4 indicates that Village 14 is the only village in which most households do not think that there is equality in access to water. Conflicts over water management are reported in both the PIU and the NIB village, with most households in both villages reporting experience of conflicts over irrigation management (see Table 12.5).

Resource management and allocation

Evidence from this study indicates that customary arrangements work well for domestic water requirements in the irrigation areas, because all the people are from one ethnic group and share the same customary beliefs and practices. Water for domestic use is required in relatively small volumes, with relatively little competition for available water resources. By contrast, large amounts of water are required for commercial irrigation, putting much greater pressure on available water resources. Ensuring a fair distribution of irrigation water requires a management system with capacity for efficient management, rule enforcement and conflict resolution. Customary water management arrangements among the Luo rely heavily on the individual's sense of duty and loyalty to his/her community. For instance, a person who decides not to let people fetch water from a spring located on his parcel of land will be shunned by the rest of the community but will not be formally prosecuted. Such a system may not provide enough certainty for commercial undertakings that require significant monetary investment. Statutory arrangements may thus be better suited because they are supported by legislation and can be enforced through statutory legal systems.

Table 12.2. Water sources and their management in the village managed and controlled by the NIB (Village 9) (from authors' analysis of the Safeguard group interview).

Source of water	Use of water	Users of water	Where do the users draw authority to use the water?	Who manages the water?	Can users transfer their rights?	Owner of land where water point is located	What forms of pollution affect the water source?	Mediating institutions
Marega River	Cooking, farming	All the villagers	Customary	None	No	NIB	Chemicals from irrigation scheme	Irrigation scheme and Government of Kenya
Nyando River	Drinking, washing		Statutory irrigation act and non-statutory	Irrigation Board for irrigation purposes	Yes		From plants and the chemical factory	Irrigation scheme and Government of Kenya
Ombeyi River			Customary	No one	Yes			Irrigation scheme and Government of Kenya
Shallow well		All	Customary	Owner of land	No	Individuals		The family; the village elders if public funds were used for construction

Table 12.3. Payment of water fees in the five Safeguard villages in the Nyando flood plain (from authors' analysis of Safeguard household survey data).

	Respondents' payments of water fees		
Village number, type of irrigation system	No	Yes	Total
9, NIB	10	17	27
10, smallholder ad hoc	1	13	14
11, smallholder ad hoc	3	7	10
13, PIU support to farmers	18	3	21
14, PIU support to farmers	1	29	30
Total	33	69	102

Table 12.4. Perceptions of equality of access to water in five Safeguard villages in the Nyando flood plain (from authors' analysis of Safeguard household survey data).

Village number, type of irrigation system	No equality of access	Equality of access	Total
9, NIB	7	23	30
10, smallholder ad hoc	10	20	30
11, smallholder ad hoc	8	21	29
13, PIU support to farmers	5	25	30
14, PIU support to farmers	17	12	29
Total	47	101	148

Table 12.5. Experience with irrigation management conflicts in five Safeguard villages in the Nyando flood plain (from authors' analysis of the Safeguard household survey data).

Village number, type of irrigation system	No experience with irrigation management conflicts	Experience with irrigation management conflicts	Missing	Total
9, NIB	2	24	4	30
10, smallholder ad hoc	1	3	26	30
11, smallholder ad hoc	3	1	26	30
13, PIU support to farmers	6	10	13	29
14, PIU support to farmers	6	24	0	30
Total	18	62	69	149

The issue of rights in an irrigation scheme goes beyond access to water. Equally important are rights to access and manage land, rights to manage labour and rights to participate in irrigation system management. Large-scale irrigated agriculture requires the organization of labour through both formal and informal arrangements. In the NIB irrigation schemes, labour is provided by tenant farmers, with the head of each household registered as a tenant of the NIB. The registered tenant-farmer makes informal arrangements with the rest of his/her household/family on how much labour each person will contribute to the irrigation field *versus* other household activities. He/she also makes infor-

mal arrangements on how the family members will divide the proceeds from the rice field.

In the initial arrangement with its tenant farmers, the NIB provided farm inputs, marketed the produce and paid the tenant-farmer the difference after recovering the cost. The farmer had no control of the processes but he was sure to get his money when funds became available. However, the informal arrangements within the household depended on the integrity of the registered tenant-farmer. Qualitative studies in two of the schemes established that many of the informal income-allocation arrangements were not honoured in some households, resulting in domestic conflicts. There are no traditional institutions to address these issues, since the public schemes are seen as a creation of the government.

One reason that the NIB schemes collapsed was the farmers' discontent with the system that gave them no room to participate in decision-making processes. These were all provisions of an act of parliament (the Irrigation Act) and could be changed only by amending the act – a long and laborious process requiring the goodwill of the government. This has discouraged many people from becoming involved. The NIB appropriated all land in the area, so that household members not involved in rice irrigation are able to engage only in other land-based livelihood strategies by hiring land outside the community. This helps explain why poverty is worst and livelihood strategies fewest in the NIB irrigation areas. With their major source of income removed, the people lack capital to invest in non-land-based livelihood strategies. Capable young people move out of the area in search of employment elsewhere, while more wealthy households purchase land outwith the scheme and move out. The result is a further concentration of poverty in the failed schemes.

By contrast, the smallholder irrigation schemes supported by the PIU are based on informal arrangements for labour allocation and sharing of produce, which are based on customary norms. Every member of the household with customary rights to land is allocated individual portions for which he/she is responsible for labour and inputs and controls the products generated on that land. This gives each person and household incentive to invest more time and money and may be part of the explanation why villages where this system is practised have a lower incidence of poverty.

Livelihood and poverty outcomes under different types of irrigation practised in the Nyando basin

One of the outputs generated by the SPM village survey is a list of all households in the village, with the 'stage of progress' currently attained by each household and the stage attained 10 years ago by that household (or its predecessor), and the stage attained 25 years ago by that household (or its predecessor). The stages are then mapped into categories of poor, not poor and relatively prosperous, using definitions provided

by each community. Fortunately, the conceptions of poverty and the stages end up being relatively similar from village to village and thus can be compiled and compared across villages.

Figure 12.3 presents a compilation of the poverty to prosperity data for all households in the 14 study villages, aggregated into three elevation zones – upper, medium and lower. The results show that poverty is generally highest in the lower-altitude zone, with about 40% of the sample households now considered poor and poverty rising by over 15% over the last 10 years. These data are consistent with the national sample and census data for Kenya, which show Nyanza province (which includes the lower Nyando basin) having the highest rate of poverty in Kenya and the highest rate of increase over the period 1994–1997. A high incidence of HIV/AIDS is one of the reasons for this overall trend in poverty: the Luo population has the highest rate of HIV/AIDS infection among both men and women in all of Kenya.

Figure 12.4 presents a breakdown of the poverty–prosperity data for villages in the flood plains. The results indicate different patterns across the three types of land tenure and water management. Twenty-five years ago, poverty rates were lowest in Villages 10 and 11, smallholder agriculture areas where the residents have long held secure land tenure. Small amounts of land in those villages are irrigated using flood, bucket and pump irrigation. Poverty has increased in those villages, slowly, until 10 years ago, and more rapidly since. The area covered by the NIB had a poverty rate of over 30% in the early years of the NIB irrigation scheme, a rate which fell just below 30% 10 years ago, then exploded to over 60% at present. This corresponds to a collapse in the NIB services to the irrigation system due to lack of financial resources, making irrigation no longer possible. Rice cultivation in the NIB village declined after 1994, and ceased in 1998. In contrast, the two villages that have been supported by the PIU since the 1980s experienced a modest decline in the rate of poverty from 25 years ago to the present time, with a current poverty rate of about 38%.

Table 12.6 lists the number of households practising different livelihood strategies at the present time in each of the five study villages in the Nyando flood plain. It also lists the total

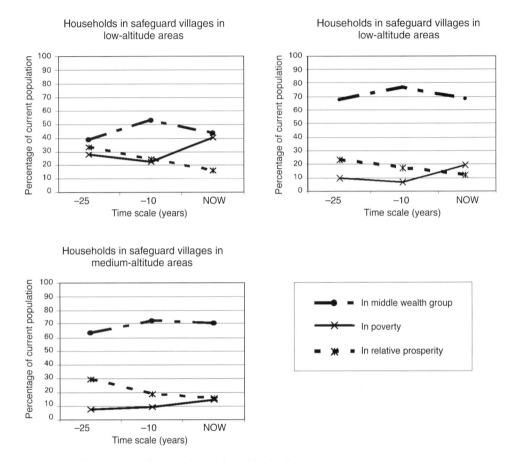

Fig. 12.3. Welfare trends in the Nyando river basin by altitude.

number of strategies listed for all households in each village, the average number of strategies reported for each household and the current rate of poverty as reported by the village representative groups. It is noteworthy that the village with the highest current rate of poverty (Village 9, 62%) has the lowest number of strategies employed per household. Households in the smallholder agriculture area are more diversified than in the irrigation areas. Village 14, which has the most functional remaining irrigation system in the area, has the highest number of households still growing rice.

Figure 12.4 indicates that, 25 years ago, the introduction of irrigation had the effect of reducing the levels of poverty in the initial stages by introducing a cash crop in a region where agriculture had been predominantly for subsistence. Irrigation was introduced at a time

when the government subsidized agricultural inputs heavily. Later policy changes reduced these subsidies, and this had the effect of reducing the profits realized by the rice farmers. In the NIB schemes the people were not involved in management and blamed the decline on the NIB. It was the beginning of discontent.

In the smallholder schemes supported by the PIU, the people understood the reasons for decline of profits and sought ways of surviving. During the period 25–10 years ago, the negative impacts of the reduced subsidies had not yet started being felt extensively and the irrigation communities were still enjoying the benefits of cash crop farming. From 10 years ago to date the situation has been made worse by liberalization. Governments in the developing world found it hard to protect their farmers from cheaply produced agricultural goods. The high cost of

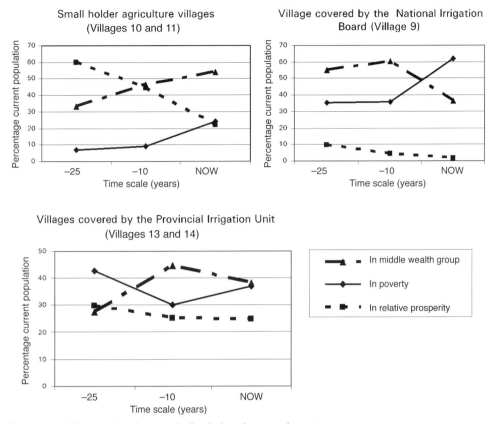

Fig. 12.4. Welfare trends in the Nyando flood plains by type of irrigation.

farm inputs and a competitive market have reduced the gains of rice irrigation. In Village 9 the collapse of the scheme is indicated by a steep increase in the number of people living in poverty, beginning 10 years ago and continuing to date. The proportion of the community living in relative prosperity stood at 10% 10 years ago, but today it has dropped to almost 1%.

The irrigation scheme in Village 13, which has been supported by the PIU, has also ceased activity. The main reason that this scheme ceased operations was the extremely high levels of siltation of the irrigation intakes and canals. Desiltation proved a high recurrent cost for the communities, a cost that they were not able to sustain over time (Ong and Orengo, 2002). The results presented in Fig. 12.4 and Table 12.6 indicate that poverty has increased in Village 13 over the past 10 years, although not nearly as much as in Village 9. Livelihood strategies are more diverse in Village 13 than in Village 9,

although not as diverse as in Villages 10, 11 or 14. Interpreted another way, households that had control of their own farmland were able to respond to the collapse of irrigated rice production with some types of agricultural enterprise, while households that did not have that control were generally forced to low-return, non-agricultural livelihood strategies. These results support the argument of Shah *et al.* (2002) that sudden withdrawal of state support in smallholder irrigation schemes is bound to lead to collapse.

Conclusions

It is clear from the study that poverty in the Nyando basin is generally concentrated in the lower parts of the basin that have the greatest potential for irrigation. This geographic clustering of poverty in the lower parts of the basin

Table 12.6. Number of households practising alternative livelihood strategies in the five villages of the Nyando flood plain.

Strategy	Village 9 – NIB	Village 10 – small-holder mixed farming	Village 11 – small-holder mixed farming	Village 13 – supported by PIU	Village 14 – supported by PIU
Beans			17	6	
Boda boda (bicycle taxi)	4	8	5	1	4
Casual labour	31	3	8	3	22
Cattle	17	29	28	20	39
Formal employment	4	6	20	11	13
Maize	15	46	61	33	71
Other skilled employment	14	13	22	14	10
Other small-scale farming	4	13	16	5	4
Rice	3	1	1	15	73
Sheep or goats	2	48	63	6	74
Sorghum or millet	1	15	37	17	70
Sugarcane		40	1		
Tomatoes	2	42	11	2	5
Trade	28	20	46	12	4
Vegetables/onion	1	51	19	1	7
Total strategies in village	126	335	355	146	396
Number of households in village	74	63	102	52	94
Average strategies/household	1.70	5.32	3.48	2.81	4.21
Current poverty status (%)	62	29	22	37	37

contrasts with the standard situation in South-east Asia and parts of South Asia, but it may be closer to the norm in East Africa. Failed irrigation development, particularly with heavy government involvement in agricultural production and local organizational arrangements (the NIB village), has been a source of impoverishment in the Nyando flood plain. Erosion in the upper reaches of the basin has also contributed to the decline of irrigation systems in the flood plain because the heavy siltation washed down by the floods has blocked the irrigation supply channels. In the context of high HIV/AIDS infections and other diseases, it is very difficult for farmers to mobilize enough labour to desilt the canals themselves, thereby increasing their dependence on government agencies for equipment in the larger-scale irrigation systems. Yet, the overall small improvement in the poverty situation in the villages supported by the PIU provides some evidence that irrigation development can contribute to welfare improvements.

One of the key lessons from this study appears to be the trade-offs between specialization and diversity in production under irriga-tion. While there may have been efficiency gains associated with specialized rice production when the NIB was functional, this specialization implied high risk in the event of failure of the irrigation system. Diversification of livelihood strategies at the household and community levels is a major source of welfare enhancement and risk minimization.

One major difference between the NIB area and the other areas included in this study lies in land tenure: in the NIB scheme, farmers are considered only 'tenants', with little tenure security. They thus have less incentive to invest in developing the land or even the irrigation facilities, and have less decision-making authority, which is needed for diversification of livelihoods. Women in the NIB village also have less control over resources than in other villages. Tenure security and resource control of both households as a whole and women in the households are important in addressing poverty reduction in the irrigated areas.

Results on water governance and gender equity in water access and management indicate that there has been some spillover of influence from irrigation management to management of domestic water sources. The village involved in the NIB irrigation scheme in particular noted the importance of the irrigation schemes and the government in mediating access to virtually all water sources in the village. In the NIB area, the government and the scheme are understood to be the owners of land on which water points are located; in the other villages the water points are located on individual land, but with relatively open access to other people living in the village. Other results from the Nyando basin show that there are, indeed, strong links between landownership and water access (Onyango *et al.*, Chapter 11, this volume).

Under the Water Act, 2002, water and irrigation management will, for the first time, be centralized in one government ministry, the Ministry of Water Resources Development and Management. This Ministry is reassessing how best to revitalize the irrigation sector and improve access to domestic water sources. In 2004–2005 the intakes and canals in some of the schemes were desilted through food-for-work schemes, and irrigation water began to flow again in some of the schemes. This research shows that landownership will be key to maintaining these gains. Farmers will be more apt to adapt to invest in land improvements and diversify their income sources if they have secure land rights.

The results from this study and experience from elsewhere in Africa indicate that markets and efficient service provision will also be important (Shah *et al.*, 2002). Different approaches to marketing may be taken. The smallholder farming approach, exemplified by the NALEP approach, would support common-interest groups around a diversity of crops, perhaps focused on vegetables sold into the Kisumu market. An alternative approach, which has proved successful in the revitalization of the Mwea irrigation scheme on the slopes of Mount Kenya (Shah *et al.*, 2002), has been to support the development of a multi-purpose farmers'

cooperative to engage in input and output marketing arrangements for high-quality irrigated rice. Such a cooperative could pay the PIU to provide irrigation services to the scheme. Another alternative that is being tried out for the first time at the NIB schemes in Ahero, a west Kano irrigation scheme, is engaging with micro-finance organizations for financing of inputs and marketing of produce. The arrangement is carried out by the farmers' organizations and is for a specific time period.

Acknowledgements

The authors acknowledge financial support from the Comprehensive Assessment of Water Management in Agriculture and the European Union. Wilson Nindo provided excellent leadership in the field activities. Peter Muraya and Maurice Baraza provided assistance with data management. We would also like to acknowledge the many useful comments provided by Barbara van Koppen and Mark Giordano.

Endnotes

[1] The process of adjudication and registration was initiated by the government to convert customary land rights to statutory land rights. After adjudication – passing judgement that a certain plot of land does belong to a particular individual according to customary arrangements – the government went ahead to survey and register the said parcel of land in that person's name and issue him or her with a title deed. Adjudication thus reaffirmed customary land claims and converted them into statutory rights (described in Onyango *et al.*, Chapter 11, this volume).

[2] Since water resources management and irrigation are both in the new Ministry of Water and Irrigation, the institutions created by the different acts have cross-cutting functions. All of the institutions created by the Water Act, 2002 will provide services for the irrigation sector.

References

Barrett, C.B. and Swallow, B. (2006) Fractal poverty traps. *World Development* 34 (1), 1–15.

Ellis, F. (2000) *Rural Livelihoods and Diversity in Developing Countries.* Oxford University Press, Oxford, UK.

Krishna, A. (2004) Escaping poverty and becoming poor: who gains, who loses and why? Accounting for stability and change in 35 North Indian villages. *World Development* 32 (1), 121–136.

Krishna, A., Kristjanson, P., Radeny, M. and Nindo, W. (2004) Escaping poverty and becoming poor in twenty Kenya villages. *Journal of Human Development* 5 (2), 211–226.

Meinzen-Dick, R. and Pradhan, R. (2002) *Legal Pluralism and Dynamic Property Pights.* CAPRI Working Paper 22, IFPRI, Washington, DC.

Narayan, D., Chambers, R., Shah, M.K. and Petesch, P. (2000) *Crying Out for Change: Voices of the Poor.* World Bank Publications, 1 November 2000.

Ong, C. and Orego, F. (2002) Links between land management, sedimentation, nutrient flows and smallholder irrigation in the Lake Victoria Basin. In: Blank, H.G., Mutero, C.M. and Murray-Rust, H. (eds) *The Changing Face of Irrigation in Kenya: Opportunities for Anticipating Change in Eastern and Southern Africa.* International Water Management Institute, Colombo, Sri Lanka, pp. 135–154.

Shah, T., van Koppen, B., Merry, D., de Lange, M. and Samad, M. (2002) *Institutional Alternatives in African Smallholder Irrigation: Lessons from International Experience with Irrigation Management Transfer.* Research Report 60, International Water Management Institute, Colombo, Sri Lanka.

Walsh, M., Shepherd, K. and Verchot, L. (2004) Identification of sediment sources and sinks in the Nyando river basin. In: Mungai, D., Swallow, B., Mburu, J., Onyango, L. and Njui, A. (eds) *Proceedings of a Workshop on Reversing Environmental and Agricultural Decline in the Nyando River Basin.* World Agroforestry Centre (ICRAF), the National Environment Management Authority of Kenya (NEMA), the Water Quality Component of the Lake Victoria Environment Management Programme (LVEMP) and the Ministry of Agriculture and Rural Development, Nairobi, pp. 27–30.

13 If Government Failed, how are we to Succeed? The Importance of History and Context in Present-day Irrigation Reform in Malawi

Anne Ferguson[1] and Wapulumuka Mulwafu[2]

[1]*Department of Anthropology, Michigan State University, East Lansing, Michigan, USA; e-mail: fergus12@msu.edu;* [2]*History Department, Chancellor College, University of Malawi, Zomba, Malawi; e-mail: wmulwafu@chanco.unima.mw*

Abstract

This chapter examines the interface between new, formal irrigation, water policies and laws and long-standing customary practices in two smallholder irrigation schemes earmarked for transfer to water user associations in Malawi. It documents how local histories and practices are shaping access to critical land and water resources in ways not anticipated by technocratic irrigation and water reform implementers. At this point, rather than creating a climate encouraging more equitable economic growth by making smallholder farmers' rights to land and water resources more secure, the opposite seems to be taking place as formalization opens the door for local elites to use diverse strategies to capture these resources.

Keywords: customary practices, irrigation management transfer, informal institutions, neo-liberal reforms, devolution, Malawi.

Introduction

Over the last decade Malawi, similar to other southern African countries, has revised most of its environmental and agricultural policies and laws. Since 1999 new irrigation, land, and water policies and supporting legislation have been approved by parliament. These new policies and laws aim to alter or formalize resource access and use practices, which once were under customary tenure, as well as to introduce new statutory laws and institutions. Customary land is to be titled under the rubric of 'customary estates', the use of water for productive purposes will require permits and fees, and government-run smallholder irrigation schemes are being turned over to users. These reforms will dramatically alter access to critical land and water resources for rural livelihoods in one of the poorest countries in the world.

This chapter focuses on the transfer of two government-run smallholder irrigation schemes in the southern region to farmers' associations in the context of the implementation of new irrigation, land and water policies, and pending laws. It provides a grounded analysis of the early effects of these reforms, drawing attention to how international and national policies and laws interact with existing informal and formal institutions and practices, and economic and political hierarchies, to yield sometimes unexpected results. Drawing on recent work by Benjaminsen and Lund (2002), Cleaver (2002), Mollinga and Bolding (2004) and others, the study examines problems inherent when blueprint reform models are implemented with little regard to context and history.

The following questions are addressed: how are reforms under way in the land, water and irrigation sectors likely to affect farmers in smallholder irrigation schemes? How do the new reforms interact with existing customary land- and water-related rights, privileges and practices? What are some of the outcomes of these interactions? Specifically, who may benefit from the transfer of the irrigation schemes to farmers' associations? Are these reforms likely to provide smallholder farmers – especially the disadvantaged – with equitable and secure rights to land and water resources as the policies espouse? Or will they create uncertainty and entrench privileged interests?

Policy Context

Malawi's economy is dependent on the export of primary agricultural products – particularly tobacco – for which the terms of trade declined significantly in the 1990s. More than 85% of the population live in rural areas, and per capita incomes have decreased significantly since the imposition of structural adjustment programmes in the 1980s. Approximately 45% of the population presently live below the absolute poverty line of US$40 per capita per annum, and 65% are considered poor by more conventional standards (Devereux, 2002a, p. 3). In the mid-1990s, Malawi had one of the poorest nutritional, health and poverty statistics of any non-conflicting country in the world, with no significant improvement in sight (Devereux, 2002a, p. 6). In 2002, the country experienced its worst famine in recent history (Devereux, 2002b). Without sufficient donor aid, 2006 may be an equally difficult year.

Deepening poverty and chronic food shortages suggest fundamental failures in development and poverty-alleviation strategies. Researchers point to the role of Washington Consensus neo-liberal economic policies and to government mismanagement and corruption in

explaining this growing impoverishment (Devereux, 2002a; Owusu and Ng'ambi, 2002). Starting in the late 1980s, new government policies promoted by the World Bank and the International Monetary Fund redefined the role of the central government and restricted state intervention in the economy. Almost all policies in the natural resources and agriculture sectors have been rewritten, and new laws are being drafted and implemented reflecting these changes. Many call for turning over management of resources to user groups or local governments. The pace of reform is staggering. Since 1998, new irrigation, agriculture, land and water policies and, in many cases, laws have been drafted and approved by Parliament, in addition to a new local government Act (Ferguson and Mulwafu, 2004). This section briefly reviews these reforms, drawing attention to common elements in them to set the stage for an examination of their effects on the smallholder irrigation schemes.

As land pressure and climate change intensify, the government is turning to irrigated agriculture as a means to increase production. Expanded irrigation is expected to boost incomes and food security and is considered to be a way of reducing poverty. The National Irrigation Policy and Development Strategy (GOM, 2000) reflects this stance. It calls for the rapid phase-out of government support to the 16 smallholder irrigation schemes and their transfer to newly created farmers' associations. The policy also advocates the expansion and intensified use of informal irrigation by small-scale farmers along stream-banks and drainage lines and in wetlands, a form of irrigation that has received little government attention (Kambewa, 2004; Peters, 2004).

Transfer of government-run irrigation schemes to farmers' associations, often referred to as irrigation management transfer (IMT), has gained popularity in southern Africa as part of neo-liberal reform models. IMT is promoted as a means of decentralizing functions of the state, reducing public expenditure and mitigating the perceived dependency syndrome by instilling a sense of local ownership and responsibility. Four conditions are identified as necessary for successful IMT: (i) it must improve the life situations of a significant number of scheme members; (ii) the irrigation system must be central to creating this improvement; (iii) the cost of self-management must be an acceptably small proportion of improved income; and (iv) the proposed organizational design must be seen to have low transaction costs (Shah *et al.*, 2002, p. 5; see also Vermillion, 1997; Vermillion and Sagardoy, 1999). The assumption is that these conditions require the introduction of new models of social organization and formal institution building, along with physical renovation of the irrigation schemes. Yet little attention has been paid to how past forms of social organization and existing customary practices may influence the creation of new institutions.

Malawi's land and water policies have also been revised. The new Land Policy (GOM, 2001) will essentially privatize customary land by creating 'customary estates'. Titling committees are to be established at the level of traditional authorities (TAs) and districts. Wetlands are to be designated as public lands under the control of TAs. Irrigation schemes are to remain public or government land, but will be leased to newly created water-user associations (WUAs). The Water Policy has been under revision since 1999 (GOM, 1999a, b). The draft 2000 Policy was approved by the Cabinet in 2002 but was subsequently rescinded, and a new Policy is about to be enacted in 2005.

The new version (GOM, 2004) calls for the creation of catchment management authorities. It embraces demand management strategies, including user and polluter pay principles, and suggests that those using water for productive purposes will be expected to obtain water user or abstraction permits. No clearly stated right to water for domestic purposes is evident in the new Policy. Instead it states that: 'The protection and use of water resources for domestic water supply shall be accorded the highest priority over other uses' (GOM, 2004, 3.3.9). Water is to be treated not only as a social good but also as an economic good, with both entrepreneurs and individuals having 'equitable' access. Communities, non-governmental organizations (NGOs) and the private sector will bear the costs of infrastructural development, maintenance and operation (Ferguson and Mulwafu, 2001; Ferguson, 2005). The new Water Law, which will formalize and clarify many of the principles in the Policy, has not yet been approved.

These reforms governing critical natural resources must be considered within the context of the local government Policy and Law (GOM, 1998a, b), which have initiated sweeping changes in how the government will operate. While line ministries will retain responsibility for policy formation, enforcement, standards and training, most administrative and political functions once concentrated in ministries at the national level are being transferred to districts and municipalities. The District Development Committee and Plan are the principal means by which integrated sectoral planning is to be achieved. Marking a significant change from the past, civil servants are now to be accountable to the populations they serve, not to their parent ministries in central government. The powers of TAs have been reaffirmed, and they are integrated into the new structure through their ex officio participation in District Assemblies, as well as serving as chairpersons of the Area Development Committees.

The new policies and laws differ significantly in the amount of public input and participation that took place during the drafting period. The new Land Policy was preceded by more than 6 years of study and broad-based consultation. In contrast, little public input into the Irrigation or Water Policies appears to have taken place. While the Land Policy was publicized on the radio, in the newspapers and through other means, information about the Water and Irrigation Policies has not been widely disseminated, and most Malawians appear to know little about them.

The new Irrigation, Water and Land Policies reflect the neo-liberal preoccupation with establishing the correct institutional framework to provide secure property rights, promote private sector investment, enact the user and polluter pay principles and decentralize and devolve ownership and management functions away from central government to local governments, communities and the private sector. Most policies and laws have been drafted at the behest of donor organizations and enacted on a sector-by-sector basis. Although they are being harmonized to resolve areas of ambiguity and conflicting clauses, questions about how the new structures will function in local contexts in relation to existing rights and practices have been largely overlooked.

Research Sites and Methods

In Malawi, irrigated land includes formal irrigation schemes operated by the government and
private estate owners, as well as lands along
stream-banks, in low-lying areas of residual
moisture and in wetlands cultivated by small-
scale farmers. The formal irrigation schemes are
often located in, and surrounded by, wetlands
and depend on the same water sources. A
recent World Bank estimate is that 28,000 ha
are under 'formal' or 'semi-formal' irrigation, of
which 6500 ha are under self-help smallholder
schemes, 3200 ha under government-run
smallholder irrigation schemes and 18,300 ha
in estates (World Bank, 2004). The common
estimate for the potential irrigated area (not
limited to wetlands) is 250,000–500,000 ha.

Our research focused on the Domasi and
Likangala watersheds in the Lake Chilwa basin
in the southern region (see Fig. 13.1). This
basin is home to six of Malawi's 16 government-run smallholder irrigation schemes, all
earmarked for transfer to farmers' associations.
The two schemes forming the basis of our study

are: Domasi irrigation scheme located on the
Domasi river, Machinga district; and Likangala
irrigation scheme on the Likangala river,
Zomba district. Domasi covers approximately
500 ha and has 1500 farmers. The Likangala
scheme is the largest in the Likangala complex,
which comprises four smaller schemes as well –
Khanda, Njala, Chiliko and Tsegula. The study
focused on the Likangala scheme itself, which is
450 ha in size and has nearly 1300 farmers.
Each plot on these gravity-fed schemes is 0.25
acre in size. Rice is grown on the schemes
during the rainy season. Rice, sweet potatoes,
maize, pumpkins, watermelons, tomatoes and
other vegetables are produced in the dry
season. Some of the plots are reassigned for
temporary use in the dry season by surrounding farmers who do not otherwise have access
to the schemes.

The study took place between 2001 and
2004 and used quantitative and qualitative
methods. In 2003, we conducted a survey of
123 farmers on the schemes to gather baseline
information on access to plots, farming and
marketing practices, water use and conflicts. We

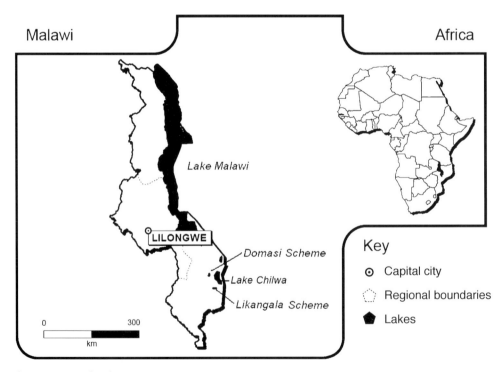

Fig. 13.1. Map of Malawi.

interviewed 63 (51%) farmers on Domasi and 60 (49%) on Likangala. An irrigation transfer survey to gather information on scheme governance and farmers' knowledge of, and participation in, the transfer process was administered to 120 of those farmers, 61 (51%) from Domasi and 59 (49%) from Likangala. All those interviewed were plot-holders.

Two field assistants were assigned to live on the schemes for the 3-year period. In addition to engaging in participant observation and writing field notes, they carried out interviews with farmers and irrigation scheme committee members on assigned topics. The qualitative research enabled us to gather information on tenure and land-use practices, conflicts over land and water, scheme governance and farmer evaluation of the handover process, which was not accessible or reliable via formal survey research. To learn about developments in the policy arena, new policy documents and reports were collected, and key actors at the national and local levels were interviewed twice a year. These included interviews with officials in the Ministry of Water Development, the Department of Irrigation and the Ministry of Lands, along with major donors, including the International Fund for Agricultural Development (IFAD), US Agency for International Development (USAID) and the World Bank. At the local level, we interviewed irrigation scheme managers, committee members, officials of the Agricultural Development Division, district authorities and project managers of the Balaka Concern Universal office responsible for training farmers on the Domasi scheme to form a WUA. Finally, we presented preliminary research findings for discussion to policy makers, project implementers and farmers through a series of workshops conducted over the research period.

History and Present Context of the Domasi and Likangala Irrigation Schemes

Historical background

Malawi's 16 government-run smallholder irrigation schemes were established in the late 1960s and early 1970s to demonstrate the methods and benefits of intensive cash crop production. The Land Act of 1965 provided for the creation of settlement schemes to foster increased peasant production (GOM, 1965). These schemes came to be closely identified with the Malawi Young Pioneers (MYP), a paramilitary wing of the ruling Malawi Congress Party (MCP) (Nkhoma and Mulwafu, 2004). Established in 1965, the MYP was originally a movement meant to integrate the youth into agricultural development. Training bases, which in some cases served as settlement schemes, were established where primary school graduates received 10 months of basic training in leadership, civics, agriculture and community development. As time went by, the MYP became involved in political matters as on-the-ground watchdogs for the country's authoritarian leadership. They were infamous for maintaining discipline and ensuring compliance with party dictates, such as the purchase of party cards.

To establish the Domasi and Likangala irrigation schemes, customary land was appropriated from TAs. State authorities consulted with chiefs, since the latter's cooperation was critical in getting the schemes off the ground. In one case, the Group Village Headman, Namasalima of the Zomba district, refused to grant land for the establishment of the scheme, citing among other factors loss of land for his subjects. The state then went to Group Village Headman Mpheta in the neighbouring Machinga district, who welcomed the idea, and the location of the scheme was changed. Villagers displaced by the schemes were resettled in surrounding villages and given plots, often in the same irrigation blocks. A small number of settlers, plus the MYP, came from outside the Lake Chilwa basin, but the majority of plot-holders were, and continue to be, from nearby villages and towns. We found that most of the farmers interviewed were born in the district where the scheme was located – 83% of respondents in the case of Likangala and 84% in the case of Domasi – while the remainder came from nearby districts.

Between the late 1960s and the 1980s, the schemes were fairly well maintained but run in a top-down, authoritarian fashion by the government (Krogh and Mkandawire, 1990). Management was formalized with the promulgation of specific rules and regulations on irrigation farming. This was followed by the creation of two departments to oversee irrigation matters in the country, the Departments of Irrigation and Settlement (GOM, 1965, 1968). As was the case with irrigation schemes elsewhere in eastern and southern Africa, farmers were supposed to comply with statutory regulations on plot allocation, use and maintenance of canals, water allocation and use, and cropping calendars (Adams et al., 1997; Bolding et al., 2004).

Initially, the Taiwanese (Chinese) Agricultural Technical Mission provided support on the technical and managerial aspects of irrigation farming. Local chiefs assisted the Taiwanese in handling sociocultural matters, such as re-settling villagers, conflict resolution and plot allocation. In the early 1970s, however, Land Allocation Committees were established to manage the schemes. Members were political appointees who often had to receive the support of MCP authorities in order to retain their positions. In fact, these committees were invariably chaired by MCP officials and chiefs, and MYP Discipline Officers were ex officio members. The Land Allocation Committees implemented statutory laws and regulations with a considerable degree of coercion (Chilivumbo, 1978). The system was highly regimented, and farmers were required to follow rules to the letter. Failure to comply often resulted in severe punishments, such as uprooting of crops or eviction. In part, this authoritarian structure explains why the performance of the schemes has been described as efficient up to the beginning of the 1990s.

That success, however, should be understood within the context of the embeddedness of the management regime in the wider national authoritarian political processes. There was an intricate fusion between statutory and socially embedded institutions on the irrigation schemes. Although the schemes were on public land, where formal rules had to be applied, informal arrangements with powerful political actors developed, and these were instrumental in the management of the schemes. While the Departments of Irrigation and Settlement represented the formal management structure, they were often overridden by chiefs, MCP officials and MYPs who acted out of the existing and workable local conditions.

The deepening economic and political crises of the 1980s, coupled with the withdrawal of Taiwanese support, forced the government to reduce its role in scheme management and upkeep. During the 1990s, in particular, physical infrastructure deteriorated. As Malawi made the transition from authoritarian rule under President Banda to a multi-party democracy in the mid-1990s, farmers often ignored cropping calendars and other rules established during the Banda regime. In 1993, the MYP was disbanded and most settlers brought in by the movement were evicted from the schemes. Although the reasons for these actions had little to do with agriculture, farmers believed that this signalled their emancipation from the previous regimental scheme management. Thus, since the early 1990s, many of the formal authority structures governing the smallholder irrigation schemes have lost legitimacy and are frequently ignored. Farmers felt that the old rules and regulations were unfair and, like the regime that imposed them, should be rejected.

Following the dissolution of the MYP, a power vacuum crept into management. The government established Scheme Management Committees (SMCs), consisting of elected members from the plot-holders as the new scheme-governing structure. The Irrigation Officer, a government employee in charge of the scheme, served as an ex officio member of the committee, but in practice retained most of the authority. The sweeping political and economic reforms in the last 10 years have reduced the number of civil service and agricultural extension posts and slashed public sector budgets. As noted, farmers themselves have resisted imposition of SMC authority. Consequently, the new structure has been highly ineffective, often laden by problems of corruption, poor governance and lack of direction, as described more fully below.

Present context

How is this legacy reflected in current efforts to transfer management of the smallholder irrigation schemes to farmers? It is against this background that the government, with pressure from donors, has opted to hand over the schemes to farmers' associations. With IMT gaining international acclaim, Malawi began the process that, to date, has resulted in more complications than solutions. From the mid- to late 1990s, experiments with IMT yielded inconclusive results. The Food and Agricultural Organization (FAO) and Danish International Development Agency (DANIDA) both funded a study of three pilot schemes, which were to be rehabilitated and handed over to farmers. DANIDA never completed its work due to political disagreements over issues of corruption with the Malawi government, which led to the suspension of aid and the closure of Denmark's embassy in 2001. When FAO completed its technical report no follow-up was made until 1999, when IFAD launched the Smallholder Flood Plains Development Programme. This project targeted physical rehabilitation, training and transfer of responsibilities to farmers' associations. However, the programme was driven by an engineering mentality with little appreciation of the complexities of the history and the social relations that had characterized the schemes during and after the departure of the MYP.

About the same time, a few social science-oriented studies revealed some deeply ingrained problems in the schemes. Kishindo (1996), Cammack and Chirwa (1997) and Chirwa (2002) showed that poor conditions existed and that human rights had, in some cases, been infringed. Their studies revealed low earnings, high rates of farmer turnover, lack of farmer participation, land dispossession, gender discrimination and autocratic administration by scheme management authorities. These findings suggested that irrigation farming was far from achieving the proclaimed goal of increasing agricultural productivity and attaining national food security. All this resonated well with the discourse on the need for IMT and provided fuel to draft and implement the new irrigation reform.

At the time of this study, the Likangala and Domasi irrigation schemes differed in the condition of their physical infrastructure, degree of farmer mobilization and source of funding for renovation and transfer of the schemes to farmers. Since its inception in 1972, Domasi has been fairly well supported by government and donor organizations, particularly the Taiwanese Agricultural Technical Mission and, most recently, IFAD. Although still in need of renova-

tion, its physical infrastructure is in better condition than that of Likangala. Domasi also is further advanced in the process of forming a WUA, adopting a constitution and by-laws and is likely to be the first smallholder irrigation scheme in Malawi to be formally handed over to the farmers' association. Likangala, in contrast, has received less government and donor support since it was established in 1969. Renovation and training have proceeded slowly, and it was not until August 2004 that preparations for establishing a farmers' association were initiated and a new constitution was adopted. Likangala was originally scheduled to be renovated with Highly Indebted Poor Country funds but, as of 2006, it appeared likely that the World Bank would become the new donor.

Our research revealed that the smallholder irrigation schemes play a vital role in both the local economy of the Lake Chilwa basin and the well-being of the farmers on them. Farmers on both schemes had diverse livelihood strategies. In addition to their irrigation plots, 93% reported having upland rain-fed fields, and 45% had either wetland or stream-bank gardens. Further, many plot-holders had sources of income in addition to farming: 40% listed casual labour, 19% marketing of crops, 23% owned small businesses and 9% had other occupations. Despite their engagement in other occupations, farmers reported that the irrigation plots constituted the major source of their household food supply and cash earnings.

The two irrigation schemes differed in important ways. There were differences in the number of years farmers had held plots, with turnover on Likangala being higher than that at Domasi. Domasi had a higher percentage of farmers (44%) who had been on the scheme for 20 years or longer as compared with Likangala (17%). When the irrigation scheme lands were originally parcelled out to farmers in the late 1960s and early 1970s, they received two to four plots, each one constituting 0.25 acre. The baseline survey revealed that the average number of plots held was greater on Domasi than on Likangala. The Domasi mean was 3.9, while on Likangala it was 2.7. Overall, 18% of the total sample reported farming five plots or more – 8% of Likangala farmers and 17% of Domasi farmers.

In order to estimate differences in wealth among farmers, a ranking of the households' assets was undertaken, with scores ranging from 7 to 1576. Households were clustered into three wealth categories. Over two-thirds fell within the lowest part of the range, 26% in the middle and 7% in the top asset group. This distribution reflects the overall distribution of poverty in Malawi and in the southern region in particular. A slightly higher percentage of Domasi than Likangala farmers had asset scores in the middle and upper clusters.

Overall, our findings indicate that the irrigation scheme constituted farmers' major source of livelihood – including food for household consumption and cash earnings. However, the differences presented above suggest that Domasi plot-holders were somewhat wealthier than those on Likangala. Irrigation scheme farmers are, on average, better off than Malawians who do not have access to dry-season irrigated fields. Many scheme farmers are able to plant twice a year or more and, consequently, are not as likely to experience food deficits as those without access to dry-season gardens. While they are not among Malawi's poorest farmers, many plot-holders remain vulnerable, as the asset profile reveals. During the January–March 2002 period, the height of the famine, the field assistants reported that some people on the schemes were consuming maize husks and grasses. Deaths, aggravated if not entirely caused by hunger, also occurred.

The Implementation of New Policies and Laws

The institutionalization of the new Irrigation, Land and Water Policies and Laws on the Domasi and Likangala irrigation schemes has just begun. In this section, we explore how this process is being shaped by local histories as well as by existing informal institutions, practices and power relations. Irrigation reform policy makers and programme implementers, for the most part, are unaware of these complexities and are often surprised when they yield unanticipated results. For example, in discussions on the status of Likangala and Domasi schemes, a member of a recent irrigation renovation project development team remarked that the donor organization would seek to avoid including schemes where there were social conflicts, unaware that this was the norm, not the exception, in most schemes.

Processes of formalization and informalization of irrigation, land and water rights produce complex situations 'which are neither regulated by predictable rules and structures nor characterized by sheer anarchy' (Benjaminsen and Lund, 2002, p. 3). While the terms formalization and informalization are often used in policy and academic literature, as Cleaver (2002) points out, they polarize reality. Her concept of institutional bricolage represents a means of framing the complex interactions among state-sponsored statutory and bureaucratic reforms and existing local institutions, practices and power relations. This approach allows us to analyse the dynamic arrangements that develop when new institutions and social relationships are adapted to existing conditions and power relations.

In this section, we demonstrate how the transfer of authority to user groups, as advocated in the Irrigation, Land and Water Policies, provides the opportunity for these groups to institutionalize not only newly established laws and procedures but also practices that were previously regarded as customary or informal in nature. This context opens the door for the powerful and well-positioned to capture resources and authority, and for others potentially to challenge them. Benjaminsen and Lund (2002, p. 1), for example, point to the importance of such everyday politics of institutionalization of rights and exclusion, noting that: '… it becomes all the more important to investigate empirically how local-level competition, conflict and power reshape social institutions and move with a distinct dynamic that does not necessarily fit with dominant discourses.' The following discussion on the implementation of the Irrigation, Water and Land Policies parallels experiences found in other countries in the eastern and southern African region (Bolding *et al.*, 2004; Juma and Maganga, 2005).

Traditional authorities and new structures of irrigation scheme governance

The rapid pace of reform has raised a number of controversial issues in the management of the smallholder irrigation schemes. One such issue concerns the role of TAs. Under the previous Land Policy and Law, the smallholder irrigation schemes were classified as public land, and they are earmarked to remain so in the new Land Policy and Law. The newly formed WUAs are to lease the schemes from the government. The new Domasi and Likangala constitutions state that chiefs are not to take part in plot allocation or dispute resolution on the schemes, although in the past they, at times, had played an unofficial but important role in these processes. The new Local Government Act and Decentralization Policy give chiefs identified roles in local administration, as Heads of Area Development Committees and ex officio members of District Assemblies, as discussed above (GOM, 1998b). In this section, we discuss some of the understandings that irrigation farmers have concerning the role of chiefs, which underscore the importance of past histories and traditions and which may result in future conflicts.

To begin with, farmers have different understandings of the tenure status of the schemes under the new legislation. Our study indicated that many were unaware that the WUAs were to receive leases from the government: 37% thought the scheme would revert to customary land; 27% thought it would become their own private property; 13% thought the WUA would be the new owner; only 16% were aware that the scheme would remain government land. This has given rise to uncertainty over the roles of chiefs in solving disputes that develop, especially between farmers on and off the schemes. When asked who solves such disputes, 57% said they were solved by the WUA Executive Committee or the Scheme Management Committee and 38% said they were solved by chiefs, while 5% said they didn't know. Because decentralization and many other processes of reform are occurring simultaneously, lines of authority are often unclear to farmers, and sometimes even to officials. This raises opportunities for multiple interpretations of rights and competing claims to land, water and other resources.

These differing interpretations regarding the tenure status of the schemes and the plots on them have given rise to conflicts. At Likangala, the widely held perception that the land will revert to farmer or customary control has opened the door to attempts by village headmen to reclaim ancestral lands. There, one village headman encouraged farmers from his village to take over plots on Blocks B and C from other farmers. His claim to these blocks was based on his assertion that these were his ancestral lands and, since the scheme was being turned back to farmers, the plots should be allocated to those from his village. He drew on past history to make these claims. This headman and many members of his village were exiled to Mozambique when former President Banda banned the Jehovah's Witnesses in the early 1970s. When they returned to the area in the early 1990s, they had very little land on which to cultivate and were refused access to all their original scheme plots (Nkhoma and Mulwafu, 2004). Other village heads have threatened that, if this headman is allowed to claim the scheme land as his village land, they will do the same.

The farmers who had been displaced in Blocks B and C reported the case to the Scheme Management Committee for resolution. When the matter had been discussed for more than a year and no binding decision was reached, it was taken to the TA as a socially recognized authority. Unfortunately, the TA also failed to resolve the matter and, at this point, referred it to the District Commissioner. Upon hearing views from both parties, he called for a meeting of all farmers in the scheme to announce his verdict. However, in a surprising turn of events, he asked the TA to pronounce the verdict, which was that those who had invaded plots should give them back to the farmers they took them from at the end of the harvest season. This particular example illustrates some of the ambiguities that exist in making claims to plots and in resolving disputes. The local government and irrigation reform processes have allowed old claims based on persisting local custom to be revisited, and have involved both customary and statutory authorities in dispute resolution.

The new irrigation policy in the local context

Drawing on previous experiences and present social location, farmers were divided in their support for the transfer of the schemes to farm-ers' associations. On both irrigation schemes there was significant opposition. Some plot-holders, particularly the wealthier ones, feared that the transfer to farmer management would potentially remove their opportunities for accu-mulation, as new rules concerning access and plot allocation could be put in place. Others were concerned that transfer would open the way for more 'outsiders' to gain access to plots. Still others, in line with the past history of irriga-tion reform, regarded rehabilitation as a government responsibility and were reluctant to take over the schemes until they had been completely refurbished. Indeed, in 2003, farm-ers at Likangala opposed the transfer because they were afraid they would inherit a dilapi-dated main canal and other structures they could not afford to fix. Many stated that they did not see how they could succeed in repairing and running the scheme when the government, with all its resources, had failed in doing so.

As mandated by the new Irrigation Policy, WUAs are being established on the Likangala and Domasi irrigation schemes, but through different means and with quite different results, reflecting local histories and interactions between formal and informal institutions and practices. At Domasi, the rehabilitation and handover process was instituted with funding from IFAD in 2000. Although the model employed called for farmers to participate in setting the conditions of the transfer process itself and for them to receive training in scheme maintenance and management, it did not occur until very late in the renovation process. Concern Universal, an NGO, was contracted in 2002 to provide farmer training at a point when most decisions regarding physical rehabilitation had already been made and renovation was well advanced. At Likangala, farmers were mobilized to supply labour for rehabilitation, but little farmer capacity building had occurred. The priority given to physical renovation reflects the backgrounds of the majority of civil servants in the Irrigation Department. Most are engineers who see renovation as a technical problem requiring little social input.

Actual rehabilitation of canals, headworks, roads and other facilities on both schemes has proceeded slowly due to numerous factors. These included delays in the arrival of funds and supplies, problems with local contractors and heavy rains that destroyed renovated struc-tures. Other factors with social and historical origins have caused delays as well, including inputs disappearing or being stolen and farm-ers' reluctance to provide labour. These delays have been great at Likangala, which has depended on government funding for renova-tions. At Domasi, delays have resulted in two postponements of the targeted date for scheme handover to the WUA. Initially, transfer was scheduled for 30 December 2002, and then for 30 September 2003. By mid-2004, govern-ment officials recognized that rehabilitation and handover would not be a single event completed by a specified date and marked by a celebration and photo opportunities for politi-cians, planners and donors. Rather, it was likely to be a phased process taking much more time and getting more complicated than originally anticipated.

At Domasi, almost all farmer capacity build-ing has focused on preparing the newly elected WUA Executive Committee members to carry out their functions, rather than on the general WUA membership. Indeed, only 15 (13%) of the farmers in the overall sample said they had received some training on handover issues. Twelve of these were from Domasi, and all of them were members of various scheme committees. Under the 'Training of Trainers' model used by Concern Universal, the assump-tion was that these farmers would inform and train their neighbours, a process that did not take place. Farmers who had undergone train-ing expected support and payment to train others and, failing this, many did little to pass on information.

Indeed, WUAs have played no real role in decision making. Most decisions concerning the Domasi scheme were made by a small group of newly elected WUA Executive Committee members and government scheme officials and were announced at general WUA meetings. Indeed, reflecting the authoritarian base of past 'elected' Scheme Management Committees, although technically elected by WUA members, most of those on the new Executive Committee were members of the previous Scheme Management Committee, composed primarily of a small group of wealthier plot-holders, their relatives and friends. Not surprisingly, such concentration of knowledge and authority in the hands of a small number of committee members mirrors the previous top-down administrative practices. It also suggests that most farmers on the schemes are poorly equipped to exercise their rights and obligations in the new governance structures. In fact, the handover survey revealed that the majority of Domasi farmers did not understand that the WUA was their membership organization: most thought it was a new name for the Scheme Management Committee.

At Domasi, a small group of relatively well-off farmers has maintained control of the WUA Executive Committee and other newly established committees. Some of these farmers owned more than ten plots each, often those with the best access to water. Actual plot ownership at the household level was greater than these figures suggest, as spouses and children had plots registered in their names as well. These farmers received extensive capacity building from Concern Universal, including the opportunity to travel and observe other WUAs being established on irrigation schemes in other parts of the country, as well as a trip to Italy for the Executive Committee President to receive training in marketing and general management of smallholder irrigation schemes. This training has helped consolidate and legitimize their present positions and their views of the future of the scheme.

At Likangala the WUA was being organized, its constitution enacted and a new SMC elected at breakneck speed in July and early August 2004. This was instigated by Zomba Rural Development Project (RDP) officials under the assumption that a WUA had to be in place to allow the scheme to qualify for newly available World Bank renovation funding. A 'participatory and consultative' meeting on problem identification and constitution building occurred at Likangala in early July 2004, involving village headmen and other TAs, scheme committee members, RDP officials and only a small number of ordinary plot-holders. One week later, RDP officials presented the constitution for ratification at a general farmers' meeting attended by fewer than 20 farmers who did not hold elected or appointed office. Few farmers knew that there was a draft constitution or that a meeting was going to take place

to discuss it, let alone the provisions contained in the document itself. At the ratification meeting the constitution was read to the farmers, who were asked to endorse it. Barely 1 week later, another meeting was held to elect a new SMC. This committee, like its predecessor, was composed of relatively wealthy and influential farmers. This top-down process echoed the way schemes were run during the Banda era and may well be challenged by Likangala WUA members in the future, if they are sufficiently organized.

In essence, the transfer of authority for scheme management to WUAs, as required by the Irrigation Policy and Law, opened the door at Likangala for the state to reassert its authority via an alliance between the RDP and newly elected SMC, while at Domasi it resulted in the further consolidation of the power of local elites in control of the WUA Executive Committee. These differences are reflected in the provisions in the two new constitutions. While they show some similarities, these constitutions also reflect differences rooted in the histories, contemporary practices and power relations existing on the two schemes, giving rise, at this point at least, to different visions of the schemes' futures.

At Likangala, the state, in the guise of the newly elected SMC, has sought to reinstate many of the rules and regulations that characterized the irrigation schemes in the 1980s. In contrast, at Domasi, the Executive Committee has adopted new provisions that formalize many informal practices that developed particularly during the 1990s, allowing greater concentration of land and other resources. We examine below how some of the key provisions of the new constitutions reflect history and context just as much as they do the new procedures and rules set out in the Irrigation Policy.

Rules limiting access

One contentious debate that took place in constitution drafting related to who should have rights of access to plots. Should it be people from surrounding villages, any person from Zomba or Machinga districts or any citizen of Malawi? When the schemes were established, as noted above, the land was converted from the customary to the public tenurial system.

Until the adoption of the new constitutions, any citizen of Malawi could technically ask for a plot by applying to the SMCs. In the immediate post-Banda period, absentee farmers and plot-seekers from urban areas increasingly began to obtain plots through informal renting and borrowing/lending arrangements and, in some cases, preferential allocation from the SMCs.

Many farmers and village headmen opposed this influx; they considered the land their ancestral territory to be inherited by their children and grandchildren. Reflecting these issues, both constitutions contain clauses limiting scheme access to residents of the area. The Likangala constitution states that access to plots is dependent on being from Traditional Authority Mwambo. The Domasi constitution contains a similar, if somewhat more vague, clause asserting that access is limited to citizens of Malawi who are residents of the area. This focus on local ownership reflects the historical tensions concerning displacement from ancestral lands, as well as more recent concerns that plots are being unjustly allocated to outsiders.

Rules concerning WUA membership criteria, plot ownership and inheritance

The two constitutions reflect quite different WUA membership criteria – clauses that are closely related to the new regulations concerning plot ownership and inheritance. At Likangala, the unit of membership was identified as the *banja* (family), consisting of husband, wife and children, while at Domasi it was identified as the individual. These variations are reflected in different provisions concerning the number of plots that can be owned. When the irrigation scheme lands were originally parcelled out to farmers, they received two to four plots, the area of each being 0.25 acre. Original by-laws on both schemes prohibited ownership of more than four plots and also forbade the practices of renting or lending plots. Many farmers believed that these rules were still in place and made efforts to disguise practices of renting, lending and plot accumulation in various ways.

The new constitutions have adopted quite different policies concerning these practices. The Likangala constitution states that families may own no more than four plots in total. It is

too early to determine whether the SMC, some of whose members have considerably more than four plots, will be willing to enforce these limits. It is worth noting, however, that the very people who have been given authority to enforce the new regulations are the ones known in the past for violating them. At Domasi, on the other hand, the constitution does not specify the number of plots an individual can have, stating rather vaguely that WUA members have a right to 'a profitable landholding size according to agreed criteria for land allocation'.

In a similar fashion, the Likangala constitution seeks to reinstate or reinforce older prohibitions on renting of plots, stating that farmers found to be renting out plots can be fined and ultimately removed from the scheme. The older regulation, that land not cultivated for 2 years reverts to the SMC, had spurred renting as a means of dealing with hardships of various kinds. Those who were unable to cultivate because they lacked inputs, did not have sufficient labour or were sick, often rented or lent their plots to better-off farmers and ended up working as labourers on their own or others' fields. At Likangala, despite its illegal status, it may be difficult to halt renting for at least two reasons – its widespread occurrence and the fact that it meets the needs of both wealthy and poor farmers. The Domasi constitution, on the other hand, is mute on the issue of renting, presumably permitting it to continue and thus formalizing what had become a common informal practice during the 1990s.

As tight control over the schemes collapsed, farmers became accustomed to leaving their plots to their spouses, children and other relatives. The new Domasi constitution states that plots can be left to a specified next of kin, who must be identified on the plot-holder's WUA membership card. The Executive Committee has the power to approve or reject this choice, as it has the authority to determine whether the next of kin meets membership criteria. The goal is to limit inheritance to one family member in good standing with the WUA. This clause may well generate opposition in the future as more farmers become aware of it, since it contradicts what has become local inheritance practice. In the opinion of many Likangala farmers, it is only when the plot-holder is unmarried and has no offspring that the plots revert to the SMC for redistribution, and then usually half goes to relatives of the deceased and the remainder to non-family members. However, the new Likangala constitution states that, upon the death of the holder, the plots are to revert to the SMC, which may redistribute them to the relatives of the deceased or to others as they see fit. In the past, the SMC had sometimes used the occasion of the death of a member to obtain plots and reallocate them, often to powerful, influential people, including members of the committee itself. Given these practices, this inheritance clause in the Likangala constitution may also generate opposition once it becomes more widely known.

Malawi's new land reform may have impacts on women's rights to irrigated land. Women's access to plots and voice in management have not been addressed directly in farmer training on the irrigation schemes to date, although the new Irrigation Policy includes normative statements supporting women's equal participation in irrigated agriculture. The Domasi and Likangala schemes are located in an area of matrilineal inheritance. Reflecting the patrilineal biases of most development planners at the time when the schemes were established, plots were registered in the names of men as heads of families. Over the years, however, many women have gained access to plots using various strategies, including appeals to cultural traditions that recognize matrilineal inheritance of land and other property. In 2003, Concern Universal estimated that, of the 1500 registered plot-holders at Domasi, 47% were women. Asked whether women should be allowed to register plots in their own names, an overwhelming 95% of the Domasi respondents said that they should, while 88% affirmed the same at Likangala. At Likangala, where the 2004 constitution limits the number of plots a family can hold to four, it is not clear what will happen to plots registered in a woman's name when the husband also has plots and the total number exceeds four.

The new Land Policy and Law propose to make inheritance more equitable by not recognizing either customary patrilineal or matrilineal inheritance practices, calling instead for children of both sexes to inherit equally from parents. It is too early to determine what the effects will be on women's land rights but, in a context where patrilineal inheritance continues to be taken as the norm by most donors, policy makers and decision makers, women in the Lake Chilwa basin and other matrilineal areas in the south and centre may lose rights to valuable irrigated land, while those in patrilineal areas may not gain more rights.

Rules on dry-season rotation of plots

Older by-laws on both irrigation schemes contained clauses requiring dry-season rotation of plots. The SMCs would reallocate plots each dry season, allowing those who did not normally have access to these valuable lands to use them temporarily. Farmers interviewed were generally supportive of this practice: 83% said it should be continued after handover for a number of reasons, including that it helped people who did not have enough food and gave access to those who did not have plots or whose lands did not receive enough water. Although farmers were supportive of this dry-season plot rotation, many criticized the way it was carried out, claiming that the SMC was corrupt and often allocated plots, not to the poor, but to better-off farmers and city dwellers. The new Domasi constitution does not mention dry-season rotation, presumably because it interferes with securing property rights and is not in the interests of those now in control of the new WUA Executive Committee. At Likangala, on the other hand, the new SMC is set to continue this practice.

Irrigation reform in the context of water reform

This section looks at the interface of the water and irrigation reforms. It shows the challenges of implementing provisions of these reforms against the backdrop of local histories and the broader political, social and economic context. Particular attention is paid to the implications of these reforms for class and gender disparities on the irrigation schemes.

Although the Water and Irrigation Policies and Laws are being harmonized to resolve areas of ambiguity and conflicting clauses, questions about how the new structures will function in local contexts in relation to existing rights, practices and resources have been largely ignored. One of these questions involves the creation of Catchment Management Authorities (CMAs), as proposed in the new Water Policy and pending Law. The policy calls for Malawi to be divided into catchment areas, which are drawn according to hydrological criteria and, in many cases, cross political–administrative boundaries. Two or more districts or TAs may fall within one CMA. While the catchment approach may make environmental sense, it creates another administrative structure that has to be negotiated and financially supported. It is unclear how CMAs will work with District Councils and other political administrative units.

In fact, this has been a significant issue in Zimbabwe, where the same organizational structure was put into place (Derman *et al.*, 2000). There, CMAs include representatives of districts, local representatives of various ministries, and major water users such as commercial farmers, smallholders, and mining and urban water user representatives. For district authorities and smallholder farmers alike, the transaction costs of participating in meetings are high, and they often lack funds to attend. Water users also have to travel long distances to CMA offices to pay fees or obtain services (Derman *et al.*, 2000; Nicol and Mtisi, 2003). In Malawi, more serious financial problems exist as well, as sustainable sources of funding for the CMAs have yet to be identified.

WUAs are likely to be represented in the proposed CMAs and expected to participate in meetings using their own resources, which will require that they raise funds from fees collected from water users, the majority of whom are poor. This may be a major limiting factor for the effective participation of WUAs in the new structure for management of water resources.

Malawi's 2000 Water Policy and 1999 Draft Law have recognized people's right to water for 'primary' purposes, defined as the provision of water for household and sanitary purposes and for the watering and dipping of stock (GOM, 1999b, part 1.1). As noted above, the 2004 Policy no longer uses this rights-based language. Instead, it states only that the protection and use of water resources for domestic water supply are to be given the highest priority over other uses. Presumably the new law, once drafted and approved, will provide clarification.

Users of water for productive purposes are likely to be required to obtain a water use permit. Current suggested guidelines in this regard indicate that an applicant will need to provide information on particulars of the land, e.g. freehold or leasehold, name or description and type of body of water from which the water is required, the point of abstraction, the amount required and the purpose of use. The payment would be made on an annual basis, with the permit renewable every 5 years. Water rentals would be determined by the amount of water to be abstracted. Again, we must await the drafting and enactment of the Water Law for clarification.

Small-scale irrigation is considered a productive use of water, and WUAs will have to obtain a water use permit, with the payment of fees passed on to plot-holders. Many scheme farmers also have other small parcels of land they farm and may have to obtain a water-use permit on their own as well. Collection of these fees promises to be an arduous task for the government, considering the high levels of poverty plus the considerable transaction costs involved in collecting fees from large numbers of smallholders unaccustomed, and often unwilling, to pay for water.

The experience from Tanzania perhaps best illustrates the seriousness of this problem. State attempts to introduce water permits and water pricing among small-scale users in the river-basin organizations in the mid-1990s ended in failure. Not only was the process of registering water users laborious and time-consuming, but the transaction costs were extremely high. Moreover, even when the users had organized themselves into WUAs, the risks of corruption and marginalization of the poor were significant. In these cases WUAs run the risk of becoming one more means through which wealthier water users advance their own interests at the expense of the poor (van Koppen *et al.*, 2004). In both Domasi and Likangala schemes, it is the wealthy and elite farmers on the WUA management committees who have so far benefited from new opportunities provided by the irrigation reform in terms of training, access to information and abilities to shape the rules governing scheme functioning.

One option for Malawi to consider is to legally recognize a smallholder's right to water for both productive and domestic purposes. This right would not require individual or collective titling or permits. It would: (i) take into account the importance that water plays in livelihood strategies; (ii) recognize that the majority of small-scale farmers lack the means to pay user fees or could best use the money for other productive purposes; (iii) acknowledge that many smallholders do not believe that water should be commoditized, and grant them a voice in deliberations over water use without having to register; and (iv) avoid having to collect fees from all of them or their organizations – a nearly impossible task in any case. Registration and collection of water permits and user fees, for the immediate future at least, can best be concentrated on large-volume water users (Electricity Supply Commission of Malawi (ESCOM), Water Boards, private estates, etc.).

Concluding Remarks

Malawi has embarked on what constitutes a radical redefinition of tenure and governance structures related to key irrigation, land and water resources. These new policies and laws draw from neo-liberal development thinking, with its emphasis on private sector initiatives, redefinition and reduction of the role of the state, and promotion of new decentralized, stakeholder-driven and community-based management institutions. The new irrigation reform, relying on IMT thinking, embodies many of these characteristics. It calls for the creation of new forms of social organization – WUAs – and formalization of rights and responsibilities, together with physical renovation of the irrigation schemes. The operating assumption to date is that these institutions and practices can be put in place without reference to history or the local context. This chapter has questioned that assumption by examining the dynamic arrangements that develop when new institutions and social relationships are adapted to local conditions and power relations on two smallholder irrigation schemes. The study raises two central questions that merit follow-up as the pace of irrigation reform accelerates.

Are the new directions likely to broaden smallholder irrigation scheme farmers' – especially disadvantaged ones' – access to the critical livelihood resources of land and water? Our findings indicate that many critical questions remain to be addressed concerning equity, poverty alleviation and strategies for pro-poor economic growth in the transfer of the smallholder irrigation schemes from the government to WUAs. At this point, it appears that the Domasi WUA Executive Committee and the new Likangala SMC have adopted different positions on equity and poverty-alleviation issues – with the Domasi Executive Committee focusing on productivity achieved through permitting greater concentration of plots and other resources and the Likangala SMC opting for more equitable distribution of them.

The constitutions of the two schemes draw on experiences from the past to reflect different visions of the future of smallholder irrigation in Malawi. The Likangala constitution seeks to reinstate the older top-down, state-sponsored rules ostensibly favouring more equal distribution of resources. It does this by limiting plot concentration, promoting dry-season plot rotation and restricting informal inheritance practices. The Domasi constitution, in contrast, seeks to strengthen tenure security and to promote the entrepreneurial spirit by formalizing informal practices dating from the late 1980s and early 1990s, including renting and increased land concentration. It remains to be seen whether these provisions will be put into practice and what their equity effects might be.

Second, one of the key driving assumptions underpinning these reforms is that people's rights to resources will be made more secure, thus spurring economic growth. Is this occurring on these two irrigation schemes? At this point, the study reveals lack of knowledge and understanding among officials and farmers alike about the irrigation, land and water reforms. Most farmers interviewed had no clear understanding of what their rights to land or water resources would be once transfer of the irrigation schemes was accomplished. No common understanding existed concerning key issues of membership requirements in the WUA: (i) tenure status of the scheme; (ii) whether plots could be bought, sold, rented, borrowed or inherited; and (iii) whether there would be a limit on the number of plots owned. Presently, rather than being more secure, farmers' rights to land and water resources appear more uncertain than they have been in the past.

Acknowledgements

This research was supported by a grant from the USAID-funded BASIS CRSP project on Institutional Dimensions of Water Policy Reform in Malawi: Addressing Critical Land-Water Intersections in Broadening Access to Key Factors of Production. We would also like to acknowledge the work carried out by the BASIS field assistants over the course of the research project: Messrs Davidson Chimwaza, Mapopa Nyirongo, Noel Mbuluma and Chancy Mulima. Messrs Chimwaza and Mbuluma lived and worked on the irrigation schemes, participating in many scheme meetings and in the daily lives of the farmers for 3 years. Two BASIS-supported graduate students, Messrs Bryson Nkhoma and Daimon Kambewa, also contributed substantially to the research project.

References

Adams, W., Watson, E. and Mutiso, S. (1997) Water, rules and gender: water rights in an indigenous irrigation system, Marakwet, Kenya. *Development and Change* 28, 707–730.

Benjaminsen, T. and Lund, C. (2002) Formalization and informalization of land and water rights in Africa: an introduction. *European Journal of Development Studies* 14 (2), 1–10.

Bolding, A., Manzungu, E. and Zawe, C. (2004) Irrigation policy discourse and practice: two cases of irrigation management and transfer in Zimbabwe. In: Mollinga, P. and Bolding, A. (eds) *The Politics of Irrigation Reform.* Ashgate Publishing Co., Burlington, Vermont, pp. 166–206.

Cammack, D. and Chirwa, W. (1997) Development and human rights in Malawi. In: Chilowa, W. (ed.) *Bwalo: Forum for Social Development, Issue 1.* CSR, Zomba, Malawi, pp. 105–121.

Chilivumbo, A. (1978) On rural development: a note on Malawi's programmes for exploitation. *African Development* 3 (2), 41–55.

Chirwa, W. (2002) Land use and extension services at Wovwe rice scheme, Malawi. *Development Southern Africa* 19 (2), 307–327.

Cleaver, F. (2002) Reinventing institutions: bricolage and the social embeddedness of natural resources management. *European Journal of Development Studies* 14 (2), 11–30.

Derman, B., Ferguson, A. and Gonese, F. (2000) *Decentralization, Devolution and Development: Reflections on the Water Reform Process in Zimbabwe.* Report. BASIS CRSP, Madison, Wisconsin.

Devereux, S. (2002a) Safety nets in Malawi: the process of choice. Paper prepared for the *Institute of Development Studies Conference 'Surviving the Present, Securing the Future: Social Policies for the Poor in Poor Countries',* Sussex, UK.

Devereux, S. (2002b) The Malawi famine of 2002. *Institute of Development Studies Bulletin, The New Famines* 33 (October), 70–78.

Ferguson, A. (2005) Water reform, gender and HIV/AIDS: perspectives from Malawi. In: Whiteford, L. and Whiteford, S. (eds) *Globalization, Water and Health: Resource Management in Times of Scarcity.* School of American Research, Santa Fe, New Mexico, pp. 45–66.

Ferguson, A. and Mulwafu, W.O. (2001) Decentralization, participation and access to water reform in Malawi. Report prepared for the *BASIS CRSP Policy Synthesis Workshop,* July 2001, Johannesburg, South Africa.

Ferguson, A. and Mulwafu, W.O. (2004) *Irrigation Reform on Malawi's Domasi and Likangala Smallholder Irrigation Schemes: Exploring Land-Water Intersections.* Final Research Report, BASIS CRSP, Madison, Wisconsin (October).

GOM (Government of Malawi) (1965) *Land Act.* Government printers, Zomba, Malawi.

GOM (1968) *Irrigation Ordinance.* Government printers, Zomba, Malawi.

GOM (1998a) *Decentralization Policy.* Decentralization Secretariat, Lilongwe, Malawi.

GOM (1998b) *Local Government Act.* Government Printers, Zomba, Malawi.

GOM (1999a) *Water Resources Policy and Strategies.* Ministry of Water Development, Lilongwe, Malawi.

GOM (1999b) *Draft Water Act.* Ministry of Water Development, Lilongwe, Malawi.

GOM (2000) *National Irrigation Policy and Development Strategy.* Ministry of Agriculture and Irrigation, Lilongwe, Malawi.

GOM (2001) *Land Policy of Malawi*. Ministry of Lands, Housing, Physical Planning, and Surveys, Lilongwe, Malawi.

GOM (2004) *National Water Policy*. Ministry of Water Development, Lilongwe, Malawi.

Juma, I. and Maganga, F. (2005) Current reforms and their implications for rural water management in Tanzania. Presented at the international workshop on *African Water Laws: Plural Legislative Frameworks for Rural Water Management in Africa*, 26–28 January, Johannesburg, South Africa.

Kambewa, D. (2004) *Patterns of Access to and Use of Wetlands: Lake Chilwa Basin, Malawi*. Report, BASIS CRSP, Madison, Wisconsin.

Kishindo, P. (1996) Farmer turn over on settlement schemes: the experience of Limphasa irrigated rice scheme, northern Malawi. *Nordic Journal of African Studies* 5 (1), 1–10.

Krogh, E. and Mkandawire, R.M. (1990) *Life as a Rice Farmer in Malawi. Socio-Economic Survey*. Smallholder Irrigation Rehabilitation Project, Bunda College of Agriculture, University of Malawi, Lilongwe, Malawi.

Mollinga, P. and Bolding, A. (eds) (2004) *The Politics of Irrigation Reform*. Ashgate Publishing Company, Burlington, Vermont.

Nicol, A. and Mtisi, S. (2003) Politics and water policy: a southern Africa example. *IDS Bulletin* 34 (3), 41–53.

Nkhoma, B.G. and Mulwafu, W.O. (2004) The experience of irrigation management transfer in two irrigation schemes in Malawi, 1960s–2002. *Journal of Physics and Chemistry of the Earth, Parts A/B/C* 29 (15–18), 1327–1333.

Owusu, K. and Ng'ambi, F. (2002) *Structural Damage: the Causes and Consequences of Malawi's Food Crisis*. World Development Movement, London.

Peters, P. (2004) *Informal Irrigation in Lake Chilwa Basin: Streambank and Wetland Gardens*. Report, BASIS CRSP, Madison, Wisconsin.

Shah, M., Osborne, N., Mbilizi, T. and Vilili, G. (2002) *Impact of HIV/AIDS on Agricultural Productivity and Rural Livelihoods in the Central Region of Malawi*. CARE International, Lilongwe, Malawi.

van Koppen, B., Sokile, C.S., Hatibu, N., Lankford, B., Mahoo, H. and Yanda, P. (2004) *Formal Water Rights in Rural Tanzania: Deepening the Dichotomy?* Working Paper 71, International Water Management Institute, Johannesburg, South Africa.

Vermillion, D. (1997) *Impacts of Irrigation Management Transfer: a Review of the Evidence*. Research Report 11, International Irrigation Management Institute, Colombo, Sri Lanka.

Vermillion, D. and Sagardoy, J.A. (1999) *Transfer of Irrigation Management Services – Guidelines*. FAO, Rome.

World Bank (2004) Irrigation, rural livelihoods and agricultural development project preparation/pre-appraisal mission, 12–24 July. Aide Memoire, World Bank, Washington, DC.

14 A Legal–Infrastructural Framework for Catchment Apportionment

Bruce Lankford[1] and Willie Mwaruvanda[2]

[1]School of Development Studies, University of East Anglia, Norwich, UK;
e-mail: b.lankford@uea.ac.uk; [2]Rufiji Basin Water Office, Ministry of Water and
Livestock Development, Iringa, Tanzania; e-mail: rufijibasin@yahoo.co.uk

Abstract

We propose a water management framework for bringing together formal and informal water rights and irrigation intake design to apportion water in catchments. This framework is based on setting and modifying seasonally applied volumetric and proportional caps for managing irrigation abstractions and sharing water between upstream irrigators and downstream users in river basins. The volumetric cap, which establishes the upper ceiling of irrigation abstractions in the wet season, relates to formal water rights and maximum intake capacities. The proportional cap, which functions in the dry season beneath the volumetric ceiling, builds on customary water negotiations and on the design and continual adjustment of intakes by users. Both caps should be viewed as being adjustable in response to dialogue between users. The analysis is informed by conditions found in the Great Ruaha river basin, southern Tanzania, where rivers sequentially provide water for irrigation, a wetland, the Ruaha National Park and for electricity generation. Consequences for catchment interventions in the face of climate, population and land use change are explored.

Keywords: water management, framework, formal, informal rights, irrigation intakes, Tanzania.

Introduction

The apportionment of water between sectors in river basins requires the resolution of three matters: (i) establishing a vision of water allocation (river basin objectives for who gets what water); (ii) creating and sustaining the physical, legal, economic and institutional means of distributing water according to this vision; and (iii) monitoring outcomes so that further adjustments can be made to both vision and means. Of these three, the most difficult is the second, requiring the deployment of a water governance architecture that: (i) utilizes various allocation devices; (ii) involves and recognizes many stakeholders; (iii) selects relevant tech-

nology and infrastructure; (iv) accommodates issues of scale and timing; and (v) is underpinned by an appropriate legal and institutional framework. With regard to the latter, the gaps, overlaps and contradictions occurring between formal and informal legal agreements that fit within that architecture pose particular problems. Arguably, this is the key challenge for integrated water resources management in Tanzania (Sokile *et al.*, 2003), and arguably for sub-Saharan Africa in the face of changes in climate and technology and land use. How this challenge might be met is the subject of this chapter.

Although theoreticians may articulate ideal legal and institutional frameworks, in reality such frameworks commonly suffer from incongruities that exist between institutional functions, practices, objectives and biogeographic properties (Moss, 2004). Water frameworks have to help achieve river basin objectives, work within the limitations imposed by inherent conditions, fit other economic and infrastructural devices and often build on existing progress made. The scope for rethinking a wholly new institutional matrix may be severely restricted. In this regard, the contribution of this chapter builds on existing legislation in Tanzania. Furthermore, systemic challenges also exist: for example, research may point to the benefits that local user agreements can play at the local level, but how do we ensure that local user agreements collectively result in large-scale and bulk-water redistribution, and how should local agreements that may operate well at the irrigation level be applied to the catchment level? If informal arrangements are not dovetailed into higher-level formalities and other allocation devices, new legislative and institutional frameworks will only partially succeed.

This chapter proposes a framework that fits together legal, institutional and infrastructural water management provisions, recognizing the synergy between different components of water management, building on present-day policy directions and acknowledging contextual properties and processes (Garduno, 2001). The framework emphasizes the division of water management into wet and dry seasons, arguing that formal water rights have a role in the wet season, and that customary or local water agreements relate better to conditions found in the dry season (though clearly a variety of ongoing discussions and consultations are required throughout the year – this is not to propose a mutually exclusive division).

The two key assumptions here are that formal rights relate to access to water quantities measured by a flow rate (e.g. l/s) and that customary agreements relate to access to water quantities described by an approximate share of the available water (e.g. 'about half of what is present in the stream'). The assumptions are valid because formal rights are denominated in volumetric terms while customary agreements in their original form (an important distinction, since customary rights can be transmuted

during formalization procedures into volumetric measures) are founded on a notion of access to an (unmeasured) quantity of water, combined with the notion that not all the water can be abstracted from a stream or irrigation channel (Gillingham, 1999; SMUWC, 2000). Therefore, customary agreements, for the purposes of this chapter, pertain to negotiations over water *shares* that theoretically range from 0% (no water is abstracted) to 100% (all the water is abstracted), with the observation that streamflows are divided by trial using proportionally based intakes rather than by measuring flow using gauges, weirs and adjustable gates.

The framework explicitly works with the wet/dry season separation to assist rather than undermine these legal pluralisms and water reallocation objectives. This fits the call by Maganga *et al.* (2003) for an approach that 'combines elements of RBM and customary arrangements at the local level' and underpins upstream–downstream transfers of water within an ecosystems services approach. The framework is not a classification as proposed by Meinzen-Dick and Bakker (2001), who examined rights associated with different water purposes. The proposal here concerns mainly agricultural productive use of surface water that also meets domestic purposes in villages within the command area. It should be emphasized that this chapter (which utilizes research from two projects – SMUWC[1] and RIPARWIN[2] – that have studied river basin management in the Great Ruaha River, part of the larger Rufiji basin) is exploratory in nature. The discussion here applies to the catchment scale[3] rather than to the larger basin scale, because it is in the former where the tensions associated with irrigation abstractions and downstream needs are most keenly felt. This chapter also briefly discusses some concerns related to the sustainability and workability of the new arrangement, particularly with respect to irrigation intake design and conceptualization.

River Basin Management Initiatives in Tanzania

The Rufiji and the Pangani are two basins that have been supported by the Ministry of Water and Livestock Development (MOWLD) and a

World Bank Project (River Basin Management and Smallholder Irrigation Improvement Project, RBMSIIP) to manage water at the river-basin scale via the establishment of river basin offices (RBOs). Although details on these projects are available elsewhere (World Bank, 1996; Maganga, 2003), two[4] key activities of the basin offices are described here.

Formal water rights

Water for irrigation is managed via the issuance of formal water rights (which will be called 'permits' according to the new Water Policy) to water users against the payment of an annual fee, that are expressed in quantitative flow units (e.g. cumecs) (Mwaka, 1999). Associated with this is the registration of users and establishment of water user associations as legal entities. Maganga (2003) outlines the new thinking in the Water Policy (MOWLD, 2002) that has been partly incorporated into the new Water Strategy (MOWLD, 2004), which aims to regulate water use on the basis of statutory legal systems. Therefore, formal water rights are the key means of achieving redistribution in Tanzania (World Bank, 1996). However, as Maganga points out, law-making to date has not recognized the role that customary agreements play at the local level, though space for customary agreements is given in the new putative legislation and, therefore, a future activity will be to incorporate customary arrangements in ways that fit the rubric of the legislation.

Recent research (van Koppen et al., 2004; Lankford et al., 2004) supports the view that customary rights have not been fully recognized and, in addition, shows that the formal statutory rights may be structurally flawed in three ways: first, payment for water is not related to volume actually used, and so they may not dampen demand as they are supposed to do, but instead help increase demand. This lack of fit relates to discrepancies between the water right abstraction rate and the designed intake abstraction rate as is explained below. Secondly, they mainly address water availabilities found in the wet season rather than in the dry season, when important redistribution objectives are equally, if not more, critical. According to the Rufiji Basin Water Office

(RBWO), there is a nominal 50% reduction in the water right during the dry season, but this too is not against measurement, and does not relate to the real decreases found in river flows, which are closer to 10% of the wet season flows. Thirdly, they demand high levels of supervision that are not commensurate with resources available to the basin authorities.

Discussions with the Ministry of Water and Livestock Development seem to indicate that there is no plan to change the policy on the use of statutory rights, and that water rights will continue to be issued. The RBWO has recently been requested by its Board to review the current status of rights already issued with a view to bringing them into line with water availability. An appropriate accommodation of customary agreements might be highly beneficial, as research shows that, in parts of the Great Ruaha basin, local users negotiate and share river flows at the irrigation system level and catchment scale (Gillingham, 1999; SMUWC, 2000).

Irrigation improvement programmes

Where identified, smallholder irrigation systems had their intakes upgraded from traditional construction (e.g. stones and mud) to that of a concrete and steel gate design using a weir to raise water levels and a sluice gate to adjust discharge (see Fig. 14.1). Theoretically, this brings water control and adjustability and makes possible the measurement of water flows – and has long been thought to raise irrigation efficiency (Hazelwood and Livingstone, 1978). This change in – or upgrading of – the intake is usually the single greatest component of the so-called 'irrigation improvement programs' (Lankford, 2004a).

However, such technological change needs to be carefully scrutinized before being termed an 'improvement'. The change can be analysed by examining two related components of the design process: (i) sizing the dimension of the intake (note that main canal sizing is part of the headworks design but, for the sake of simplicity, the discussion here refers to the intake design); and (ii) configuring the operability of the intake – its ability to be operated, adjusted and understood in terms of a volumetric or

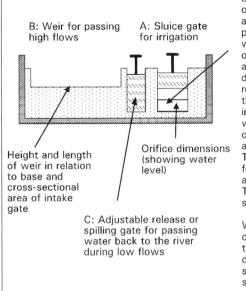

B: Weir for passing high flows

A: Sluice gate for irrigation

Height and length of weir in relation to base and cross-sectional area of intake gate

Orifice dimensions (showing water level)

C: Adjustable release or spilling gate for passing water back to the river during low flows

Dimensions and adjustability of the undershot orifice gate (or sluice gate) determine flow rates and the ability to adjust flows. When water passes over the weir (the water level is below the weir here) water undershoots through the gate opening (or orifice) of a known cross-sectional area. This flow is driven by a head (water level) difference, termed a discharge/head (Q/H) relationship. The head difference is affected by the water level over the weir and water level inside the orifice, as affected by the height of the weir lip and its length, and level of the base of the orifice and cross-sectional area of intake, the latter also adjusted by lowering or raising the sluice. These dimensions determine flow-through intake for a given height of water passing over the weir and set the *maximum capacity* of intake, Q max. This allows engineers to design gates to give a specific flow.

With this design, it is difficult to guess the flow division. Undershot orifice gates do not assist in transparency of water division because the discharge rate through the gate changes with several factors, and disproportionately so (gate setting and height of water over weir).

Fig. 14.1. Schematic of commonly used design for an improved intake.

proportional division of the incoming river flow into two outgoing flows – the intake flow and the downstream river flow.

Regarding the dimensions of the intake, as a process, upgrading follows standard procedures for irrigation infrastructure design – the selection and setting of the crop and irrigation system water requirement. This procedure and its rationale are explained in more detail in Lankford (2004b), but it can be summarized as a formulation of a fixed peak water supply to meet a given command area, crop type, climate and efficiency. The key point is that, without better recognition of the total and frequently changing catchment demand, this fixed peak amount becomes *physically embodied* as the maximum discharge rate of the intake, rendering future claims to adjust the share of water between the intake and downstream that more difficult.

As shown in the chapter, it is this maximum discharge design that overrides other considerations such as the amount of legal water right. The maximum discharge when orifices are fully open is one of the most important design parameters, because users tend to default to this setting – meaning that improved gates are normally

opened to their maximum. This is the reason that, when the rivers are in peak flow, intakes tend to take the maximum flow possible, and that in the dry season intakes can abstract all the available water. In theory and ideally, the legal water right should be the same as the maximum discharge (although frequently it is a different value) and, moreover, both should be adjustable in the light of new circumstances.

With respect to intake operability, in many cases problems have arisen, suggesting that this component of design is worth further scrutiny. Undershot orifice gates (see explanation in Fig. 14.1) obscure the ability to guess the proportionality of flow division – termed here transparency. Since in all cases water measurement is lacking (Gowing and Tarimo, 1994; Lankford, 2004a), the lack of proportional division (explained later in the chapter) makes it difficult for users to negotiate fairer shares of available water. The current gate and weir model is designed mainly for the wet season, allowing flood flows to be throttled back so that fields are not surcharged with excess water. However, such events are in the minority and, on the whole, headworks are largely unable to affect water management and efficiency *within* the irrigation system and therefore should not be designed with the wet season solely in mind. Instead, it is the pattern of water sharing and associated conflicts during the dry season that require more attention, because current structures allow the very small flows found in this season to be completely tapped without allowing any downstream flows (as well as being more likely to be washed away, according to custom, traditional intakes could not be built to block the whole river (Gillingham, 1999)).

We argue that, more than 'paper' water rights, it is the concrete and metal forms of irrigation intakes and the design process leading to them that determine the actual water taken throughout the hydrological year, affecting the share of water between irrigation and downstream sectors, and influencing how easy it is to adjust that share. Incorrectly assigned water rights that do not match intake capacities add complications. Upgrading of intakes and improvement of water control at the intake are commendable objectives, and are desired by farmers; however, it is the end purpose that should be rethought. As this chapter argues, there is a case for improving intakes so that they work more in harmony with water rights across both seasons within a dynamic catchment, rather than solely, in a rather static manner, for the irrigation system in question. If water rights are to be a key means to allocate water, and formal and informal rights are to be used together, then it is the design process of the mediating irrigation infrastructure that needs to be held to account.

Case study description

The Great Ruaha River basin is found in southern Tanzania (*see* Fig. 14.2). Previous articles, to which the reader is referred, describe in detail the geography of the area (Baur *et al.*, 2000; Franks *et al.*, 2004).

Some of the conditions relevant to this analysis of river basin initiatives are as follows:

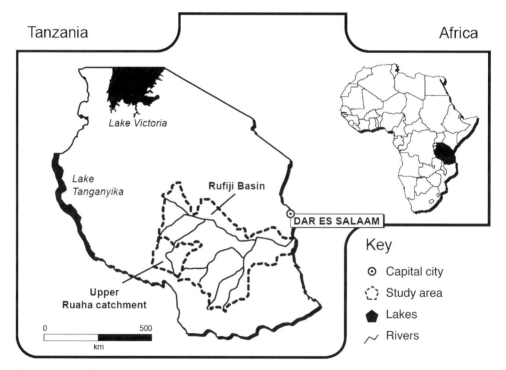

Fig. 14.2. Map of Tanzania.

1. The size of the sub-basin (68,000 km^2) poses logistical problems for managing water by formal rights alone that require monitoring and policing. To reduce these costs and to manage conflicts at the catchment scale requires robust forms of subsidiarity.

2. The basin effectively experiences a single rainy season (of about 600–1000 mm average depending on climate and altitude). Rivers swell during this period, but shrink dramatically during the dry season between May and November, a period that suffers from water stress and conflict. This considerable dissimilarity in water availability and associated dynamics suggests that wet and dry seasons need different forms of management and, in particular, the dry season necessitates special care.

3. The area lacks an aquifer or any large-scale storage that can support irrigation (although the downstream hydropower has storage). Irrigation has to rely on run-of-river supplies, and this points to the need to manage surface water resources carefully without the benefit of storage buffering.

4. There is competition between upstream irrigation and downstream; a RAMSAR wetland, the Ruaha National Park and hydropower. This competition exists in both wet and dry seasons, but not on the scale of the competition envisaged by RBMSIIP (Machibya *et al.*, 2003). In addition, the policy for the river – 'restoring the all-year-round flows'[5] – presents a goal by which river basin management can be tested. During a normal year, competition is found mainly during the dry season, arising from downstream needs for domestic use, animal watering and ecological functioning provided by in-stream flows, which support aquatic and terrestrial wildlife. Management during the dry season is affected by large wet-season abstractions that make it more difficult to throttle demand during the dry season. This, combined with the changeable climate that brings shortages during the wet season, means that water management is required throughout the year. Furthermore, the authors argue that purposive decisions over upstream–downstream allocation should replace the ad hoc unplanned change in distribution that has arisen within the last 30 years and that may continue in the future.

The challenges ahead

Reviewing the discussion above, we see that there are a number of concerns for water management in the basin:

- To build on the water rights currently provided so that they help achieve river basin objectives.
- To improve the system that caters to both the wet and dry seasons, and that manages the switch in water availability and demand between the two seasons.
- To draw up an arrangement that incorporates without incongruities both formal and customary agreements.
- The necessity of drawing together the water rights and the infrastructural works so that these match and, together, fit the hydrology, water demands and social make-up of the catchments in question.
- That the National Water Policy is implemented effectively, especially with regard to its institutional framework.

This chapter aims to answer these concerns and the call by Moss (2004, p. 87) for 'creating better fit' between institutions and other components, and is a contribution to the request in the National Water Policy (MOWLD, 2002, pp. 28–29): 'Thus the legislation needs to be reviewed in order to address the growing water management challenges.' It should be emphasized that this chapter does not propose an actual distribution of water but aims to show how available water might be shared between sectors. In addition, the framework described here is relevant in other closing and closed river basins, such as the Pangani in northern Tanzania.

Upstream–Downstream Water Allocation

Definitions and theory

Because irrigation is the major upstream water abstractor in the basin, it is the main determinant explaining the share of water between this sector and downstream sectors. Simply put, water for downstream (for domestic, livestock, fishing and wildlife purposes) is the remainder after irrigation abstraction has occurred (following the observation that return flows of drainage water are a minor proportion of abstracted flow or are accounted for). This relationship is captured in Fig. 14.3 and is explained here.

The abstraction flow-rate to feed a single irrigation scheme is a function of four factors (see Eqn 1): (i) the design of the intake capacity; (ii) the number of irrigation intakes feeding that system; (iii) any operation of these intakes that adjusts their discharge; and (iv) the flow of water in the river that affects the head of water at the intake. Intake design incorporates a discharge-head relationship between intake flow, orifice size and head of water at the weir so that for most intakes, without adjustment, intake flow increases as the river flow increases. As has been shown by Lankford (2004b), the intake rate is a function of river discharge rather than of response to changes in irrigated area or of crop water demand, except when intakes are throttled to safeguard fields from rare, extreme damaging floods (see Fig. 14.1 for a further brief explanation of how standard intakes and weirs work and, in addition, web sites or text books on canal irrigation engineering provide additional information, e.g. Kay, 1986).

Q (single irrigation system) = f (intake design, intake number, intake operation, flow in river)   (1)

Where Q is discharge expressed as a volume of water per time unit (e.g. l/s). By simple mathematical balance, the flow for downstream irrigators is the remainder of the river flow once upstream intake abstraction has occurred (see Eqn 2 and Figs 14.3 and 14.4).

Q (individual downstream irrigator intake) = (Q river supply – Q upstream intake)   (2)

The flow of water being abstracted into the whole irrigation sector (a summation of all intakes within a catchment) is a result of the river supply and the total intake capacity combined with any cumulative effect of operational decisions (Eqn 3):

Q (total irrigation) = f (all intakes design, number of intakes, cumulative operation, river flow)   (3)

When applied to 'between sector' computations (Eqn 4), it is the cumulative upstream irrigation abstraction in a catchment that determines the water available for downstream users:

Q (downstream) = (Q river supply – Q total upstream irrigation intake)   (4)

Over 1 year, abstraction fluctuates as a result of the four factors (intake design, intake number, intake operation, supply in river), creating an abstraction hydrograph (see Fig. 14.4), which follows the river supply hydrograph with greatest abstraction during the wet season and lower abstraction in the dry season. Via mathematical continuity, the downstream hydrograph will be a function of the upstream irrigation abstraction. Figure 14.4 is further explained in the discussion below on volumetric and proportional caps. We can now determine a simple indicator of river basin management, the 'irrigation allocation ratio' (IAR) of irrigation abstraction to total supply (Eqn 5), a measure of the equity of distribution between irrigation and other sectors. A proportion of about 50% indicates that water is evenly divided between irrigation and other sectors, while an IAR of 90% tells of a highly skewed supply to irrigation.

Irrigation allocation ratio, IAR = (irrigation abstraction)/(upstream supply flow)   (5)

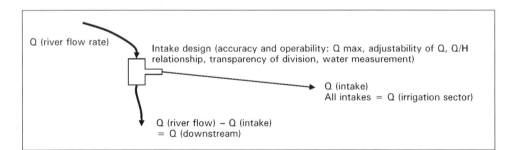

Fig. 14.3. Irrigation abstractions establishing downstream allocation.

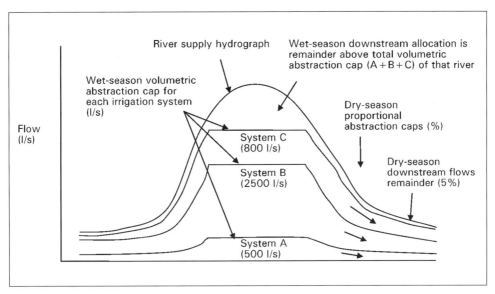

Fig. 14.4. Supply, abstraction and allocation hydrographs.

Introduction to volumetric and proportional caps

To manage the irrigation allocation ratio in Eqn 5 requires an understanding of volumetric and proportional caps. Figures 14.4 and 14.5 and the worked example below show how setting two types of 'caps' (equivalent to 'ceilings' 'maxima' or 'thresholds') affects the irrigation allocation ratio (IAR). As explained in the next section on legal–infrastructural framework for catchment apportionment (LIFCA), these two caps relate closely to the properties of intake structures and to the season.

The volumetric cap is determined by the maximum volumetric capacity of the intake, or 'Q max'. This cap, it is argued, applies during the main part of the wet season when river flows are larger. Figure 14.4 shows this as a *fixed plateau* on each intake hydrograph where the maximum intake capacity stops more water from being abstracted. Note that the height of the cap plateau is only set from the zero on the Y axis for the first intake but, for the others, the level is set by counting up from the previous intake plateau. Figure 14.4 is a stylized rather than an exact representation of the worked example given below. The volume of the water for downstream during the wet season is the area of the graph between the river hydrograph and the uppermost plateau of intake C.

The proportional cap is determined by the design features of the intake that function when the river flow is lower than Q max. These design features are discussed in greater detail below. More to the point, proportional caps in Fig. 14.4 can be seen as sloping lines, denoting a *constant fraction* (but reducing quantity) being apportioned to intakes A, B and C. The volume of the water for downstream during the dry season is the area of the graph between the river supply hydrograph and the sloping line of intake C.

Worked example

A worked example in Table 14.1 demonstrates the effects that adjusting volumetric and proportional caps have on water apportionment in a catchment (see also Figs 14.4 and 14.5). Three intakes feeding irrigation systems, A, B and C, are located in a single catchment. The current design allows a maximum of 500, 2500 and 800 l/s, respectively, giving a total abstraction of 3800 l/s. During the dry season when this flow is not exceeded, the share for A, B and C is 15, 50 and 30%, respectively, providing 5% for downstream sectors. Under a new modified arrange-

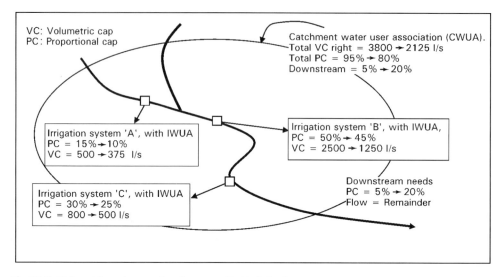

Fig. 14.5. Volumetric and proportional caps applied to irrigation systems.

Table 14.1. Existing and modified settings for volumetric and proportional caps (worked example).

Case	Volumetric cap units (m³)		Proportional cap units (%)	
	1 (Wet season, existing)	2 (Wet season, modified)	3 (Dry season, existing)	4 (Dry season, modified)
Irrigation system A cap	0.5	0.375	15	10
Irrigation system B cap	2.5	1.250	50	45
Irrigation system C cap	0.8	0.500	30	25
Volumetric (cumecs) and proportional cap (%) for irrigation	3.8	2.125	95	80
Remainder for downstream	Remainder	Remainder	5	20

ment, the volumetric caps are reduced, leaving 375, 1250 and 500 l/s, respectively (giving a total catchment right of 2125 l/s) and, when water does not exceed this volume, the proportional shares of A, B and C are also reduced to 10, 45 and 25%, respectively, providing more water (20%) downstream.

The volumetric outcome can be seen in calculations given by Lankford and Mwaruvanda (2005). In annual volumetric terms, the amount of water diverted for irrigation decreased by 29,352 MCM (million cubic metres), from 75,062 to 45,710 MCM, a drop of 39%, giving an extra 29,352 MCM to downstream. Calculation of the irrigation allocation ratio (IAR) shows that the revised caps decreased irrigation impact on the hydrology of the catchment from 56 to 36%. Furthermore, the downstream share benefited considerably from only slight reductions in each irrigation system's abstraction. This is particularly notable in the dry season and was a result of the relatively low starting fraction given to downstream needs, combined with three intakes releasing water. Each intake needed to give a 5–10% compensation to result in 15–30% total extra water flowing downstream.

Application to the Great Ruaha River basin

In 2000, SMUWC found that about 45 cumecs were the maximum total intake capacity of irrigation. Once this was exceeded, water went to the Usangu wetland, the Ruaha National Park and hydropower stations. In the future, the total intake capacity could be revised to manage the balance of water heading downstream. This means bringing in a new volumetric cap, which might be determined on the basis of observations and modelling, and might be set at 50 cumecs – which amount was the estimate of abstraction for the year 2005. During the dry season, the improvement of intakes in the last 25 years in the Great Ruaha basin has resulted in some taking all of the dry-season flow. From observations (SMUWC, 2000), the proportional cap in the dry season was about 90–100% – in other words, until the abstraction capacity was exceeded by flood water, nearly all the water was taken by irrigation in those catchments with irrigation. In the future, should LIFCA be applied, catchments would have their dry-season irrigation abstraction altered according to local circumstances, perhaps ranging from 70 to 90%.

Legal–Infrastructural Framework for Catchment Apportionment

Introduction

Having discussed concepts underlying the allocation of water in a catchment, it is possible to propose a synergistic framework of water management, design and legislative dimensions. This LIFCA is presented in Table 14.2. Each column represents either the wet or dry season. For each season a water management arrangement is proposed. This multi-layered arrangement coheres with: (i) the type of water threshold decision to be made (volumetric or proportional); (ii) the design of the maximum capacity; (iii) the operability of intakes; (iv) the type of property right (formal or informal); (v) the level of stakeholder decision making (river or irrigation user association); and (vi) the nature of water payment made. LIFCA is described first before detailing the technology required to support the framework.

Following Table 14.2, in the wet season, to distribute water between irrigation and downstream sectors, first, a maximum cap on abstraction is required. This cap is physically designed in by sizing the maximum apertures of the intakes so that no more water than this cap can be abstracted. This cap is underpinned by the formal water rights sold by the government

(requiring the current system of volumetric water rights to be improved so that this cap is set accurately and legally). In turn, the legal right relates to either individual water user associations that represent irrigation systems or to the catchment water user associations (CWUAs) that represent the irrigation sector within that catchment. If the latter occurs, then the CWUA can divide up the volumetric right to its various constituent intakes, represented by irrigation water user associations (IWUAs). Either way, the individual intake or total intake capacity should be expressly and accurately related to the formal rights and managed both at the individual and catchment level. Water basin officials would then be interacting with representatives of both individual intakes and the whole catchment iteratively to ensure coherence between these water volumes.

In the dry season, (see Table 14.2), arrangements switch because the designed-in maximum capacity for abstraction is now above the river supply; thus, the meagre river supply needs sharing between irrigators, and between irrigation and downstream sectors. This requires a maximum threshold on the share provided to irrigation. This allocation is more likely to be implemented by the regulation (partial throttling) of gated adjustable intakes but would benefit from being 'designed in' using proportional weir-type structures (see next section for further information on different types of structures). Since the 'rights' to these dry-season flows are below the flow rates set by the formal rights, the dry-season shares (or 'rights') have to be negotiated informally as customary rights between all users in and below the catchment and then backed up by a mixture of intake design and adjustment. These latter rights would have to be articulated, not in the form of flow rates (l/s) but as proportions of the water, for example 'an intake would receive 20% of the river flow water'.

The role of the river basin official would change in the dry season when the formal rights are no longer 'active'. Greater emphasis would be placed on conflict-resolution services to assist the CWUA in sharing more equitably the available water, altering the proportions of water according to changing circumstances or encouraging stakeholders to permit more water to remain for downstream environmental and domestic flows.

Table 14.2. LIFCA – a framework of seasons, caps, intake design, rights and WUAs.

	Wet season	Dry season
Type of cap	Total volumetric abstraction cap, the flow rate in l/s or cumecs (m³/s) abstracted	Proportional abstraction cap, in proportions (%) of flow abstracted
Intake design required	A proportional intake design can accommodate both wet-season volumetric caps and dry-season proportional caps. Proportional intakes can be designed to have a maximum capacity, beyond which no extra water is tapped and below which water is abstracted proportionally; a well-designed intake can be adjustable and transparent, assisting users in knowing the division of water between the intake and downstream	
Part of intake design most closely associated with this season and cap	The design of the maximum capacity of the intake is a critical step, and will generally establish the maximum volumetric cap. Accuracy of design sizing is important here, as is future adjustability if maximum intake capacity is also to be adjusted in the future. Maximum flow (Q), determined from flume dimensions and main canal specifications. Excess flow can be returned to the river. (Q max focus = l/s)	Design can be used to implement proportional divisions (using fixed or adjustable proportional gates). Accuracy of sizing the proportional division important. In addition, design allows on-off shutters for time schedules (focus = % of division)
Part of intake operability most closely associated with this season and cap	Advised to rely on Q max rather than on throttling because gates are opened to maximum setting. Thus the accurate design of the maximum abstraction is very important (see row above). Although Q max will be a flow rate, users should be able to discern the division of water, and therefore transparency will also be key	Incremental adjustments of intakes or on–off adjustments to schedule water are advised. Further alteration of the intakes may be necessary to reflect ongoing negotiations, but if fixed proportional dividers are well designed this need not be a regular or onerous activity. Adjustability and transparency required here
Type of rights most closely associated with this season	Formal water right (volumetric). The design of the maximum flow through the intake matches the flow rate of the water right. In addition, the total intake capacity of all irrigation in the catchment matches the total rights disbursed to the catchment	Customary agreements and rights (proportional, or time schedule basis). These are expressed and negotiated in terms of shares (e.g. 40% of the available water) or time (e.g. 2 days for taking the total supply) or a mixture of the two
Role of irrigation water user association (IWUA) or catchment water user association (CWUA)	Water right to CWUA and division of right to irrigation WUA (IWUA) representatives	Division of river supply agreed between users or irrigation WUAs and agreed apportionment of water between total irrigation and downstream users
Institutional connections	Basin Office to facilitate and mediate catchment water user association negotiations of the total formal water right	Intake to intake representatives of irrigation water user associations plus RBWO mediation explore customary water rights
Payment structure	Fixed payment for formal water right	No payment envisaged for proportional share, though might be possible

With regard to payments for water, in the current legislation, payments for the water right are pegged to the allocated amount rather than to the actual measured amount. This same arrangement could be applied to this framework, which therefore does not, at least in the initial stages, envisage a volumetric basis to determine a water charge, although this would be a future goal that various stakeholders might wish to explore. A more efficient and appropriate step would be to ensure that maximum intake capacity (max Q) is the same as the water right (either for an individual intake or for the whole catchment) so that payment, the right and the maximum amount that could be taken are the same. Following this, it would be necessary only for occasional flow measurement or for stakeholders to report unsanctioned changes to the amount abstracted. The agreements over the dry-season shares do not involve financial transactions, but result from discussions held within the catchment users' organization, mediated by the basin authority.

In summary, the framework can be expressed within five objectives:

- To match formal water rights with maximum water flows abstracted, at both the intake and catchment level so that the volumetric cap is built in.
- To make the gate design facilitate water sharing during the dry season when flows are meagre, to match customary water rights and build in the proportional cap.
- To bring adjustability and flexibility so that users may frequently adjust flows and turn them on and off.
- To enhance transparency so that users may know how the flows are being divided, either volumetrically or proportionally.
- To empower local users in managing water at the catchment level, including building and adjusting intakes that meet their requirements and wider, downstream allocation objectives. It is proposed that the framework would function best when all five objectives are brought simultaneously together in a coordinated fashion.

Infrastructural design to support LIFCA

As proposed above, because irrigation is upstream of other demands on the plains of Usangu, it is the presence and type of irrigation intake structures that 'hard-wire' in the apportionment of water and its adjustment. The discussion here explores how this infrastructure, particularly proportional intakes, might solve the five objectives of the LIFCA. To alter water apportionment (or IAR) in both wet and dry seasons requires three parts or functions of intake design and operation to be understood. These are accuracy, operability and operation. All three parts work simultaneously and interrelate and, when carefully considered, support the objectives encapsulated in LIFCA. To meet these objectives, an intake or series of intakes would be accurately sized, fully adjustable, highly transparent and well understood by local users. To explain how the appropriate design of the three parts embody the objectives of LIFCA, the reader is referred again to Fig. 14.1 for the common but problematic design used in current improvement programs, and to Figs 14.6 to 14.8, showing a selection of proposed designs of proportional intakes that better encapsulate LIFCA objectives.

Intake accuracy

The first part or function is to 'build in' accurate intake dimensions so that the size of the intakes assists in two ways. First, having an accurate 'Q max' means that the maximum flow rate closely equals the water right and matches the volumetric cap. This can be achieved for both individual intakes and by adding up all intakes in a catchment, the total cap for irrigation abstraction. Second, the accurate dimensions of the proportional ratios of the cross-sectional areas of the proportional flumes then match informal water rights proposed by catchment stakeholders and, in combination with other proportional intakes, accurately set the total proportional cap of water abstracted by irrigation during the dry season.

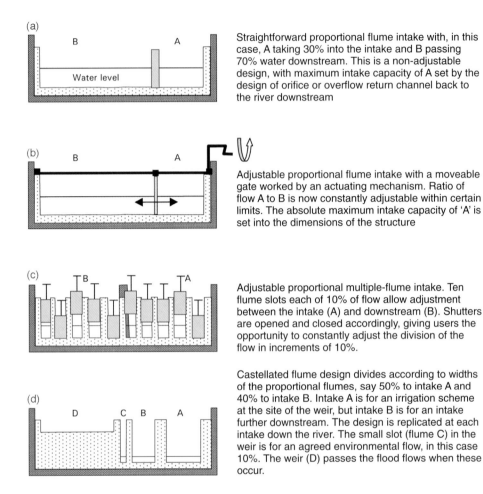

(a)

Straightforward proportional flume intake with, in this case, A taking 30% into the intake and B passing 70% water downstream. This is a non-adjustable design, with maximum intake capacity of A set by the design of orifice or overflow return channel back to the river downstream

(b)

Adjustable proportional flume intake with a moveable gate worked by an actuating mechanism. Ratio of flow A to B is now constantly adjustable within certain limits. The absolute maximum intake capacity of 'A' is set into the dimensions of the structure

(c)

Adjustable proportional multiple-flume intake. Ten flume slots each of 10% of flow allow adjustment between the intake (A) and downstream (B). Shutters are opened and closed accordingly, giving users the opportunity to constantly adjust the division of the flow in increments of 10%.

(d)

Castellated flume design divides according to widths of the proportional flumes, say 50% to intake A and 40% to intake B. Intake A is for an irrigation scheme at the site of the weir, but intake B is for an intake further downstream. The design is replicated at each intake down the river. The small slot (flume C) in the weir is for an agreed environmental flow, in this case 10%. The weir (D) passes the flood flows when these occur.

Fig. 14.6 (a to d). Proportional flume intakes; various designs.

Intake operability

The second part is to build in better *operability* of allocation so that intelligent gate adjustments can be made. Operability depends on three factors: the *adjustability*, *water measurement* and *transparency* of the gate flows. 'Adjustability' is designed by considering how the gate orifice (opening) can be set at partial settings and how any head-controlling structure such as a weir can also be adjusted. The actual operation of this intake structure is then the adjustment of the intake flow either by closing and opening the orifice gate, or by increasing or decreasing the height of any weir structure. It is this adjustability that also explains why, in the wet season, farmers will throttle down their intake when very high floods threaten their systems and why, in the dry season, negotiations between upstream and downstream farmers can be physically transformed into gate adjustments that release water downstream. The current improved gate technology chosen in Usangu does enable flows to be adjusted (see Fig. 14.1), but the same technology is not very transparent for reasons described below.

Without much difficulty, as seen in Figs 14.6b, c and 14.7, proportional gates can be made adjustable. The adjustment of the cross-sectional area ratio between A and B is either actuated by a constantly moveable dividing plate (see Fig. 14.6b) or by an array of on–off shutters giving incremental steps (see Fig. 14.6c). With respect to the fifth LIFCA objective

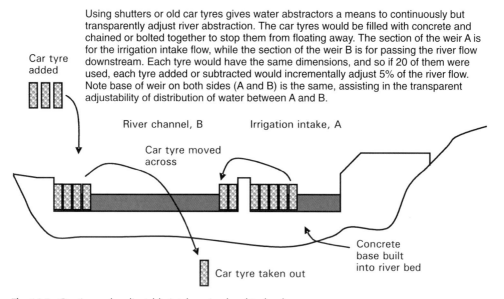

Using shutters or old car tyres gives water abstractors a means to continuously but transparently adjust river abstraction. The car tyres would be filled with concrete and chained or bolted together to stop them from floating away. The section of the weir A is for the irrigation intake flow, while the section of the weir B is for passing the river flow downstream. Each tyre would have the same dimensions, and so if 20 of them were used, each tyre added or subtracted would incrementally adjust 5% of the river flow. Note base of weir on both sides (A and B) is the same, assisting in the transparent adjustability of distribution of water between A and B.

Car tyre added

River channel, B   Irrigation intake, A

Car tyre moved across

Concrete base built into river bed

Car tyre taken out

Fig. 14.7. Continuously adjustable intake using local technology.

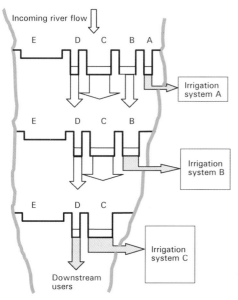

Incoming river flow

E   D   C   B   A

Irrigation system A

E   D   C   B

Irrigation system B

E   D   C

Irrigation system C

Downstream users

A castellated weir design is replicated down the river at each intake. There are three irrigation systems to be supplied, A, B and C from three flumes, A, B and C. There is also a small flume (D) for downstream users (e.g. cattle keepers) and a flood weir to pass high flows (E).

Each system receives a flow in proportion to its cropped area or other negotiated agreement, which in turn gives rise to the designed-in ratio of cross-sectional areas of the flumes apportioned to A, B, C and D. Regardless of the incoming flow (except in flood periods), the four flumes divide a percentage of the flow consistently, which in this case might be A, 20%, B, 30%, C, 40% and D, 10%.

The benefit of castellated weirs is enhanced transparency of water apportionment; irrigators from systems C and B can walk up the river to the first weir and observe that their water is coming through their particular flume without being tampered with.

The flumes' cross-sectional areas could be fixed or adjustable.

Fig. 14.8. Castellated weirs for proportional distribution of water.

of local water management, Fig. 14.7, showing a 'local technology' concept, is worth explaining here. It is an example of incremental steps embodied by the movement of old car tyres into and out of the two parts of the weir, A and B. The use of car tyres here is conjectural and is proposed as a potential example of how stake-

holders might use local artisans, discussions and material in arriving at a satisfactory structure. The key point with the car tyre concept is that the *functions* of flexible intakes can be captured (accuracy, operability and transparency) without being fixated on *form* – the use of old tyres rather than concrete and metal.

The car tyre intake specifically endorses local construction, knowledge and ongoing adjustments of water allocation to match rapidly changing catchment circumstances.

However, stakeholders might decline adjustability, so that the intakes are more 'tamper-proof'. Proportional intakes can be fixed (see Figs 14.6a, d and 14.8). Here, the dimensions of the flumes would be agreed with representatives of the irrigation and downstream users – but it is possible that such fixed dimensions may in time not represent the share of the demand due to ongoing growth of water use in the catchment.

'Water measurement' is made possible by having specialized structures, such as gauging plates and recorders. Measurement is a large topic (see Kraatz and Mahajan, 1975), and is not discussed or proposed here because it is seen by the authors as a future option, complicated, currently lacking in Usangu and not immediately integral to the functioning of LIFCA. In the future, however, as concerns over water escalate, local users may end up requesting water-measurement structures to arbitrate in disputes. Robust and simple water-flow measurement is possible in such cases so that users can compare each other's shares. However, enhancing transparency of comparison would be a satisfactory precursor or alternative to water measurement.

'Transparency' is supported by having the dimensions of the weir and intake relate directly to the proportion of flow division. Transparency must be considered in the absence of water measurement so that the intake dimensions (adjustable or not) and their resulting discharge outcomes are closely connected and transparent. Transparency of water division is part of gate operability, even if this simply supports, in the absence of possible adjustment, observation of the water division between the intake and downstream river. This is because visual clues should be given to the operator that his or her adjustment results in an increase or decrease of the intake flow by a given and knowable amount. Transparency negates the need for water measurement, but brings intelligent purposive operation of the intake.

The current design of intakes (see Fig. 14.1) obscures knowledge of flow division, because the incoming river flow is not simply divided between two flows. The intake divides the flow, with one flow going over a weir which is long and high up and the other going through an orifice which is set lower down. Also, incremental adjustments to the sluice gate do not bring pro rata changes in flow; the changing head difference and changing cross-sectional area combine to bring unpredictable flow changes (hence the need to have a gauging plate with such structures, a device that is missing from nearly all intakes in Usangu). In contrast, the designs in Figs 14.6 to 14.8 employ 'flume'-type gates (a flume is a small channel with two parallel straight sides and is open at the top in contrast to orifices that, by definition, are enclosed either as round or square orifices).

Constructed carefully, flume gates bring enhanced proportionality and transparency because the open top of the flume, combined with its straight sides and equal base height with the weir, make the relationship between water height (H) and water discharge (Q) more linear – that is, with an increase in water depth comes an increase in discharge. For any given incoming river flow the division between flow into the intake (given as 'A') and the flow passing downstream ('B') is a function of the ratio of cross-sectional area between A and B; it is this proportion between the area of A and B and the simplicity of the division design that provides the advantageous visual clues and feedback to the irrigators.

In Fig. 14.7, transparency is assured by using car tyres of the same dimensions, so that each tyre acts to block or open a set and known fraction of the total weir length across the intake and river sections (A and B). In Fig. 14.8, we explore another level of transparency provided via a novel arrangement of proportional flumes, termed 'castellated weirs' (whose details are given in Fig. 14.6d). Each castellated weir is replicated down the river, so that users from both upstream and downstream areas may come together at any given weir to observe that their particular portion of water is being passed down without being obstructed. More description of the weir system is given in Fig. 14.8.

Intake operation

The third factor, arising out of accuracy and operability, is to rely on the *operation* of intakes by water users to frequently regulate any infrastructure so that these adjustments generate the intended outcomes. Regulation involves adjusting and closing intakes so that downstream flows are altered, or open and closed completely so that flows are scheduled against time windows (e.g. '3 days all river flow for this intake, 3 days for the next intake and 1 day sending the river flow downstream'). Scheduling, over a given time period, is therefore another way upstream or downstream users share the available water, and is an alternative expression of a customary right.

In addition, operation of the existing undershot orifice gates is another way of adjusting the dry-season proportion taken by irrigation so that, rather than have proportional gates achieve this, users manipulate the sluice gate to arrive at mutually agreed divisions of water. The aim of LIFCA, however, is to have these shares built in and more transparent by employing different designs of intakes. The nature of the operation (opening and closing of shutters, turning a valve or moving car tyres) is an outcome of the design process of the operability and accuracy of the gate, and hence it is the latter that needs careful thought if operation is to support LIFCA. The foremost aim of an improved design process is to give all catchment users, not just upstream irrigators, intakes that meet the five objectives expressed in LIFCA.

Discussion

Although theoretically the framework resolves the contradictions of how formal and informal rights can operate together by splitting them into different seasons, in reality this may present some problems. It is difficult to foresee all complications, but some are identified here.

The setting of thresholds

Setting the caps will inevitably create winners and losers, as shares increase for some and drop for others. The process by which the caps are set would benefit from being participative and informed by good-quality hydrology and observations of current patterns of water use. Incremental adjustments might be advisable

during different parts of the river hydrograph; indeed, for the very lowest and driest part of the year, local users might agree that all water should be kept in the river with only domestic (rather than productive) quantities tapped.

Sharp-eyed readers will have noticed that, by definition, the wet season begins once the total abstraction capacity of all intakes in the catchment has been exceeded by river flows, and that a dry season is that time period when the river is lower than this threshold. The dry season is, by definition, the period when the river flow no longer exceeds intake capacity, and that negotiated customary agreements need to interject. This can be realized by setting conditions with the rights that recognize these negotiations. These definitions do not follow other ways of naming the two seasons (start of rains, based on long-term records or related to other farming activities). It follows that, the higher the abstraction capacity the shorter the wet season, until the point where total abstraction might grow to exceed all but the highest peak flows, in which case throttling and adjustment are necessary nearly all the time. Clearly, the thresholds and resulting design modifications have to be set so that expectations of irrigators and other sectors match the hydrology and climate of the area. Other ways of adjusting the caps to take into account varying flows from one year to another can be built into the intakes, with the maximum cap being adjusted by the addition or subtraction of a special shutter or plate to the intake gate.

Information transparency

The test of the arrangement will be the switch from the wet season to the dry season, a transition period of care and attention. The switch will not happen automatically – though it could be very much assisted by a combination of appropriate intake infrastructure, sharing information up and down the catchment and, in the future, water measurement. Problems might arise when a river flow has exceeded the capacity of the uppermost intake but not the capacity of all the intakes combined. The upper irrigators will probably feel, on observing 'good flows', that it is their right to tap this water with their gate set at maximum, even though this will skew their proportion above that agreed. Key to this transition, and to the management of the arrangement as a whole, will be water measurement or transparent water division (structures that split water without the need for measurement).

Allowing flexibility and change

It would be mistaken to impose this arrangement on water users without allowing them to bring their own ideas and suggestions (even rejecting it!). Each catchment has its own properties and dynamics, necessitating a flexible, situational response. In addition, the system should be allowed to change over time responding to shifts in demand, problems arising and possibly changes in supply. It is possible that in the future, the volumetric and proportional caps might be traded between intakes and sectors, a facility now recognized in the new water legislation, expressed as tradable water rights. Flexibility is a key part of the framework, acknowledging how rapidly both the demand and supply of water have changed in the recent past and may continue to change in the future.

Institutional ownership and sustainability

It would be a truism to argue that the arrangement would depend on all stakeholders meaningfully agreeing to the constraints and benefits imposed by it. However, some significant factors that promote institutional sustainability might be:

- The four concerns above (process of setting thresholds, information needs, the role of design, allowing flexibility) are important.
- The river basin office would need to focus on delivering a variety of services, including conflict resolution, resetting the caps and ensuring follow-up modifications to infrastructure.
- The chapter has focused on the question of 'share management' rather than 'supply management' (in the usual sense of augmenting supply), or 'demand management' (persuading farmers to be more efficient so that intake flows can be reduced).

Although demand and supply management are often connected, the success of managing shares via abstraction flow reduction for a particular user would depend on whether their productivity of water can be raised, which research in the area suggests it can (Mdemu *et al.* 2003).

Retuning river basin infrastructure

Central to the success of the framework is a commitment to revising the existing intake infrastructure in each catchment. Many objectives of the Legal Infrastructure Framework for Catchment Apportionment would not work without intake infrastructure being rethought. A redesign programme could, in promoting the manageability of river basin allocation via the framework, draw on an extensive literature based on irrigation designs (e.g. Yoder, 1994).

Moreover, intake design should move from being the domain of irrigation engineers to being a negotiated process with and by local representatives of the total catchment. Each individual intake would have to be designed so that it relates iteratively to a number of factors: area of irrigation, crop types, renegotiated shares, population density and so on. Deriving irrigation intake designs on the basis of crop water requirements would appear to be an anachronistic methodology in this highly dynamic multi-user environment (Lankford, 2004b). Being able to adjust the maximum cap to account for hydrological and demand-side changes would benefit the workability of the arrangement and fit the principle of continuous and flexible adjustment that this framework is built upon.

Conclusions

This chapter shows how two decisions – setting the maximum volumetric cap and maximum proportional cap – embodied in flexible intake designs determine the allocation of water in a river basin characterized by an order of abstraction and the presence of wet and dry seasons. These decisions allow us to think of ways how (if irrigation is upstream of wetlands and hydroelectric plants) irrigation abstractions could be managed and modified by both intake design and operation. Moreover, this analysis provides possible means to rationalize the interface between formal water rights (that establish and relate to the volumetric cap) and customary agreements (that relate to negotiations over shares of the in-stream water). These coordinated arrangements are termed here 'legal–infrastructural framework for catchment apportionment' (LIFCA). Thus, with respect to the latter, this chapter demonstrates how, if strengthened and supported, local customary negotiations combined with water management interventions, might help set and relate to the proportional cap of water abstraction that applies during the dry season.

Furthermore, this chapter argues that the design of irrigation intakes, in terms of maximum capacity, adjustability and transparent proportional capability, needs to be revisited and retuned so that the intakes fit and help support any newly modified caps and their associated sharing arrangements. At the heart of this framework is the belief that intakes should be designed to encourage and facilitate the continuous negotiation of intake settings so that their iterative and frequent adjustment is an ongoing part of water allocation at the catchment and basin scales.

These conditions, which invoke this framework as an option, are found in the wider Rufiji basin, and in parts of the Pangani basin. The latter also suffers from considerable conflicts that have arisen due not only to increasing demand but also to the imposition of a formal water rights structure that has yet to be further refined. Although one option is given here, various possibilities include managing the status quo, an outright return to customary rights, constructing storage or building in volumetric water measurement to charge for water used. Substantively, the authors therefore call for further discussions on how a more equitable allocation is to be effected and made relevant to the issues found at the catchment scale. We believe that solutions to water shortages in a sub-Saharan Africa affected by climate change and population growth cannot be met only by storage or institutional reform, but by combining those synergistically with the apportionment infrastructure to foster catchment manageability.

Acknowledgements

This chapter is an output of the RIPARWIN research project (Raising Irrigation Productivity and Releasing Water for Intersectoral Needs), funded by DFID-KAR (R8064) and the International Water Management Institute (South Africa Office). It is co-managed by the Overseas Development Group (ODG, University of East Anglia, UK), the Soil Water Management Research Group (SWMRG, Sokoine University of Agriculture, Tanzania) and the International Water Management Institute. The authors gratefully acknowledge the discussions held and inputs made by N. Hatibu, H. Mahoo, S. Tumbo, D. Merrey, B. van Koppen, H. Sally, H. Lévite, M. McCartney, D. Yawson, C. Sokile, K. Rajabu, R. Kadigi, J. Kashaigili, J. Cour, M. Magayane and R. Masha. The authors would particularly like to thank reviewers of the earlier chapter for their comments.

Endnotes

1 SMUWC – Sustainable Management of the Usangu Wetland and its Catchment, a natural resources research and development project funded by DFID during 1999–2001.
2 RIPARWIN – Raising Irrigation Productivity and Releasing Water for Intersectoral Needs, a research project funded by DFID KAR during 2001–2005.
3 In the Great Ruaha basin, the term actually used is 'sub-catchment' but, for the sake of simplicity, 'catchment' is used here.
4 A third activity is the monitoring of river flows in selected sites using automatic gauging stations, although some of these are now non-functional. Although this is a vital part of river basin management, such measurements are not related to demand or management of water, and consequently users have no stake in this information being collected and distributed.
5 Up to 1993/94, the Great Ruaha was a perennial river flowing through the Ruaha National Park. Since that date, the river has dried up for between 2 and 8 weeks each year during the tail end of the dry season. The main explanation for this is continuing abstraction into irrigation intakes for a variety of productive, domestic and non-productive purposes. RIPARWIN and RBWO (and other stakeholders) share a common vision of water

distribution, which can be distilled down to the need to return the Ruaha river to year-round flow by 2010. This directly relates to the statement by the Prime Minister of Tanzania, Frederick Sumaye, in London (6 March 2001), made with UK Prime Minister Blair for the Rio+10 Summit: 'I am delighted to announce that the Government of Tanzania is committing its support for a programme to ensure that the Great Ruaha River has a year-round flow by 2010. The programme broadly aims at integrating comprehensive approaches towards resources planning, development and management so that human activity does not endanger the sustenance of the Great Ruaha ecosystems.' Achieving year-round flow would be, from a number of perspectives, a marker of success in achieving integrated water management in the basin.

References

Baur, P., Mandeville, N., Lankford, B. and Boake, R. (2000) Upstream/downstream competition for water in the Usangu basin, Tanzania. *Seventh National Hydrology Symposium, BHS National Hydrology Symposium Series*, University of Newcastle, UK. British Hydrological Society, London.

Franks, T., Lankford, B.A. and Mdemu, M. (2004) Managing water amongst competing uses: the Usangu wetland in Tanzania. *Irrigation and Drainage* 53, 1–10.

Garduno, H.V. (2001) *Water Rights Administration, Experiences, Issues and Guidelines.* FAO Legislative Study 70, Food and Agriculture Organization of the United Nations, Rome, pp. 17–19.

Gillingham, M.E. (1999) *Community Management of Irrigation.* Working Paper for SMUWC, DFID Project: 'Sustainable management of the Usangu Plains and its catchment', Ministry of Water and Livestock Development, Dar es Salaam, Tanzania.

Gowing, J.W. and Tarimo, A.K.P.R. (eds) (1994) Influence of technology and management on the performance of traditional and modernised irrigation schemes in Tanzania. *Proceedings of XII World Congress on Agricultural Engineering*, International Commission of Agricultural Engineering (CIGR), Milan, Italy, August–September 1994, pp. 360–368.

Hazelwood, A. and Livingstone, I. (1978) *The Development Potential of the Usangu Plains of Tanzania.* Commonwealth Fund for Technical Co-operation, The Commonwealth Secretariat, London.

Kay, M. (1986) *Surface Irrigation: Systems and Practice.* Cranfield Press, Cranfield, UK.

Kraatz, D.B. and Mahajan, I.K. (1975) *Small Hydraulic Structures.* FAO Irrigation and Drainage Papers Nos 26/1 and 26/2, Food and Agriculture Organization, Rome.

Lankford, B.A. (2004a) Irrigation improvement projects in Tanzania; scale impacts and policy implications. *Water Policy* 6 (2), 89–102.

Lankford, B.A. (2004b) Resource-centred thinking in river basins: should we revoke the crop water approach to irrigation planning? *Agricultural Water Management* 68 (1), 33–46.

Lankford, B.A. and Mwaruvanda, W. (2005) A framework to integrate formal and informal water rights in river basin management. In: *International Workshop on 'African Water Laws: Plural Legislative Frameworks for Rural Water Management in Africa'*, 26–28 January 2005, Gauteng, South Africa.

Lankford, B.A., van Koppen, B., Franks, T. and Mahoo, H. (2004) Entrenched views or insufficient science? Contested causes and solutions of water allocation; insights from the Great Ruaha River Basin, Tanzania. *Agricultural Water Management* 69 (2), 135–153.

Machibya, M., Lankford, B. and Mahoo, H. (2003) Real or imagined water competition? The case of rice irrigation in the Usangu basin and Mtera/Kidatu hydropower. In: *Tanzania Hydro-Africa Conference*, Arusha, Tanzania, 17–19 November 2003.

Maganga, F.P. (2003) Incorporating customary laws in implementation of IWRM: some insights from Rufiji River Basin, Tanzania. *Physics and Chemistry of the Earth, Parts A/B/C*, 28 (20–27), 995–1000.

Maganga, F.P., Kiwasila, H.L., Juma, I.H. and Butterworth, J.A. (2003) Implications of customary norms and laws for implementing IWRM: findings from Pangani and Rufiji basins, Tanzania, *4th WaterNet/Warfsa Symposium: Water, Science, Technology and Policy Convergence and Action by All*, 15–17 October 2003. WaterNet. (http://www.waternetonline.ihe.nl/aboutWN/4thProceedings.htm).

Mdemu, M.V., Magayane, M.D., Lankford, B.A., Hatibu, N. and Kadigi, R.M.J. (2003) Conjoining rainfall and irrigation seasonality to enhance productivity of water in large rice irrigated farms in the Upper Ruaha river basin, Tanzania. *WARFSA/WATERnet Symposium Proceedings* 15–17 October 2003, Gabarone, Botswana, pp. 97–101.

Meinzen-Dick, R. and Bakker, M. (2001) Water rights and multiple water uses: framework and application to Kirindi Oya irrigation system, Sri Lanka. *Irrigation and Drainage Systems* 15, 129–148.

Moss, T. (2004) The governance of land use in river basins: prospects for overcoming problems of institutional interplay with the EU Water Framework Directive. *Land Use Policy* 21 (1), 85–94.

MOWLD (Ministry of Water and Livestock Development) (2002) *National Water Policy.* MOWLD, Dar es Salaam, Tanzania.

MOWLD (2004) *National Water Sector Development Strategy (Circulation Draft).* MOWLD, The United Republic of Tanzania, Dar es Salaam, Tanzania.

Mwaka, I. (1999) *Water Law, Water Rights and Water Supply (Africa): TANZANIA – Study Country Report.* Cranfield University, Silsoe.

SMUWC (Sustainable Management of the Usangu Wetland and Its Catchment) (2000) *Supporting Volume A on Water Management.* For the Directorate of Water Resources, Ministry of Water, Government of Tanzania, The SMUWC Project, Mbeya Region, Tanzania.

Sokile, C.S., Kashaigili, J.J. and Kadigi, R.M.J. (2003) Towards an integrated water resource management in Tanzania: the role of appropriate institutional framework in Rufiji Basin. *Physics and Chemistry of the Earth, Parts A/B/C* 28 (20–27), 1015–1023.

van Koppen, B., Sokile, C.S., Hatibu, N., Lankford, B., Mahoo, H. and Yanda, P. (2004) *Formal Water Rights in Rural Tanzania: Deepening the Dichotomy?* Working Paper 71, International Water Management Institute, Johannesburg, South Africa.

World Bank (1996) *River Basin Management and Smallholder Irrigation Improvement Project* (RBMSIIP). Staff Appraisal Report, Washington, DC.

Yoder, R. (ed.) (1994) *Designing Irrigation Structures for Mountainous Environments: a Handbook of Experiences.* International Irrigation Management Institute, Colombo, Sri Lanka, 206 pp.

15 Intersections of Law, Human Rights and Water Management in Zimbabwe: Implications for Rural Livelihoods

**Bill Derman,[1] Anne Hellum,[2] Emmanuel Manzungu,[3]
Pinimidzai Sithole[4] and Rose Machiridza[5]**

[1]*Department of Anthropology, Michigan State University, USA and Department of
International Environment and Development Studies (NORAGRIC), Norwegian
University of Life Sciences, Norway; e-mail: derman@msu.edu or
bill.derman@umb.no;* [2]*Faculty of Law, University of Oslo, Oslo, Norway;
e-mail: anne.hellum@jus.uio.no;* [3]*Department of Soil Science and Agricultural
Engineering, University of Zimbabwe, Harare, Zimbabwe;
e-mail: manzungu@mweb.co.zw;* [4]*Centre for Applied Social Sciences, University of
Zimbabwe, Harare, Zimbabwe; e-mail: spinimidzai@yahoo.com;* [5]*Department of Soil
Science and Agricultural Engineering, University of Zimbabwe, Harare, Zimbabwe;
e-mail: roma877@yahoo.co.uk*

Abstract

As poverty has increased in Zimbabwe and elsewhere in Africa, the importance of water for smallholder agriculture has intensified. This chapter draws attention to the human right to water adopted in General Comment 15 by the Committee on Economic, Social and Cultural Rights, supplanting the Dublin Principles, which have too often been understood in the African context to mean water with the 'right' price. The chapter relates this human rights framework for law and policy, embedded in international and regional African instruments, to the history of national water legislation in Zimbabwe and its recent water reform. We ask how the historically evolving component of 'Primary Water Rights' tallies, or not, with international human rights approaches. It also traces the implications for rural livelihoods of the recently introduced obligation to pay fees for any water use that exceeds these 'primary water uses'. Further, the international human rights approach to water and the national notion of 'primary water uses' are compared with the multiple ways in which men and women share and manage land and water, including local norms and practices within a broader right to livelihood. Field research in Zimbabwe suggests the existence of a right to water and livelihood in local water management that can respond better to poverty and gender inequities. We suggest that a right to livelihood could be used for an active research programme to examine integration of local norms and practices within water management laws and policies and small-scale irrigation as an alternative to the overemphasis upon large-scale commercial agriculture.

Keywords: human rights, rights to water and livelihood, small-scale agriculture, local norms, gender discrimination, Zimbabwe.

Introduction

Water forms part of a broad 'right to life' that underlies rural livelihoods in Zimbabwe. It is expressed in the Romwe Catchment in southern Zimbabwe as 'water is life' (*hupenyu*) (Nemarundwe, 2003), in Shamva district as 'drinking water should be for everyone' (Matondi, 2001) and in Mhondoro Communal area as 'one can't deny drinking water to anyone' (Derman and Hellum, 2003). This right endures despite efforts by both colonial and independent governments to redefine rural citizens' relationship to water. The newly enunciated international human right to water accords well with the practices and norms within most, if not all, of Zimbabwe's communal and resettlement areas, but does not fit with either the colonial past or the current focus of water reform efforts. The idea that to deny water is to deny life indicates a profounder truth that there can be no human life without water. To deny people water denies them life.

The United Nations has included 'a right to water' in the International Covenant on Economic, Social and Cultural Rights (ICESCR), 1966 in its development policy.[1] In its global report on water, *Water for People, Water for Life*, the United Nations Educational and Scientific Organization (2003) emphasizes the right to water explicitly. This right is implicitly recognized in the Convention on the Rights of the Child (CRC), 1989, in the Convention on the Elimination of All Forms of Discrimination Against Women (CEDAW), 1979 and in the General Comment on the Right to Health, 2000. The previous global consensus around the Dublin Principles[2], with its emphasis upon water as an economic and social good, seems to be receding in the face of a growing movement toward recognizing a human right to water as an axiom for development and poverty-elimination policies. The Millennium Development Goal aimed at halving the number of people without clean drinking water emphasizes the critical importance of clean water. The World Bank, which had been in the forefront of arguing that water was not a human right but an economic good that required proper financing (World Bank, 1993, 2002, 2003a), has shifted toward examining human rights and equity (Salman and McInerney-

Lankford, 2004). It would seem that the many elements of the global system are catching up with villagers.

National water legislation and recent reform involve how a nation's waters are managed and understood. In this respect it is important to understand the colonial roots of water legislation in Zimbabwe, which protected and developed water resources in the interest of the colonial settlers at the expense of Africans, whose access to water was minimal and therefore largely falling under 'primary water uses' (see below). Entrenched inequities were factually perpetuated under Zimbabwe's water reforms in the 1990s that were enacted principally with the four Dublin Principles, rather than the human rights' framework, in mind. The principles fit better with the long-standing state biases toward large-scale commercial agriculture, and are at variance with what happens in the communal areas where the majority of the people live. In these areas, residents cultivate small plots, drawing upon local norms and practices and often resisting unsympathetic state policies. A common feature of local norms and practices, as observed in a wide range of contemporary studies of natural resources management in Zimbabwe's rural areas and decentralization, is the emphasis on resources that are vital for livelihood, such as food and water.[3]

In this chapter we identify local principles underlying access to water and land, and we have been surprised at the strength of fundamental norms despite a literature that emphasizes contestation and overlapping and conflicting spheres of authority. In turn, this has led us to examine if and how these normative local frameworks are consonant with some principles of the right to livelihood and right to water now embodied in a range of international human rights instruments, as well as within the current national legislation. This chapter connects researchers' observations on the principle of a 'right to water' in rural Zimbabwe with how that right could be considered within the broader context of a 'right to livelihood'. We suggest that the conceptual division made between land and water does not fit with local conceptions of livelihoods or the growing evidence of the importance of the land–water interface, which includes 'natural' wetlands and irrigation systems. We have chosen to probe these issues in Zimbabwe due to a long history of colonial state support for irrigation for white farmers, the difficulties in establishing small-holder irrigation and the contemporary processes of water reform in light of studies investigating water management.[4]

This chapter does not include in any depth the medium- and long-term implications of the current fast-track land reform programme (FTLP) underway for the 'right to water' and the 'right to livelihood' (Derman and Hellum, 2003; Hammar *et al.*, 2003; Manzungu, 2003; Hellum and Derman, 2004, 2005). It is too early to speculate on what directions Zimbabwe will take after President Mugabe leaves office. However, one can observe that there has been a dramatic increase in the numbers of Zimbabwe's poor, a direct consequence of the FTLP. Zimbabwe has one of the highest rates of inflation in the world, combined with a shrinking economy. In the past few years, it has fallen from a medium human-development nation to a low one (Human Development Report, 2003), and was ranked 147th out of 177 in the global human development index. High rates of education (now dropping rapidly) keep Zimbabwe from the bottom. Due to HIV/AIDS, population growth is projected to be at 0.2% of the annual growth rate. Life expectancy at birth has fallen from 56.0 to 33.1, perhaps the most powerful indicator of failed policies.

In this context, there needs to be a much greater coordination between water policies and poverty alleviation strategies. The question that can be asked is whether the global water agenda, with its emphasis on commercialization, effectively engages with local realities in a collapsing country like Zimbabwe, whose political leadership first brought in neo-liberalism for a time but then rejected it (Manzungu, 2002).[5] In the context of Zimbabwe, attention to the local gains is of increased importance, since villagers are far more reliant upon their own resources than before.

This chapter proceeds as follows: in section 1 we detail the emergence of the right to livelihood and the right to water in the United Nations system, the African Union and other international and national forums. We do this to examine how a human right can be constructed on the basis of other human rights. If and how

the human right to water becomes accepted and implemented remains to be decided, based upon many factors known and unknown. Section 2 considers Zimbabwe's water history, water reform and water management. In section 3 we examine the contemporary water reform programme, which was intended to address the inequalities produced by settler rule and realities of contemporary integrated water management. How local norms and practices respect rights to livelihood and water forms the substance of section 4. While we note how little the new laws have affected these, we propose greater attention to those elements of local practice that are best conserved. In the conclusions, section 5, we examine how human rights – with its obligations to protect, respect and fulfil – set new responsibilities for states to accomplish. This is a significant challenge in contemporary Zimbabwe, with its divergence from internationally accepted human rights standards. See Fig. 15.1 for a map of Zimbabwe.

Water as a Part of the Human Right to Livelihood

When Zimbabwe passed its new water acts the human right to water had not been explicitly recognized, although it had been included in some international conventions.[6] In more general terms, the human right to water derives from the right to life, the right to livelihood and the right to health. It has evolved through piecemeal international, regional and national law-making. It is recognized in Article 24 of the Convention on the Rights of the Child, explicitly stating that the child has a right to clean drinking water. Article 14.2h of the Convention on the Elimination of all forms of Discrimination Against Women states that rural women have a right to 'enjoy adequate living conditions, particularly in relation to housing, sanitation, electricity and water supply, transport and communications' on an equal basis with men. Article 15 of the Protocol to the African Charter on Human and Peoples' Rights on the Rights of Women in Africa on the right to food[7] obliges states partly to 'provide women with access to clean drinking water, sources of domestic fuel, land and the means of producing nutritious food'. The human right to water is also recognized in the United Nations Convention on

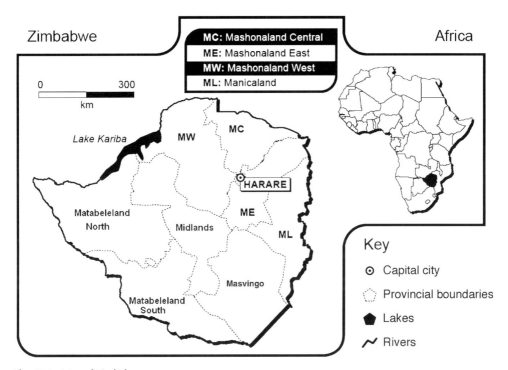

Fig. 15.1. Map of Zimbabwe.

the Law of Non-Navigational Uses of Water-courses.[8]

The Southern Africa Development Community (SADC) Protocol on Shared Water Course Systems of 1995 emphasizes equitable utilization of shared watercourses applying existing international law, existing practices and community interest taking into account, among other things, the environmental, social and economic needs and the impact of intended uses of the watercourse (Article 2).

Safe, adequate and available water

A major shift in underlining the significance of a right to water was the General Comment No. 15 of July 2002 by the UN Committee on Economic, Social and Cultural Rights, whereby the Committee concluded that there is a human right to water embedded in Article 11 in the Convention on Economic, Social and Cultural Rights (CESCR), defining the right to livelihood as 'including adequate food, clothing and housing'. The term 'including', as understood by the

Committee, indicates that the catalogue of rights encompassing the right to livelihood is not exhaustive but must be adapted to changing social and economic concerns such as the global water crisis (Eide, 2001). Concluding that water is a human right, the Committee (2002) emphasizes the interdependence between human rights in general and between access to water and the right to health in Article 12,1, the right to food in Article 11 and the right to life and human dignity enshrined in the International Bill of Human Rights (1948).

Recognizing that water is required for a range of different purposes that are essential for human life, the Committee on Economic, Social and Cultural Rights (2002) signalled three elements: (i) water must be adequate for human life; (ii) it must be safe and available; and (iii) it must be available on a non-discriminatory basis. Adequate water, according to the CESCR, is a far broader concept than just clean drinking water, since it encompasses water for personal and domestic uses and the necessary water resources to prevent starvation and disease. The scope and extent of the human right to water are defined through its link to the right to life, the right to health and the right to food. In the view of the CESCR, especially important is that sustainable access to water resources for agriculture is necessary to realize the right to adequate food (General Recommendation No. 12, 1999). Disadvantaged and marginalized farmers (women and men) should have equitable access to water and water management systems, including sustainable rain-harvesting and irrigation technology.

State obligation to respect, protect and fulfil

The obligation to respect, protect and fulfil rights cuts across urban and rural water supplies and services. The obligation to respect includes a duty to refrain from interfering arbitrarily with customary or traditional arrangements for water allocation, unlawfully polluting water or destroying water services and infrastructure during armed conflicts (General Recommendations 15, 23 and 24). Taking note of the duty in Article 1, paragraph 2 of the Covenant (1966), which provides that people cannot 'be deprived of their means of subsistence', states parties should ensure that there is adequate access to water for subsistence farming and for securing the livelihoods of indigenous peoples. This aspect of the human right to water is also expressed in the Statement of Understanding accompanying the United Nations Convention on the Law of Non-Navigational Uses of Watercourses (A/15/869 of 11 April 1997), which affirms that, in determining vital human needs in the event of conflicts over the use of watercourses, 'special attention is to be paid to providing sufficient water to sustain human life, including both drinking water and water required for production of food in order to prevent starvation'.

The obligation to protect as part of all human rights treaties and conventions requires state parties to prevent individuals, groups, corporations or other agents acting under their authority from interfering with the right to water. States parties are under an obligation to prevent private water service operators from compromising the right to equal, safe and affordable water in terms of regulatory systems, including independent monitoring, public

participation and penalties for non-compliance (General Recommendations 15, 23 and 24).

As regards the duty to fulfil, states parties must, to ensure that water is affordable, adopt measures including: (i) use of a range of appropriate low-cost techniques and technologies; (ii) appropriate pricing policies such as free or low-cost water; and (iii) income supplements. Any payment for water services has to be based on the principle of equity, ensuring that these services, whether privately or publicly provided, are affordable for all, including socially disadvantaged groups. Equity demands that poorer households should not be disproportionately burdened with water expenses compared with richer households (General Recommendations 15, 26 and 27). This has implications for the implementation of the 'user pays' principle, which has become ubiquitous in both urban and rural settings.

Taking the human right to water beyond the nation state, the Committee on Social and Economic Human Rights in General Recommendation 15 also recommends that United Nations agencies and other international organizations concerned with water – including all United Nations' organizations (World Health Organization, etc.) – should cooperate effectively with state parties in relation to the implementation of the right to water.

The Committee also recommends that the international financial institutions, notably the International Monetary Fund (IMF), the World Bank, the African Development Bank, etc. should take into account the rights to water in their lending policies, credit agreements, structural adjustment programmes and other development projects. The emerging literature on the human right to water by the World Bank and the World Health Organization (WHO) suggests a paradigmatic change (WHO, 2003; Salman and McInerney-Lankford, 2004).

Non-discrimination

State parties are also obliged to ensure that the right to water is enjoyed without discrimination on the grounds of sex, class, colour, religion or political opinion. State parties are to ensure that new laws, policies and programmes do not deny this right either de jure or de facto to selective portions of the population. Inappropriate resource allocation can lead to indirect discrimination. Investment should, according to Comment 15, not disproportionately favour expensive water supply services and facilities that are available only to a small percentage of the population.

The CEDAW and Protocol to the African Charter on Human and Peoples' Rights on the Rights of Women[9] in Africa substantiate the principle of non-discrimination in relation to water, land and food security. Simply having gender-neutral laws and policies in a situation where resources (time, money, land, water, for example) are unevenly distributed between men and women is insufficient. To ensure substantive equality CEDAW and the Protocol oblige state parties to take measures to eliminate both direct and indirect discrimination.[10] Indirect discrimination points to the unintended effects of seemingly gender-neutral laws and policies. It is defined as 'any distinction, exclusion or restriction made on the basis of sex which has the effect that they impair or nullify, on a basis of equality between men and women, human rights in the political, economic, social, cultural, civil or any other field (CEDAW Article 1)'.

The concept of indirect discrimination encompasses development policies and programmes that, on face value, are gender-neutral but in practice are biased against large groups of female users in comparison with male water users (Hellum, 2007). Policies, programmes and plans for improvements and investments in water that are based on a division between domestic and productive water use will often have a discriminatory effect both in terms of class and gender. Women farmers' hand irrigation of small vegetable gardens has too often, for example, been seen as unproductive by conventional economic standards. Seemingly gender-neutral investment policies targeted towards productive water uses have, as a result, often disproportionately favoured larger or more expensive water supply services controlled by men. States parties to the CEDAW and the African Charter are obliged to take measures to eliminate this form of discrimination. In accordance with Article 26 of the Protocol to the African Charter on the Rights of Women in Africa, state parties are obliged to undertake to 'adopt all necessary measures and

in particular provide budgetary and all other resources for the full and effective implementation of the rights'.

In the next section we trace how the human right to water, as defined internationally today, relates to the historical developments in water legislation in Zimbabwe and to the recent reform, in particular with regard to small-scale productive uses in rural areas.

Water Legislation and Smallholder Irrigation in Zimbabwe

Developments in water legislation and smallholder irrigation in Zimbabwe are closely linked with the country's socio-political history. In its simplified form this can be summarized in three phases. Between 1890 and 1980, the colonial state machinery favoured white settler political, social and economic interests at the expense of the black majority population. The attainment of independence in 1980 saw the post-colonial state seeking to redress the historical race, class and gender imbalances. The social expansion of health, education, agricultural extension and resettlement between 1980 and 1990 proved to be unsustainable as it was not supported by a strong economic base. This led to the IMF/World Bank-inspired Economic Structural Adjustment Programme (ESAP) of the 1990s. ESAP promoted economic deregulation to the extent that the anticipated economic and social gains were not achieved. From 1997 the economic malaise gave birth to complex political, economic and legal crises, which have resulted in the state being unable to deliver on political and economic rights. We provide below a synopsis of the major developments with regard to water.

The early years: 1890–1927

In the early years, the colonial state's preoccupation with mining was reflected in the water sector. The interest changed to agriculture when it was apparent two decades later that mining was not going to be a profitable venture, as was the case in South Africa. In this endeavour there was scant regard paid to the rights of the black population. For example, in the allocation of water rights a frontier mentality was displayed with such claims as: '… being a new country, Southern Rhodesia is unhampered by the pernicious common law relating to riparian ownership' (Manzungu and Senzanje, 1996). In this way the water rights of the black population that predated the settler claims were disregarded. But this is not to suggest that there was harmony within the settler community. In fact, conflicts over water within the settler community did not take long to develop. The frequent and often costly litigations between rival claimants to the use of water culminated in the Union Irrigation Act of 1912. This made provisions for the control, apportionment and use of water. The Act was based on the common law as evolved and expounded by the courts (McIlwain, 1936). In 1913, the Water Ordinance was passed as a way of comprehensively dealing with problems of rights to water (McIlwain, 1936).

The South African connection had a strong influence on some aspects of water management. For example the settlers, encouraged by the British South Africa Company (BSAC), used the Roman Dutch Law that had been brought from Holland and then in use in South Africa – by then the English riparian rights had already been adopted. This was regarded as unsuited to the water resources and production of the region (McIlwain, 1936). There was, however, continued use of the riparian rights doctrine in interpreting access rights and differentiated water-use types. The 1920 Water Ordinance explained that: 'If a farmer has land well suited for irrigation and there is a stream that can be economically utilized, he can acquire the right to use the whole of the water for irrigation even though it may leave others without water except for primary purposes.'

While there were disagreements between the settlers over which was the better legal model to guide water allocation, the situation was worse for the black population. Land appropriation disadvantaged communal area residents since it left them downstream of white settlers with less ability to access water. In sum, the legal system was set against them, and compounded by a shortage of the necessary finance.

The concept of primary water use was provided for from the early years. Historically, primary water was a concept adopted from the earliest South African water laws. The first regulation of water in Zimbabwe was by the Order in Council, 1898, Section 81 pertaining to the British South Africa Company. It required the company to ensure that the natives or tribes had a fair and equitable portion of springs or permanent water. Primary water use was water for human and farm livestock use and was set at 50 gallons (~ 228 l)/person/day. This was quite generous because it could be used in and around the homestead, including gardening. Water for 'secondary purposes' was for irrigation and watering of stock other than farm stock. 'Tertiary purposes' included the needs of the mines and railways. While in theory there was nothing that stopped the black indigenous population benefiting from primary rights, a combination of lack of information and the dry terrain they were forced to inhabit did not help matters.

Agriculture-based water law: 1927–1980

With increasing water use by white farmers, new water laws were required to establish the 'rules of the game'. In 1927, for the positive requirement of fair and equitable availability of primary water for Tribal Trust Lands (TTL, what are now the Communal Areas), water use was changed. The new act specified that changes in primary water for TTL residents be approved by the Board of Trustees for Tribal Trust Lands (Hoffman, 1964). However, participation by 'tribal' and later communal area residents in water decision making was nil. By the Act of 1927 the priority right to water, granted to the mining industry within the Gold Belt areas, was modified in favour of irrigation (Government of Southern Rhodesia, undated). Therein were a number of clauses that disadvantaged the black population. First of all, water rights were attached to land, which disadvantaged most black Zimbabweans who had been dispossessed of land and placed in the reserves where they did not enjoy full rights. Rights to land in the reserves were registered with Communal Area bodies (formerly known as Tribal Trust Lands) and not with individuals. Natives could therefore only apply for water rights as a community, and through government officials. Even then the District Administrator or Minister of Water Development held the water right on the behalf of TTL residents. There was, however, provision for the appointment of representatives of 'native interests' in the Irrigation Boards and in the Water Courts. Not much is known about whether or not they were actually represented. On the other hand, settlers could individually apply for water rights because they owned land in their own private capacity.

Another problem was that water rights were issued based on the priority date system; this meant that rights were granted on a first-come, first-served basis. The black indigenous people were disadvantaged because they had not applied for water rights (Derman, 1998; Manzungu, 2001). When they later understood this, most of the water was already committed to rights held by the settlers. Water rights were also issued in perpetuity, which meant that a water right once issued could not be revoked except in special circumstances, such as the declaration of a drought or when someone else applied for the same water and was willing to pay compensation. By virtue of the fact that settlers applied for the rights way before the indigenes, most of the water was committed. It should be noted that, although racial water allocation was provided for in the 1927 Water Act, it was only in the 1940s that massive transfers of water to the whites actually occurred because of cheap finance made available for both dam building and irrigation. This emphasizes the argument that it is not necessarily changes in water legislation that determine (lack of) access to water.

The 1947 Water Amendment Act had loose allowances for primary water users, especially for gardens and riparian users. The Act also defined *vleis* (seasonally flooded wetlands or wetlands in depressions), springs and streams that lay outside public management because they were defined as 'private water'. This changed later with restrictions on *dambo* (wetlands) cultivation, mainly because of fear of degradation, which had been noticed in the white farms. The Act also identified new water uses such as fish farms and conservation activities that were a result of new commercial interests.

The Water Act of 1976 upheld the principles of the 1927 Water Act, i.e. rights to water were linked to land, the priority date system of allocating water and granting a water right in perpetuity. The Act also provided for catchment outline plans to be prepared for the development and use of surface water. Three types of water were recognized: public water, private water and underground water. The Act, under a 1984 amendment, also provided for some stakeholder participation in such institutions as River Boards. The participation was, however, restricted to holders of water rights. The Act also required applicants for water rights to put in place water measuring devices for a water right to be confirmed as permanent. This explains why most water rights in the native areas were temporary – the natives could not afford to put in the requisite measuring devices.

In summary, it can be said that the system of water allocation in the colonial period was based on the matrix of ideas of efficiency and modernity and rooted overall in European power. The process was held almost entirely in male hands, with an extension of racial and patriarchal notions toward the great African majority who were denied both adequate land and water (Campbell, 2003). Campbell (2003) further argues that the settler state's planning mechanism was organized around the concept of the scarcity of water. Politicians, agricultural extension officers, water resources managers, hydrologists, engineers, planners and economists propagated the concept of water scarcity when, in reality, the problem of water availability for black Zimbabweans was distribution, not scarcity. This was reinforced by the myth that only whites could have an efficient and productive agriculture. African farmers, despite their early successes (Ranger, 1985; Phimister, 1988), were excluded from access to water.

Smallholder irrigation: 1980–1997

The legal framework that was put in place in the colonial era was by and large upheld by the new nationalist government that was elected in 1980, after 10 years of war. In the resettlement programme that was established after liberation the government supplied drinking water, but only in a few instances did it include irrigation. Irrigation systems existing on commercial farms transferred to smallholders or resettlement farmers were not maintained or protected. Another government initiative to elicit increase in water accessibility by reserving 10% of water in government dams was a complete failure, as no measures to make the water available were taken (IFAD, 1997). Once again, the problem was not availability of water but its delivery.

In general, government-sponsored and/or -funded irrigation schemes did not do nearly as well as expected. Several evaluation studies have suggested that smallholder irrigation schemes (initiated and constructed by the government, which may be community- or government-managed) have poor performance and are not sustainable. Problems identified include poor water utilization, in terms of its timeliness and adequacy to the field, and poor water application to the field (Pearce and Armstrong, 1990; Donkor, 1991; Makadho, 1993). Crop yields have been low and way below those achieved in the commercial farming sector. The poor agricultural performance has translated into poor financial and economic viability, thereby necessitating heavy government subsidies, up to 75% in some cases. The Rukuni Commission (Rukuni, 1994) found that the irrigation subsector in the communal and resettlement areas was dramatically under-budgeted by the state and required change.[11] Pointing out the interdependency between land and water policies, the commission made a series of recommendations[12] to increase the efficiency of the agriculture sector. These recommendations were neither accepted nor systematically incorporated into state policy.

The largest area under smallholder irrigation remained informal, with little or no economic support. It continues to use a mixture of indigenous and introduced technologies. This sector, estimated to cover at least 20,000 ha in the late 1990s, was said to be more productive than the formal sector (IFAD, 1997). Indigenous irrigation has therefore been undervalued to the extent that it does not feature in official statistics and policies, despite the fact that it contributes significantly to rural livelihoods and sustainable resources management. Bolding *et al.* (1996) have commented on the merits of indigenous irrigation. In assessing irrigation systems in the Eastern Highlands they noted why these systems were effective in contributing to food security and rural wealth. Based upon detailed empirical studies, there were several factors that led to their efficacy. These include, for example,

the use of locally available materials, a norm of equity that minimized conflicts over water quantities, flexibility in adjusting to rainfall variability, fewer expenses to build and maintain and demonstrated sustainability over many years. It appears that, although this was not addressed in the original research, all those who sought irrigated land would receive some.

These local principles did not enter the central government's policy frameworks. This has meant that small-scale irrigation and local water resources management principles and practices that could have been used to support food security at the household level have not been valued or made part of government policies.

As noted previously, a different approach had been presented to the government. Of particular significance was the fact that the recommendations made by the Rukuni Commission were very much in the spirit of Article 11 of the International Covenant on Economic, Social and Cultural Rights. This emphasized the obligations to take steps to ensure the realization of the right to livelihood by 'reforming agrarian systems in such a way as to achieve the most efficient development and utilization of natural resources'. Among the Commission's recommendations to increase the efficiency of the agriculture sector, the main recommendation was increased investment in water in communal and resettlement areas. Other recommendations as to how agricultural production could be increased in communal and resettlement areas included legally secure tenure, improved credit and financial services and comprehensive agricultural support institutions. It appeared that the Zimbabwean government had little interest in smallholders. It did, however, have a new and deep interest in water management.

In the next section we turn our attention to how Zimbabwe's water reform responded to global discourse on water reform in general and how the new water policy might address small black farmer needs.

Zimbabwe's 1998 Water Reform: Addressing the Colonial Legacy

The core of Zimbabwe's water reform, initiated in the mid-1990s, rested on increasing access to water by black Zimbabweans while ensuring the productive use of water.[13] New participatory structures were created to increase access to water management decision making. These are called Catchment and Sub-catchment Councils, and are based in Zimbabwe's seven hydrological zones. In addition, a new parastatal was established – the Zimbabwe National Water Authority (ZINWA) – to shift water management expenses from government to users and to increase the productive use of Zimbabwe's water. Prior to the Water Act of 1998, large-scale commercial farmers controlled Zimbabwe's water through a 'water rights' system – first in time, first in line. This made it very difficult for new appropriations to be made to Black small-scale farmers, who had great difficulty in finding the resources to obtain water rights and to negotiate with the bureaucracy to secure those rights.

The Water Act of 1998

Under the Act all water is vested in the President and no person can claim private ownership of any water. In presenting the first reading of the new draft Water Bill, Attorney General (now Minister of Justice) Patrick Chinamasa emphasized the following: 'What the existing legislation has done is that the water is the President's water but the President then put in legislation to give permission to people to exploit it and that is what is peculiarly known as the water right' (Zimbabwe Parliamentary Debates 1998, p. 1566).

In defending the abolition of the concept of private water, Chinamasa also asserted the common Zimbabwean understanding of water: 'Water is a public resource. It is a gift from God. None of us here are rainmakers, and that includes commercial farmers. The rainmaker is God. He provides His people and that water forms part of the hydrological cycle' (Zimbabwe Parliamentary Debates, 1998, pp. 1562–1563).

This is consistent with Zimbabwe's history as a centralized state while appearing to incorporate new global water management policies (Derman *et al.*, 2001). The 1998 water legislation transferred most national planning functions from the Department of Water Development to the new parastatal ZINWA with oversight from the Ministry of Water Development and Rural Resources. ZINWA is funded through the sale of water collected behind government dams, the

provision of water to cities and the levying of water charges to large-scale users. Management of Zimbabwe's water is to be shared with the new stakeholder organizations of Catchment Councils and Sub-catchment Councils.[14]

Commercial and primary water: the continued colonial legacy?

Zimbabwe's water is still divided into two categories – commercial water and primary water. Primary water is defined in the Water Act of 1998 as water used for: (i) domestic human needs in or about the area of residential premises; (ii) animal life; (iii) making of bricks for private use; and (iv) dip tanks.[15] In sum, it is not restricted to drinking water but seen as an integrated part of livelihood necessities such as food and housing in the communal areas. The state is obliged to respect and protect the right to primary water as embedded in the Act. What is meant by 'domestic human needs in and about the area of residential premises' is, however, not clear. The Water Act (51.1) asserts the importance of primary water: 'No permits granted by a catchment council, other than permits for the use of water granted to a local authority for primary purposes, shall have the effect of depriving persons of the use of water for primary purposes.'

This makes provision for ensuring that primary water users will not lose any further water. However, to actualize this right means knowing how further water abstractions would affect primary users. No catchment in Zimbabwe knows the amount of primary water used because the catchment planning exercises, which were to make accurate estimates, have yet to be completed. In addition, there has been a loss of information on commercial water use. The implication that primary water use had priority over commercial water use has not been asserted to the knowledge of the authors since the implementation of the Water Act. No volumes have been provided for primary water use. While a general national estimate has been made of rural primary water use, amounting to 1% (Zimbabwe Government, 2000c), there are no detailed empirical estimates of actual use. It has been assumed that the amount of use has not justified registration in comparison with commercial water use (see below).

New innovative forms of commercial cropping emerging within the common property regimes in the communal lands, such as gardening for consumption and sale, represent a challenge in how Catchment Councils, when issuing water permits, draw a dividing line between commercial and primary water uses. These uses render problematic the division between commercial and primary water. Under the new Water Act of 1998, it is only water used for commercial purposes that requires a permit in terms of Section 34. The definition of commercial water depends upon use – water used for purposes including agriculture, mining, livestock, hydroelectric power, etc. It follows from the ZINWA Act, Section 41, that only permitted water is subject to the user pays principle in terms of the new water levy.[16] Thus rural primary water users do not have to do so.

One Catchment Council, the Mazowe, debated what constitutes the difference between commercial and primary water use. The Council Chairman suggested a technological answer: if the water is moved by hand it is primary water, if it is moved by machine then it will be considered commercial. The Catchment Manager from ZINWA present at the meeting indicated that, as yet, ZINWA had not decided what the guidelines should be in deciding whether water use was primary or commercial.[17] Villagers from Bangira, in Mhondoro Communal Lands, who argued that they would refuse to pay for water moved by a pump to provide their vegetable gardens with water, contested this view. A couple who had worked hard to establish funding for the local dam in order to raise their living standards and those of other families argued that, since the surplus from the gardens was used for livelihood essentials such as clothes, school fees or medicine, the water use should not be seen as commercial.

Research investigating different catchment councils demonstrates that the intention of the new Water Act to 'ensure that the availability of water to all citizens for primary purposes' was not realized (Derman, *et al.*, 2001; Dube and Swatuk, 2002; Mtsi and Nicol, 2003; Manzungu, 2004b). Indeed, the emphasis was upon catchments and sub-catchments to raise revenue for them and for ZINWA. In the Mzingwane catchment, which had limited commercial water to levy, there was a suggestion of levying a charge

for every herd of cattle. In the Save Catchment levies were proposed for any water use where some income was realized.

This lack of conceptual and policy clarity also applies to the thousands of boreholes currently used in Zimbabwe. ZINWA's policy was to register all boreholes and then charge borehole owners for water because, like all water, it belongs to the government. On the other hand, the Presidential Land Review Committee under the Chairmanship of Dr Charles Utete recommended that levying water from boreholes should be stopped because 'it discourages investment in water resource development and the enhancement of production on farms through irrigation' (Utete, 2003, vol. 1, p. 177). No attention was paid in this report to the scale and intensity of borehole water use.

The Water Act and human rights

These different conceptualizations do not sit well with the definition of water of the Committee on Economic and Social, and Cultural Rights as a part of the right to livelihood as stated in General Recommendation 15. This recommendation emphasizes that the sustainable access to water resources for agriculture is necessary to realize the right to adequate food. Local management systems as described earlier cut across the commercial/primary division. In our view, an approach based on human rights calls for a clearer definition of primary water uses that transcends clean drinking water and includes the legitimate concerns of poor small-scale farmers. While such legal clarifications may be undertaken by the stroke of the pen, the CESCR also obliges states to take positive steps to fulfil the human right to water. Such positive steps call for long-term economic commitments, implying that internal and external economic resources invested in infrastructure are beneficial for the poor in all of Zimbabwe's rural areas, not just in the newly resettled ones.[18]

Despite the emphasis on equality of access in the initial phases of water reform, most attention has been devoted to increasing the number of commercial water users. Zimbabwe's new water management system was based on the premise that fees for commercial water use would be used for the development of water resources. The areas under irrigation in Zimbabwe have diminished greatly, since the irrigation systems on the former commercial farms have not been sustained and older government-sponsored irrigation schemes have been unable to continue in light of the harsh macro-economic climate following the fast-track land reform. According to Manzungu (Utete Report, 2003, Vol. 2, p. 89) the total number of hectares under irrigation has fallen from 186,600 to 120,410. This loss of 66,000 ha under irrigation has primarily been in the formerly large-scale commercial farm sector.

The institutional separation of water supply through Rural District Councils from water resources management issues through Catchment Councils in the communal lands is another factor that has inhibited water development. Under the water reform, Catchment and Sub-catchment Councils had reasonable sources of funding where Rural District Councils are underfunded and have too many obligations. Questions of water supply are also linked to borehole provision for combined domestic and productive use, especially in the communal and resettlement areas. Under the new water policy, the Integrated Rural Water Supply and Sanitation Program remains separate from the above while continuing to be tasked with providing safe, protected drinking water supplies for all rural water users and to ensure that every household had at least an improved, partially enclosed latrine. This separation has tended to alienate primary water users, who are the vast majority of Zimbabwe's water users (Manzungu, 2004a, p. 13).

Water supply programmes have been especially vulnerable to government service shrinkage and donor withdrawal. It is poorer women who rely heavily on water sources that are free of charge, such as borehole water for their gardens, and who find themselves caught in the gaps and mismatches between these different policies and institutional structures. Their water needs fall outside the scope of both the water and sanitation programme and the water reform policy aimed at larger-scale users. In our view, better coordination and linkages between Catchment and Sub-catchment Councils with Rural District Councils would have been a better strategy, although admittedly difficult. The involvement of both water institutions could have provided appropriate incentives for the participation of small-scale users.

The new Zimbabwean water policy seeks to have a single uniform water management system in place. The regulations framing the new water management system are, by and large, moulded on a large-scale commercial farming model giving little attention to the potential of local irrigation systems and methods developed by communal-area farmers. Local irrigation methods and principles as described above and analysed in the natural resources management literature have not entered the central government's policy frameworks (Bolding *et al.*, 1996). This has meant that small-scale irrigation and local water resources principles and practices that could have been used to support food security at the household level have not been valued or made part of government policies.

In communal areas and resettlement schemes both men's and women's access to water still rely heavily on use rights embedded in local norms and practices (Pinstrup-Andersen, 2000, p. 13). These local use rights have, in part, been protected, as described above, by the concept of primary water. In the next section we explore whether there might be an explicit or implicit recognition of 'a right to livelihood', at least with respect to access to water for livelihood purposes in Zimbabwe's rural areas.

A Rights Perspective on Water and Livelihood

Derman, Hellum and Sithole have, since 1999, been studying water management in the three villages of Bangira, Murombedzi and Kaondera in the chieftainship of Mashamayombe in Mhondoro Communal Land (Derman and Hellum, 2003; Hellum and Derman, 2005).[19] This local qualitative study was part of a wider study of national water reform in Zimbabwe that was undertaken by the Center for Applied Social Studies (CASS) at the University of Zimbabwe. We chose this area due to a rapid and recent increase in tobacco growing, a relatively large number of private wells and the existence of a dam project. Apart from dry-season vegetable gardens located along streams, rivers, vleis and boreholes, agriculture in this area remains primarily rain-fed maize with an expansion of irrigated tobacco. Because of these trends in commercialization, we expected to find a decreasing open access to the area's water resources. We made the assumption that, because the deep and open wells were located on homesteads and that there was a great increase in tobacco production, that these wells would become increasingly 'private'.

Water for life: the right to safe drinking water

Our study in Mhondoro suggested that at the local level, as in human rights law, there is a right to clean drinking water. Villagers demonstrated a surprising degree of consistency over time and space in upholding the norm that no one can be denied clean drinking water (Derman and Hellum, 2003). The obligation to share drinking water extended to wells, which were privately dug, and on basically private land. In one village a private borehole, paid for by one household, rapidly became a village source of drinking water. In another village a borehole built by the Zimbabwe Tobacco Association for irrigating tobacco seedlings became an important drinking source for the entire village. In a third village in the study area, the private well of a widow served as a source of drinking water for almost the entire village.

Based on the norm and practice of sharing, access to drinking water extended to boreholes constructed for principally commercial, dedicated or private use. The duty to share increased rather than decreased during drought periods. Such sharing cut across kinship and village borders, and it has been upheld during the accelerating economic and political crisis. Water users and well owners reported that they had never paid or received money or given gifts. To breach the norm of providing drinking water meant risking sanctions or being the target of witchcraft.[20] Universal access to drinking water in Mhondoro points to a morally based duty rather than a negotiable and reciprocity-based notion of property, often pointed to as a characteristic feature of African customary laws (Berry, 1993).[21] Applicable to men and women, insiders and outsiders, it also points to a notion of equality and non-discrimination.

These findings are consistent with our readings of a series of Zimbabwean monographs on natural resources management including water, wetlands, forests and land (Cleaver, 1995; Derman, 1998; Sithole, 1999; Matondi, 2001; Nemarundwe, 2003; Walker, undated, unpublished paper). All the empirical records from communal areas in Shamva, Mutoko, Chiduku, Dande, Masvingo, Guruve and Matabeleland suggest that water for drinking can, and should, be made available to all. Nemarundwe, in her doctoral thesis, reports from the Romwe catchment in the Chivi district, South Zimbabwe that drinking water is made available to all no matter what the source of water is. Available water sources include boreholes, river bed wells, rivers, wells, collector wells and dams. No matter the tenurial status, whether publicly or privately owned, the water sources are available for drinking water. In a powerful and clear manner she writes: 'Because water is considered hupenyu (life), there has been no case of denying another village access to water during drought, although rules of use are enforced more stringently during drought periods' (Nemarundwe, 2003, p. 108).

The study points to actual incidents where this general ideal was challenged. One example is a well owner who prevented others from accessing his well. Two days after he locked the gate to the well he found a dead dog. In response to this the well owner later unlocked the gate (Nemarundwe, 2003, p. 113). In a similar vein Prosper Matondi, who carried out his research in an area of resettlement farmers and two irrigation schemes in the Shamva district near Bindura, the provincial capital of Central Mashonaland Province, found that drinking water remained available for all despite growing scarcity of both land and water resources. In a parallel fashion, Bevlyne Sithole's research in Mutoko and Chiduku communal areas in eastern Mashonaland and Manicaland Provinces, respectively, summarizes farmers' views on water as follows: 'Water should be available to all, rich or poor, but the person who impounds the water is the one who makes the river dry' (Sithole, 1999, p. 195). Frances Cleaver's study in the Nkayi communal land in Matabeleland suggests that water user rules that limit poor people's access to water are invalid. She observed that poor women got away with breaking the rules that limited water resource to certain individual users (Cleaver, 1995, p. 357).

Water for livelihood: the right to garden?

Rural people in Zimbabwe see land and water as closely interconnected in fulfilment of livelihood needs. Livelihoods are no longer just about access and use of land and water in rural areas.[22] Access to basic livelihood resources such as health services, food and housing also depends on cash. Like many rural southern African residents, Zimbabweans are dependent upon remittances from kin in cities or abroad, or reliant upon their own engagement with paid jobs or market activities. Households and families are quite different and even in one rural area there are significant differences between them in terms of reliance upon land and water. Yet, within the context of this mixed rural livelihood structure, dambo cultivation has particular significance since wetlands have grown in importance due to the unpredictability of Zimbabwe's rains, increased reliance upon cash crops and the possibilities of hand irrigation.

Almost every family in the three villages in Mhondoro had gardens when we began our study in 1999. A quantitative survey of water management in the area demonstrated that 90% of households had some form of dry-season garden requiring hand irrigation.[23] The larger and more productive gardens tended to be close to, or in, wetlands, but there were significant gardens at the homestead if there was a borehole or productive well close at hand.

The family gardens were usually the main responsibility of the women. The crops in the gardens are covo, rape, onions, tomatoes, beans, groundnuts, maize, sugarcane and cabbage. There are also fruit trees including banana, papaya and mango. These rely heavily on the common pool water resources including rivers, boreholes, deep wells and shallow wells. Gardens are often situated on land that is either seasonally flooded or holds water from the rainy season long into the dry season. The gardens are as much a source of income as of food for the family. The income is often used for meeting household needs including food,

education, clothing and medicine. In the recent years of drought and economic hardship, the produce from women's gardens has been an essential source of livelihood.

Dambo cultivation in Mutoko and Chiduku in eastern Mashonaland and Manicaland coincided with the establishment of mission schools and hospitals in the mid-20th century (Sithole, 1999, p. 140).[24] As in Mhondoro, the major garden crops came first from large-scale commercial farms and then from agricultural extension officers during and after the colonial period. Three Mhondoro elders told us they were the first villagers to start gardening in the 1950s. They were taught to grow vegetables by an agricultural extension officer in the colonial administration. Women especially expanded their gardens after independence to provide green vegetables for their families. Gardening increased in the 1990s as the rate of inflation rose. Government construction of boreholes, cement wells and some small dams facilitated further garden expansion. The Zimbabwean government began withdrawing from rural areas during the 1990s under the combined policies of structural adjustment and decentralization. People in Mhondoro, as local communities elsewhere, have since been left to find alternative economic sources for expanding water supply for drinking water, watering cattle and irrigation. The CASS survey indicated that more than 70% of the households in the three villages had invested work and money in water, including private wells and other water resources.

The mixed character of the principles that derive from this agricultural practice is neither traditional nor modern, demonstrating that rural people in their livelihood strategies draw on a wide variety of sources. Our study of gardens in Mhondoro suggests that the right to water as part and parcel of rural livelihoods extends beyond the right to clean drinking water (Hellum, 2007). Households that needed garden lands were allocated appropriate land.[25] Everyone we interviewed in one village stated they had obtained the headman's explicit or implicit approval to access land for gardens on vleis or close to rivers. The gardens, the *Sabhuku* (village headman) said, were important sources of livelihood and self-reliance. For this reason he had not taken action when people allocated themselves gardens without his permission. Another reason was fear of revenge from *ngozi* (bad spirits). This suggests the existence of an underlying norm of sharing. A similar pattern was observed in another village, where people's gardens were moved from the wetlands to communal gardens close to a newly constructed dam. Everyone was granted land for gardens in this area. If the land allocated for the communal gardens was insufficient, the headman saw it as his duty to allocate more land. None of the villagers we interviewed had paid for the land. This suggests a wider right to livelihood that is not limited to clean drinking water, but extends to access to garden lands with available water sources.

The right to use available water for gardens, however, appears to be subject to greater contestation than a right to drinking water. For example, Nemarundwe (2003) provides a short illustrative case of water conflict at a small dam between richer and poorer, women and men and livestock owners and non-livestock owners. During a drought year the dam committee chairman sought to stop villagers from planting gardens until it was clear that there was enough water for livestock. Garden project members protested, indicating that such a move would disadvantage poor farmers who, after all, did not own livestock and depended on the irrigated plots for their livelihoods (Nemarundwe, 2003, p. 166). The dam chairman proceeded to seal off (with the assistance of two other villagers) all outlet valves at the dam so that no water could flow to the gardens. As a result, villagers challenged him publicly. The dam chairman then let out all the water, until it was below the outlets. The resolution of disputes required external authorities to help sort out the conflict. The dam chairman was subject to a tribunal organized by the Rural District Council and the NGO supporting the project. He was reprimanded and the villagers called for him to resign from the dam committee. However, he apologized to the project members and promised to cooperate with other farmers in conserving water resources.

Our reading of Sithole and Matondi suggests that, unlike the right to safe drinking water, the right to garden lands with available water was, in the final analysis, limited to kin. Prosper Matondi's study from Shamva focused on the growing scarcity of arable land near water (Matondi, 2001). As is the case throughout Zimbabwe, dambo gardens are located near the streams dissecting the vleis that are also used as grazing areas. However, over time they have been used more for gardens than for grazing. With the presence of livestock, gardens have to be fenced to prevent animals from eating the produce and drinking from the well. The fencing of vegetable gardens along rivers or on wetlands is common practice all over Zimbabwe. This suggests that, once the land is allocated for gardening, the land and the water available for irrigation become family property. Access to both land and water thus may be restricted on the basis of kinship ties. In the same vein, Sithole (1999) documents increased desiccation of dambo areas in Mutoko (Mashonaland East Province) and Chiduku (Manicaland Province) communal areas, and thus increased difficulties in using the water from dambos for small-scale irrigation.

The main mechanism for sharing scarce livelihood resources under these conditions is subdividing the land among kin within the broader family. This suggests that, in situations of scarcity of common pool resources, the norm of sharing is placed on the kin. Often, this scarcity has been created by the unequal divisions between land and water and between the commercial farm sector and the communal and resettlement areas. The pattern was that rather than deny some families or households access to dambo land, the gardens were subdivided into smaller areas. It remains to be seen whether this situation has been altered by the Fast Track Land Reform.

While access to gardens with available water resonates with villagers' deep concern for livelihood it is, unlike the right to safe drinking water, not available on a universal and non-discriminatory basis. Outsiders do not have access and the land is, in principle, allocated to the male head of the household on behalf of the family. Yet livelihood concerns crosscut the male status rule so as to make land available to single and childless women, widows and divorcees. While married women, due to these formalities, have been seen as landless, Sithole (1999, p. 80) observed that women seem to be acknowledged by most men as owners of the garden.[26] This strongly suggests that ownership within the family is not acquired through rules concerning family representation but by actual use and work on the land.

While accepted within and amongst local communities, these norms are frequently overlooked and disregarded in development policies, projects and practices. In one of the largest resettlement projects in a communal area in the Zambezi valley, Derman (1997) reports that women farmers could no longer maintain their dambo gardens since they were moved away from streams and rivers. Boreholes were provided for drinking water and watering livestock. There was no broader concern for livelihood as people were left to dig their own garden wells to supply water for vegetables. Some women continued walking long distances to keep up their gardens, while other families invested in private wells. For many women the only solution was to use the scarce borehole water for irrigating vegetables. Because of the very dry conditions and livestock water requirements there is great pressure upon the functioning boreholes which, in turn, has meant that many women have had to give up or reduce their gardens.

Conclusions and Reflections on the Right to Water and Livelihood

In sum, these practices from different parts of Zimbabwe point toward the existence of a set of interrelated norms of sharing of land and water that are essential for livelihood. Both clean drinking water and access to land with available water are shared between and within village households on a day-to-day basis. This norm underlies trouble-free cases (cases where agreement is reached through everyday practice without involvement in any dispute resolution), but it is also confirmed by ideal statements from villagers (what people say) and, more importantly, by trouble cases from Nemarundwe's, Sithole's, Matondi's and our own research. However, three interrelated processes threaten these norms and practices: (i) the broad economic and social crises that have coincided with the Fast Track Land Reform and have altered the rural landscapes; (ii) the fiscal crises of ZINWA and Catchment Councils, leading them to want to increase their sources of water

revenues; and (iii) pressures upon water resources due to drought or conflicts over use between mining or livestock and gardens.

The widespread acceptance of these norms emerges as vital in the ways that local communities handle poverty and food security. These local norms and practices interconnect with emerging human rights law that considers water as part of the right to livelihood. This includes clean drinking water and adequate access to water for subsistence farming and for securing livelihoods. The current multi-level and multi-layered political and economic crises in Zimbabwe pose challenges to the use of human rights as a framework for reform. Because international human rights are considered to be incompatible with the current Africanist directions of the Government of Zimbabwe, the Covenant on Economic, Social and Cultural Rights, the Convention on the Rights of the Child and the Protocol to the African Charter on the Rights of Women (among others) have been deemed irrelevant to the government's policies. Our research suggests that this dichotomous perception of African culture and human rights is false insofar as the rights to water and livelihood are concerned. It shows that prevailing norms and practices in communal areas and the emerging human right to water and livelihood provide common ground for a new framework facilitating active and direct support to small-scale (and often poor) farmers.

Primary water can be a starting point for national legislation and policies to include a 'right to water' and a 'right to livelihood'. The idea of a priority *right* to primary water for basic human needs, including domestic, animal and house-building functions, is unique in the region. It has meant that such water in principle has been protected from the growing demand for 'user pay' which, according to the Water Act, is restricted to commercial water. However, the pressure upon a more privatized water sector, led by the Zimbabwe National Water Authority, to be self-financing in the context of a national economic crisis demonstrates the need for greater legal and political clarity for primary water. Indeed, the 'goal' of water reform appears to be increasing the amount of water that can be labelled 'commercial' rather than 'primary'.

In our view, then, priority could focus on how to use primary water for socially beneficial and development purposes other than simply expanding commercial water use. Primary water enables the concept to be developed in the light of local concerns and the wider regional and international human rights laws. However, we do not think under current circumstances that Zimbabwe can, in practice, achieve the broadening of such rights. Rather, given the growing scarcity of resources we can easily envision ZINWA or Catchment Councils whose members are attempting to obtain revenues by defining these small gardens as commercial ventures, in which case they will be said to be using agricultural water, which attracts a price.

Another related problem is where Catchment Councils label all water as stored, which attracts a higher price compared with what is called normal flow. Basically, this means that smallholder farmers will, in principle, have to pay for the irrigation water. Once again, rural people's decision making seems highly responsive and sensible in the light of changing survival requirements and should guide laws and policies. A better approach would be to assist smallholders to increase their use of water and therefore production, with the likelihood of increased nutrition and decreased illness, especially if sanitation is improved simultaneously.

Local discourses and practices of distribution and management of water speak of the emerging notion of water as a human right. Older dambo cultivation and more recent gardens have been utilized under a principle of a right of access to both land and water for livelihood purposes. The concept of livelihood, as locally understood, has responded to a changing social and economic environment by including sale of produce but with the understanding that it is for socially understood purposes, including education of children, health expenses, clothing, house repair, etc., along with the consumption of garden products. It cuts across a narrow distinction between commercial and primary water. From the perspective of a local livelihood, it makes little sense to make a distinction between garden products that are directly consumed by the family and products that are sold to provide for medicine, food or clothes.

Neither the Zimbabwean land reform nor the water reform addresses how to assist those engaged in small-scale irrigation. The priority has been given to commercial water and to redeveloping irrigation systems in what had been the large-scale commercial farming sector. In Zimbabwe, most communal area irrigation is outside of formal irrigation schemes. Neither the Zimbabwe water acts nor recent policy documents make any mention of how to support informal irrigation carried out in Zimbabwe's communal areas and, increasingly, in the former commercial farmlands. This has to do with the division between the development functions for communal and resettlement areas tasked to Rural District Councils and central government, water management functions given to Catchment Councils, ZINWA and the Ministry of Water Development and the rural water-supply functions that are separate from the new institutions of water reform.

Lastly, given the importance of women, a grounded human rights analysis would greatly strengthen efforts to identify potential discriminatory effects and to suggest policies to increase women's production. One problem is that water sources used by female small farmers, for example irrigation of vegetable gardens by borehole water, have been seen by conventional economic standards as unproductive. As a result of the gendered character of land and water uses, seemingly gender-neutral investment policies have often disproportionately favoured expensive water supply services controlled by men. This may lead to indirect discrimination, in terms of both CEDAW and the Protocol of the Rights of Women to the African Charter.

Acknowledgements

Drs Manzungu and Machiridza would like to thank the Challenge Programme, Project Number 47: 'Transboundary water governance for agricultural and economic growth and improved livelihoods in the Limpopo and Volta basins – towards African indigenous models of governance', without whose financial support the writing of this paper would not have been possible. Bill Derman was supported in his research by a Fulbright-Hays Research Grant, a Wenner-Gren Foundation Grant for Anthropological Research and the BASIS CRSP for Water and Land Research in Southern Africa. Anne Hellum has been supported by the Ministry of Foreign Affairs/Norwegian Research Council Program Development Grant for the Institute of Women's Law at the Faculty of Law, University of Oslo and the NORAD-funded cooperation between the Institute of Women's Law (University of Oslo) and the Southern and Eastern African Center of Women's Law (SEACWL) at the University of Zimbabwe. Mr Pinimidzai Sithole's research has been supported by the BASIS Mentors' Program and the BASIS CRSP for Water and Land Research in Southern Africa.

Endnotes

[1] In general, Conventions are instruments passed by the UN General Assembly. Conventions are made binding for state parties by ratification. Two conventions are formally termed covenants. These are the Covenant on Civil and Political Rights, 1966 and the Covenant on Social, Economic and Cultural Rights, 1966. General recommendations/general comments are the interpretations of the human rights treaty bodies that, in accordance with the respective conventions, have the power to make such recommendations. The general recommendations are not directly binding for the state parties to the conventions, like the conventions themselves. They are sources of interpretation accorded weight by international and national courts.

[2] The four Dublin Principles are: (i) fresh water is a finite and vulnerable resource, essential to sustain life; (ii) water is an economic and social good; (iii) water development and management should be based on a participatory approach involving users, planners and policy makers at all levels; and (iv) women play a central part in the provision, management and safeguarding of water. The thinking behind these principles has been incorporated into policy documents authored by the World Bank and other donor organizations (World Bank, 1993, 2002, 2003a, b).

[3] Southern African rural scholarship and practice have focused upon livelihood strategies, access to resources and the necessary institutional changes to support rural livelihoods. See for example Scoones, 1996; Sithole, 1999; Benjaminsen *et al.*, 2002; Nemarundwe, 2003.

[4] Water reform has been part of the general process of decentralization. The argument runs that, if natural resources are managed at the local level, then they will be looked after better and more efficiently, resulting in improved opportunities for

sustainable livelihoods (SLSA Team, 2003a, p. 3). There was, however, no discussion of the local practices and norms that can influence or even determine whether decentralization will be successful.

[5] There continues to be an important debate about the origins of fast track and its relationship to the parliamentary elections of 2000. One line of thinking views the land invasions as politically motivated to win the elections (Sachikonye, 2003, 2005; Hellum and Derman, 2004, among others), while the other perspective contends that fast track was an unplanned response to pressures from landless people (Moyo and Yeros, 2005).

[6] In Africa, the right to water had been incorporated into national instruments in the region. For example, the right to water is embedded in the Bill of Rights in Section 27 (1) (b) of the South African Constitution. It states that everyone has the right to have access to sufficient water. Article 12 of the Zambian Constitution maintains that the state shall endeavour to provide clean and safe water. According to Article 90 of the Ethiopian Constitution, every Ethiopian is entitled, within the country's resources, to clean water. The preamble to the Namibian Sixth Draft Water Resources Management Bill of 2001 states that the government has overall responsibility for and authority over the nation's water resources and their use, including equitable allocation of water to ensure the right of all citizens to sufficient safe water for a healthy and productive life and the redistribution of water.

[7] The Protocol was adopted by the 2nd Ordinary Assembly of the African Union, Maputo, 11 July 2003 and entered into force in 2006.

[8] The statement of understanding states that: 'In determining vital human needs in the event of conflicts over the use of water courses, special attention is to be paid to providing sufficient water to sustain human life, including both drinking water and water required for production of food in order to prevent starvation.'

[9] The introduction to the Protocol states that Articles 60 and 61 of the African Charter on Human and Peoples' Rights recognize regional and international human rights instruments and African practices consistent with international norms on human and people's rights as being important reference points for the application and interpretation of the African Charter.

[10] This obligation is embedded in Article 1 of the CEDAW and in Article 2 in the Protocol to the African Charter on the Rights of Women.

[11] The Commission of Enquiry into Appropriate Agricultural Land Tenure Systems (referred to as the Rukuni Commission 1994).

[12] Recommendations 8.8.1, 8.8.3, 8.8.4 and 8.8.5, Rukuni Commission (1994).

[13] There is a substantial literature on different dimensions of Zimbabwe's water policies and water reform, including Derman *et al.*, 2001; Dube and Swatuk, 2002; Derman and Gonese, 2003; Mtisi and Nicol, 2003; Bolding *et al.*, 2004; Hellum and Derman, 2005, among others.

[14] These are the Sanyati, Manyame, Mazowe, Save, Runde, Mzingwane and Gwayi.

[15] Water Act 1998 section 32 (1).

[16] In accordance with section 41 in the ZINWA Act, the Minister may, in consultation with the approval of the Minister responsible for finance, by statutory instrument, impose a water levy on any person holding a permit issued in terms of the Water Act (Chapter 20, p. 24).

[17] Derman Research Notes, February 2000. At a Mazowe Catchment Council meeting there was a discussion on whether to ask the Centre for Applied Social Sciences to suggest a definition for commercial water. This discussion ended when the Council's Chair suggested the technological definition.

[18] There has been a large decline in support to communal areas due to the emphasis upon land acquired during the Fast Track Land Resettlement Programme.

[19] Mhondoro Communal Land is situated in the Chegutu district, which is made up of commercial farm, small-scale commercial, communal, resettlement and urban areas 120 km west of Harare. The major river that flows through this high plateau area is known as the Mupfure. It is part of the larger Sanyati River Catchment south-east of Harare and flows through communal and commercial land, including the city of Chegutu.

[20] The norms of sharing and potential sanctions exist in those areas of the three catchments where the CASS water research team has been working.

[21] There is an intense debate on the degree and extent to which access to land can be obtained through kin ties and networks and the extent to which it is being concentrated and access controlled by an emergent property class (Berry, 2002; Peters, 2004). Increasing land concentration and control will have significant consequences for access to water.

[22] The process of decreasing dependence upon agriculture alone has been called de-agrarianization by Deborah Bryceson, 1999.

[23] CASS BASIS survey data, CASS 2000–2001.

[24] As noted earlier, *dambo* or wetlands cultivation is quite old, but dry-season gardens are recent.

[25] 'Informal irrigation' land constitutes the vast majority of irrigated lands in Zimbabwe's communal areas. Yet the Irrigation Strategy of 1994, which was carried out in preparation for water reform, focused only on government-sponsored formal irrigation schemes that comprised only 2000 ha at that time (Zimbabwe Government, 1994).

[26] This is not straightforward. Sithole (1999, p. 80) writes: 'It seemed impossible for women and men for that matter to think about ownership in terms of this belonging to this one or that one.'

References

Benjaminsen, T., Cousins, B. and Thompson, L. (eds) (2002) *Contested Resources: Challenges to the Governance of Natural Resources in Southern Africa.* Programme for Land and Agrarian Studies, University of the Western Cape, Cape Town, South Africa.

Berry, S. (1993) *No Condition Is Permanent: the Social Dynamics of Agrarian Change in Sub-Saharan Africa.* University of Wisconsin Press, Madison, Wisconsin.

Berry, S. (2002) Debating the land question in Africa. *Comparative Studies in Society and History* 44 (4), 638–668.

Bolding, A., Manzungu, E. and van der Zaag, P. (1996) Farmer-initiated irrigation furrows: observations from the Eastern Highlands. In: Manzungu, E. and van der Zaag, P. (eds) *The Practice of Smallholder Irrigation: Case Studies from Zimbabwe.* University of Zimbabwe, Harare, Zimbabwe, pp. 191–218.

Bolding, A., Manzungu, E. and Zawe, C. (2004) Irrigation policy discourse and practice: two cases of irrigation transfer in Zimbabwe. In: Mollinga, P. and Bolding, A. (eds) *The Politics of Irrigation Reform: Contested Policy Formulation in Asia, Africa and Latin America.* Ashgate, Aldershot, UK, pp. 166–206.

Bryceson, D. (1999) *Sub-Saharan Africa Betwixt and Between: Rural Livelihood Practices and Policies.* Working Paper 43, African Studies Center, Leiden, Netherlands.

Campbell, H. (2003) *Reclaiming Zimbabwe: the Exhaustion of the Patriarchal Model of Liberation.* David Philip Publishers, Johannesburg, South Africa.

Central Statistical Office (2002) *Preliminary Report, National 2002 Population Census.* Government Printers, Harare, Zimbabwe.

Chakaodza, A.M. (1993) *Structural Adjustment in Zambia and Zimbabwe: Reconstructive or Destructive?* Third World Publishing House, Harare, Zimbabwe.

Chidenga, E.E. (2003) Leveraging water delivery: irrigation technology choices and operations and maintenance in smallholder systems in Zimbabwe. PhD thesis, Wageningen University, Wageningen, Netherlands.

Cleaver, F. (1995) Water as a weapon: the history of water supply development in Nkayi District, Zimbabwe. In: Grove, R. and McGregor, J. (eds) *Environment and History Journal: Special Issue Zimbabwe* 1 (3), 313–333. The White Horse Press, Cambridge, UK.

Derman, B. (1997) How green was my valley! Land use and economic development in the Zambezi valley, Zimbabwe. In: Isaac, B. (ed.) *Research in Economic Anthropology* 18, 331–380. Greenwood, Greenwich, Connecticut.

Derman, B. (1998) Balancing the waters: development and hydropolitics in contemporary Zimbabwe. In: Donahue, J. and Johnston, B. (eds) *Water, Culture and Power*. Island Press, New York, pp. 73–94.

Derman, B. and Gonese, F. (2003) Water reform: its multiple interfaces with land reform and resettlement. In: Roth, M. and Gonese, F. (eds) *Delivering Land and Securing Livelihood: Post-Independence Land Reform and Resettlement in Zimbabwe*. Centre for Applied Social Sciences, University of Zimbabwe, Harare and Land Tenure Center, University of Wisconsin, Madison, Wisconsin, pp. 287–307.

Derman, B. and Hellum, A. (2003) Neither tragedy nor enclosure: are there inherent human rights in water management in Zimbabwe's communal lands? In: Benjaminsen, T. and Lund, C. (eds) *Securing Land Rights in Africa*. Frank Cass, London, pp. 31–50.

Derman, B. and Hellum, A. (2007) Land, identity and violence in Zimbabwe. In: Derman, B., Odgaard, R. and Sjaastad, E. (eds) *Citizenship and Identity: Conflicts over Land and Water in Contemporary Africa*. James Currey, London.

Derman, B., Gonese, F. and Ferguson, A. (2001) *Decentralization, Devolution and Development: Reflections on the Water Reform Process in Zimbabwe*. Centre for Applied Social Sciences, University of Zimbabwe, Harare, Zimbabwe.

Dube, D. and Swatuk, L. (2002) Stakeholder participation in the new water management approach: a case study of the save catchment, Zimbabwe. *Physics and Chemistry of the Earth* 27, 867–874.

Eide, A. (2001) The right to an adequate standard of living including the right to food. In: Eide, A., Krause, A. and Rosas, A. (eds) *Economic, Social and Cultural Rights: a Text-Book*. Kluwer International, Dordrecht, Netherlands.

Government of Southern Rhodesia (undated) *Water Act of 1927*. National Archives of Zimbabwe, Harare, Zimbabwe.

Hammar, A., Raftopolous, B. and Jensen, S. (2003) *Zimbabwe's Unfinished Business. Rethinking Land, State and Nation in the Context of Crisis*. Weaver Press, Harare, Zimbabwe.

Hellum, A. (2001) Towards a human rights based development approach: the case of women in the water reform process in Zimbabwe. *Law, Social Justice and Global Development, University of Warwick Online Journal*. Available at http://elj.warwick.ac/uk/global/issue/2001-1/hellum.html

Hellum, A. (2007) Human rights encountering gendered land and water uses: family gardens and the rights to water in Mhondoro Communal Land. In: Hellum, A., Stewardt, J., Ali, S. and Tsanga, A. (eds) *Human Rights: Gendered Women's Realities and Plural Legalities*. Weaver Press, Harare, Zimbabwe.

Hellum, A. and Derman, B. (2004) Land reform and human rights in contemporary Zimbabwe: balancing individual and social justice through an integrated human rights framework. *World Development* 32 (10), 1785–1805.

Hellum, A. and Derman, B. (2005) Negotiating water rights in the context of a new political and legal landscape in Zimbabwe. In: von Benda-Beckmann, F., von Benda-Beckmann, K. and Griffiths, A. (eds) *Mobile People, Mobile Law: Expanding Legal Relations in a Contracting World*. Ashgate, Aldershot, UK and Burlington, Vermont, pp. 177–198.

Herbst, J. (1990) *State Politics in Zimbabwe*. University of Zimbabwe Publications, Zimbabwe and University of California Press, Berkeley, California.

Hoffman, H.J. (1964) *Water Law in South Rhodesia*. The Government Printer, Salisbury, Rhodesia.

IFAD (International Fund for Agricultural Development) (1997) *Smallholder Irrigation Support Programme*. Formulation Report, Harare, Zimbabwe.

Kujinga, K. (2002) Decentralising water management: an analysis of stakeholder participation in the management of water in Odzi subcatchment area, Save Catchment. In: Jonker, L., Beukman, R., Nyabeze, W.R., Kansiime, F. and Kgarebe, B.V. (eds) *Integrated Water Resources Management: Theory, Practice, Cases. Journal of Physics and Chemistry of the Earth* 27 (11–22), 897–905, Pergamon Press, UK/USA.

Kujinga, K. and Manzungu, E. (2004) Enduring contestations: stakeholder strategic action in water resource management in the Save Catchment Area, eastern Zimbabwe. *Eastern Africa Social Science Research Review* 20 (1), 67–91.

Makadho, J.M. (1993) An approach to quantifying irrigation water delivery performance. Paper presented at the University of Zimbabwe/AGRITEX/IFPRI workshop: 'Irrigation Performance in Zimbabwe, Juliasdale, Zimbabwe, 4–6 August 1993.

Makarau, A. (1999) Zimbabwe's climate: past, present and future. In: Senzanje, A., Manzungu, E. and van der Zaag, P. (eds) *Water for Agriculture in Zimbabwe: Policy and Management Options for the Smallholder Sector*. University of Zimbabwe Publications, Harare, Zimbabwe, pp. 3–16.

Manzungu, E. (1999) Strategies of smallholder irrigation management in Zimbabwe. PhD thesis, Wageningen University, Wageningen, Netherlands.

Manzungu, E. (2001) A lost opportunity: the case of the water reform debate in the Fourth Parliament of Zimbabwe. *Zambezia* XXVIII (i), Zimbabwe.

Manzungu, E. (2002) Global rhetoric and local realities: the case of Zimbabwe's water reform. In: Chikowore, G., Manzungu, E., Mushayavanhu, D. and Shoko, D. (eds) *Managing Common Property in an Age of Globalisation: Zimbabwean Experiences*. Weaver Press, Harare, Zimbabwe, pp. 31–44.

Manzungu, E. (2003) Of science and livelihoods strategies: two sides of the commercialization debate in smallholder irrigation schemes. In: Bolding, A., Mutimba, J. and van der Zaag, P. (eds) *Interventions in Smallholder Agriculture: Implications For Extension in Zimbabwe*. University of Zimbabwe Publications, Harare, Zimbabwe, pp. 110–130.

Manzungu, E. (2004a) *Water for All: Improving Water Resource Governance in Southern Africa*. Gatekeeper Series No. 113, International Institute for Environment and Development (IIED), London.

Manzungu, E. (2004b). Public institutions in smallholder irrigation in Zimbabwe. In: Moll, H.A.J., Leeuwis, C., Manzungu, E. and Vincent, J. (eds) *Agrarian Institutions between Policies and Local Action: Experiences from Zimbabwe*. Weaver Press, Harare, Zimbabwe.

Manzungu, E. and Kujinga, K. (2004) Enduring contests: Stakeholder strategic action in water resource management in Save Catchment Area, Eastern Zimbabwe. *Eastern Africa Social Science Research Review* XX (1) (2004), 67–92.

Manzungu, E. and Senzanje, A. (1996) A political-economy approach to water reform in Zimbabwe's agricultural sector. Paper presented at the University of Zimbabwe/ZIMWESI Workshop: 'Water for Agriculture: Current Practices and Future Prospects', Mandel Training Centre, Marlborough, Harare, Zimbabwe, 11–13 March 1996.

Matondi, P. (2001) The struggle for access to land and water resources in Zimbabwe: the case of Shamva district. Doctoral thesis, Swedish University of Agricultural Sciences, Uppsala, Sweden.

McGregor, J. (1995) Conservation, control and ecological change: the politics and ecology of colonial conservation in Shurugwi, Zimbabwe. In: Grove, R. and McGregor, J. (eds) *Environment and History: Special Issue Zimbabwe* 1 (3), 257–279. The White Horse Press, Cambridge, UK.

McIlwain, R. (1936) Water law in Southern Rhodesia. *Rhodesia Agricultural Journal* 33, 788–801.

Moore, D.S. (1995) Contesting terrain in Zimbabwe's eastern highlands: the cultural politics of place, identity and resource struggles. Doctoral thesis, Stanford University, Palo Alto, California.

Moyo, S. and Yeros, P. (2005) Resurrecting the peasantry and semi-proletariat: a critique of the new Marxist analyses of land occupations and land reform in Zimbabwe: towards the national democratic revolution. In: Moyo, S. and Yeros, P. (eds) *Reclaiming the Land: the Resurgence of Rural Movements in Africa, Asia and Latin America*. Zed Books, London and New York, pp. 165–205.

Mtisi, S. and Nicol, A. (2003) *Caught in the Act: New Stakeholders, Decentralisation and Water Management Processes in Zimbabwe*. Sustainable Livelihoods in Southern Africa Paper Series, Research Paper 20, Sustainable Livelihoods in Southern Africa Research, Brighton, UK.

Nemarundwe, N. (2003) Negotiating resource access: institutional arrangements for woodlands and water use in southern Zimbabwe. Doctoral thesis, Swedish University of Agricultural Sciences, Uppsala, Sweden.

Pearce, G.P.R. and Armstrong, A.S.B. (1990) *Small Irrigation Design, Nyanyadzi, Zimbabwe: Summary Report of Studies on Field-Water Use and Water Distribution*. Report OD 98, Hydraulics Research, Wallingford, UK.

Pinstrup-Andersen, P. (2000) Introduction. In: Bruns, B.R. and Meinzen-Dick, R. (eds) *Negotiating Water Rights*. Vistaar, New Delhi, India.

Peters, P. (2004) Inequality and social conflict over land in Africa. *Journal of Agrarian Change* 4 (3), 269–314.

Phimister, I. (1988) *An Economic and Social History of Zimbabwe, 1898–1948*. Longman, London and New York.

Ranger, T. (1985) *Peasant Consciousness and Guerilla War*. James Currey, London.

Rukuni, M. (1993) Irrigation issues in Zimbabwe. Paper presented at the University of Zimbabwe/AGRITEX/IFPRI Workshop: 'Irrigation Performance in Zimbabwe', Juliasdale, Zimbabwe, 4–6 August 1993.

Rukuni, M. (1994) The evolution of agricultural policy: 1890–1990. In: Rukuni, M. and Eicher, C. (eds) *Zimbabwe's Agricultural Revolution*. University of Zimbabwe Publications Office, Zimbabwe, pp. 15–39.

Sachikonye, L. (2003) From 'growth with equity' to 'fast-track' reform: Zimbabwe's land question. *Review of African Political Economy* 96, 227–240.

Sachikonye, L. (2005) The land is the economy: revisiting the land question. *African Security Review* 14 (3), 3331–3344.

Salman, S. and McInerney-Lankford, S. (2004) *The Human Right to Water. Legal and Policy Dimensions*. World Bank, Washington, DC.

Scoones, I. (1996) *Hazards and Opportunities: Farming Livelihoods in Dryland Africa: Lessons from Zimbabwe*. Zed Books, London.

Sithole, B. (1999) Use and access to *dambos* in communal lands in Zimbabwe: institutional considerations. A thesis submitted in partial fulfilment of the requirements for a PhD at the Centre for Applied Social Sciences, University of Zimbabwe, Harare, Zimbabwe.

SLSA (Sustainable Livelihoods in Southern Africa) Team (2003a) *Decentralisations in Practice in Southern Africa*. Programme for Land and Agrarian Studies, University of the Western Cape, Cape Town, South Africa.

SLSA Team (2003b) *Rights Talk and Rights Practice: Challenges for Southern Africa*. Programme for Land and Agrarian Studies. University of the Western Cape, Cape Town, South Africa.

UNDP (United Nations Development Programme) (2003) *Human Development Report: Millennium Development Goals: a Compact among Nations to End Poverty*. Oxford University Press, New York and Oxford, UK.

UNESCO (United Nations Educational, Scientific and Cultural Organization) (2003) *Water for People, Water for Life: the United Nations World Water Development Report*. UNESCO and Berghahn, Barcelona, Spain.

Vincent, L.F. and Manzungu, E. (2004) Water rights and water availability in the Lower Odzi watershed of the save catchment. In: Moll, H.A.J., Leeuwis, C., Manzungu, E. and Vincent, J. (eds) *Agrarian Institutions between Policies and Local Action: Experiences from Zimbabwe*. Weaver Press, Harare, Zimbabwe.

World Bank (1993) *Water Resources Management: a World Bank Policy Paper*. Washington, DC.

World Bank (2002) *Bridging Troubled Waters*. Operations Evaluation Department, Washington, DC.

World Bank (2003a) *Water Resources Sector Strategy: Strategic Directions for World Bank Engagement*. World Bank, Washington, DC.

World Bank (2003b) *World Development Report: Sustainable Development in a Dynamic World*. World Bank, Washington, DC and Oxford University Press, New York.

WHO (World Health Organization) (2003) *Right to Water*. Health and Human Rights Series No. 3, World Health Organization, Geneva.

ZCTU (Zimbabwe Congress of Trade Unions) (1996) *Beyond ESAP: Framework for a Long-term Development Strategy in Zimbabwe*. ZCTU, Harare, Zimbabwe.

Zimbabwe Government (1994) *Commission of Enquiry into Appropriate Agricultural Land Tenure Systems* (3 Vols). Government of Zimbabwe Printers, Harare, Zimbabwe.

Zimbabwe Government (1998a) *Water Act, No. 31/1998*. Government of Zimbabwe Printers, Harare, Zimbabwe.

Zimbabwe Government (1998b) *Zimbabwe National Water Authority Act, No. 11/1998*. Government of Zimbabwe Printers, Harare, Zimbabwe.

Zimbabwe Government (2000a) *Statutory Instrument 33*. Water (Catchment Councils) Regulations, Harare, Zimbabwe.

Zimbabwe Government (2000b) *Statutory Instrument 47 of 2000*. Water (Sub-catchment Councils) Regulations, Harare, Zimbabwe.

Zimbabwe Government (2000c) *Towards Integrated Water Resources Management*. Government of Zimbabwe Printers, Harare, Zimbabwe.

Zimbabwe Government (2003) *Report of the Presidential Land Review Committee under the Chairmanship of Dr. Charles Utete* (2 Vols). Government of Zimbabwe, Harare, Zimbabwe.

Zimbabwe Human Development Report (2003) *Redirecting Our Reponses to HIV and AIDS: Towards Reducing Vulnerability – the Ultimate War for Survival*. Produced by the Poverty Reduction Forum, Institute of Development Studies, University of Zimbabwe, Zimbabwe.

Zimbabwe Independent Newspaper (2004) Muckraker Column, 26 November 2004.

Zimbabwe Parliamentary Debates (1998) 25 (26), Tuesday, 3 November 1998.

Index

Note: Page references in *italic* refer to tables in the text, those in **bold** to figures and the letter n indicates material in the endnotes